Artificial Intelligence and Big Data Applications

Artificial Intelligence and Big Data Applications

Editors

Amar Ramdane-Cherif
Ravi Tomar
Thipendra P Singh

Basel • Beijing • Wuhan • Barcelona • Belgrade • Novi Sad • Cluj • Manchester

Editors

Amar Ramdane-Cherif
Versailles Systems
Engineering Laboratory,
University of Versailles
Versailles
France

Ravi Tomar
School of Computer Science,
University of Petroleum &
Energy Studies
Dehradun
India

Thipendra P Singh
School of Computer Science
& Technology
Bennett University
Greater Noida
India

Editorial Office
MDPI
St. Alban-Anlage 66
4052 Basel, Switzerland

This is a reprint of articles from the Special Issue published online in the open access journal *Information* (ISSN 2078-2489) (available at: https://www.mdpi.com/journal/information/special_issues/X1L23OGC1T).

For citation purposes, cite each article independently as indicated on the article page online and as indicated below:

Lastname, A.A.; Lastname, B.B. Article Title. *Journal Name* **Year**, *Volume Number*, Page Range.

ISBN 978-3-7258-0555-6 (Hbk)
ISBN 978-3-7258-0556-3 (PDF)
doi.org/10.3390/books978-3-7258-0556-3

Contents

About the Editors

Amar Ramdane-Cherif

Amar Ramdane-Cherif received his Ph.D. from Pierre and Marie Curie University in Paris in 1998. In 2007, he obtained his HDR degree from Versailles University. From 2000 to 2007, he was an associate professor at the University de Versailles and worked in the PRISM Laboratory. Since 2008, he has been a full professor at the University of Versailles—Paris-Saclay and works in the LISV Laboratory. His research interests include software ambient intelligence: semantic representation of knowledge, modeling of the ambient environment, multimodal interaction between people/machines and machines/the environment, the system of fusion and fission of events, and ambient assistance; software architecture: software quality, quality evaluation methods, functional and non-functional measurement of real-time, and reactive and software embedded systems. He has written 7 book chapters and 50 international journal articles and has participated in 130 international conferences. He has supervised more than 20 doctoral theses (18 supported/3 in progress).

Ravi Tomar

Ravi Tomar works as a senior architect at Persistent Systems, India. Dr. Tomar has been an experienced academician and professional with a history in the higher education industry for a decade. He is skilled in computer networking, stream processing, Python, the Oracle database, C++, Core Java, J2EE, RPA, and CorDApp. His research interests include wireless sensor networks, image processing, data mining and warehousing, computer networks, big data technologies, and VANET. He has authored 100+ papers in different research areas, filed 4 Indian patents, edited 5 books, and authored 4 books. He has delivered training to corporations nationally and internationally on Confluent Apache Kafka, stream processing, RPA, CordaApp, J2EE, and the IoT to clients such as KeyBank, Accenture, Union Bank of the Philippines, Ernst and Young, and Deloitte. Dr. Tomar is officially recognized as an instructor for Confluent and CordApp. He has taken part in various international conferences in India, France, and Nepal. He was awarded the Young Researcher in Computer Science and Engineering by RedInno, India, in 2018, the Academic Excellence and Research Excellence Award by UPES in 2021, and the Young Scientist Award by UCOST, Dehradun.

Thipendra P Singh

Thipendra P Singh currently works as a dean (academic affairs) and professor of computer science at Bennett University, Greater Noida, NCR, India. Prior to this, he was associated with UPES University and Sharda University. He holds a doctorate in Computer Science from Jamia Millia Islamia University, New Delhi. He carries with him more than 27 years of rich experience. Under his guidance, one research scholar has completed her Ph.D., and currently, 5 others are working towards their doctoral degrees. Of these students, two scholars are being jointly supervised by French and UK universities. He is a widely traveled academician and has participated in various platforms across the world including the UK, France, the UAE, and Singapore. He is the editor/author of 20 books on various relevant topics in the field of computer science. He has also authored more than 50 research papers in various highly reputable journals indexed in the Web of Science and Scopus. Dr. Singh is a senior member of the IEEE and a member of various other professional bodies including the IEI, ACM, EAI, ISTE, IAENG, etc., and also serves on the editorial/reviewer panel of different journals. He has been a Fellow of the IETA—India since 2019. He is also on the board of studies of different universities in India and abroad. He has taken part in many conferences throughout India and abroad as a chair, organizing committee member, TPC member, session chair, etc.

Preface

The 21st century has witnessed an unprecedented surge in technological advancements that have reshaped the very fabric of our existence. Artificial intelligence (AI) and big data are two technologies that have not only revolutionized different industries but also permeated into the core of our daily lives. It is within this context that this reprint of the Special Issue *Artificial Intelligence and Big Data Applications* emerges as a worthy compilation to navigate through some interesting details about these groundbreaking technologies.

As we delve into the realms of AI and big data, it becomes apparent that their synergy holds the key to unlocking a myriad of possibilities. Artificial Intelligence, with its ability to mimic human intelligence and perform complex tasks, has transcended the realm of science fiction to become an integral part of our reality. Simultaneously, big data, fueled by the exponential growth of digital information, has emerged as the lifeblood that nourishes the algorithms of AI. This compilation endeavors to unravel the symbiotic relationship between these two forces and explore their combined potential to shape the future.

The pages that follow encapsulate a holistic exploration of AI and big data applications across diverse domains. From healthcare and finance to manufacturing and education, the impact of these technologies is both profound and far-reaching. Real-world case studies and original research papers pepper the narrative, providing readers with tangible examples of how AI and big data are reshaping industries and enhancing decision-making processes.

Furthermore, this reprint of the aforementioned Special Issue places a strong emphasis on the development of algorithms in various application domains.

In addition to catering to technologists and data scientists, this reprint is designed to be accessible to a broader audience. Whether you are a business leader navigating digital transformation, a student seeking to understand the future employment landscape, or simply a curious mind eager to comprehend the technological forces shaping our world, this reprint is crafted to be your companion on the journey of discovery.

In conclusion, this reprint of the Special Issue on *Artificial Intelligence and Big Data Applications* aims to serve as a compass, guiding readers through the intricate terrain of these transformative technologies. As we stand at the crossroads of innovation and ethical considerations, this book invites you to embark on a journey of exploration and understanding, where the realms of artificial intelligence and big data converge to redefine the boundaries of what is possible in our rapidly evolving digital era.

Amar Ramdane-Cherif, Ravi Tomar, and Thipendra P Singh
Editors

 *information*

Article

Predicting COVID-19 Hospital Stays with Kolmogorov–Gabor Polynomials: Charting the Future of Care

Hamidreza Marateb [1], Mina Norouzirad [2], Kouhyar Tavakolian [3], Faezeh Aminorroaya [4], Mohammadreza Mohebbian [5], Miguel Ángel Mañanas [1,6], Sergio Romero Lafuente [1,6], Ramin Sami [7] and Marjan Mansourian [1,4,*]

[1] Biomedical Engineering Research Centre (CREB), Automatic Control Department (ESAII), Universitat Politècnica de Catalunya-Barcelona Tech (UPC), 08028 Barcelona, Spain; hamid.reza.marateb@upc.edu (H.M.); miguel.angel.mananas@upc.edu (M.Á.M.); sergio.romero-lafuente@upc.edu (S.R.L.)
[2] Center for Mathematics and Applications (NOVA Math), NOVA School of Science and Technology (NOVA SST), 2825-149 Caparica, Portugal; m.norouzirad@fct.unl.pt
[3] School of Electrical Engineering and Computer Science, University of North Dakota, Grand Forks, ND 58202, USA; kouhyar.tavakolian@und.edu
[4] Epidemiology and Biostatistics Department, School of Health, Isfahan University of Medical Sciences, Isfahan 81746-73461, Iran; faeze96amini@gmail.com
[5] Department of Electrical and Computer Engineering, University of Saskatchewan, Saskatoon, SK S7N 5A9, Canada; mom158@usask.ca
[6] CIBER de Bioingeniería, Biomateriales y Nanomedicina (CIBER-BBN), 28029 Madrid, Spain
[7] Department of Internal Medicine, School of Medicine, Isfahan University of Medical Science, Isfahan 81746-73461, Iran; r.sami@med.mui.ac.ir
* Correspondence: marjan.mansourian@upc.edu; Tel.: +34-671-314-185

Citation: Marateb, H.; Norouzirad, M.; Tavakolian, K.; Aminorroaya, F.; Mohebbian, M.; Mañanas, M.Á.; Lafuente, S.R.; Sami, R.; Mansourian, M. Predicting COVID-19 Hospital Stays with Kolmogorov–Gabor Polynomials: Charting the Future of Care. *Information* **2023**, *14*, 590. https://doi.org/10.3390/info14110590

Academic Editors: Amar Ramdane-Cherif, Ravi Tomar and TP Singh

Received: 28 August 2023
Revised: 9 October 2023
Accepted: 23 October 2023
Published: 31 October 2023

Abstract: Optimal allocation of ward beds is crucial given the respiratory nature of COVID-19, which necessitates urgent hospitalization for certain patients. Several governments have leveraged technology to mitigate the pandemic's adverse impacts. Based on clinical and demographic variables assessed upon admission, this study predicts the length of stay (LOS) for COVID-19 patients in hospitals. The Kolmogorov–Gabor polynomial (a.k.a., Volterra functional series) was trained using regularized least squares and validated on a dataset of 1600 COVID-19 patients admitted to Khorshid Hospital in the central province of Iran, and the five-fold internal cross-validated results were presented. The Volterra method provides flexibility, interactions among variables, and robustness. The most important features of the LOS prediction system were inflammatory markers, bicarbonate (HCO_3), and fever—the adj. R^2 and Concordance Correlation Coefficients were 0.81 [95% CI: 0.79–0.84] and 0.94 [0.93–0.95], respectively. The estimation bias was not statistically significant (p-value = 0.777; paired-sample t-test). The system was further analyzed to predict "normal" LOS $\leq$ 7 days versus "prolonged" LOS > 7 days groups. It showed excellent balanced diagnostic accuracy and agreement rate. However, temporal and spatial validation must be considered to generalize the model. This contribution is hoped to pave the way for hospitals and healthcare providers to manage their resources better.

Keywords: COVID-19; Kolmogorov–Gabor polynomials; length of stay; hospital capacity; regularized least squares; validation studies

1. Introduction

The fast spread of the SARS-CoV-2 coronavirus has placed immense strain on healthcare systems across the globe. As infected individuals surged, the demand for hospital admissions grew accordingly [1]. Past outbreaks have demonstrated that limited bed capacity and hospital resources significantly contribute to higher infectious disease mortality rates [2]. Hence, guidelines for prioritizing patients and determining who should be admitted for essential care are instrumental in addressing resource limitations. Neglecting this could jeopardize the lives of COVID-19 patients [3].

Nine to eleven percent of COVID-19 hospitalizations required enhanced life-support interventions [4]. However, the ICU faced challenges accommodating these needs due to limited beds and shortages in monitoring equipment, life-sustaining machinery, and skilled staff crucial for top-tier care [5]. In a study encompassing 183 nations in 2021, Sen-Crowe et al. [2] reported that high-income areas registered the highest average ICU beds at 12.79 and 402.32 hospital beds for every 100,000 individuals. On the other hand, regions with upper-middle income showed dominance in average acute-care beds, numbering 424.75 per 100,000 inhabitants. This is not the case for low- and middle-income countries, where the number of ICU beds is often insufficient, and the equipment is often old and poorly serviced. This number was five beds per one million people in Africa [6].

Challenges in managing hospital capacity throughout this pandemic spanned various phases, including testing, treatment, and preparation for future patients. As a result, there is a pressing need to accurately predict and prioritize patients based on the likelihood of their condition escalating in severity. It is part of the pandemic preparedness action plan.

In predicting hospital length of stay (LOS), the overarching challenges introduced by COVID-19 cannot be overlooked. The pandemic has significantly strained hospital capacities, potentially altering standard care pathways and discharge protocols. Furthermore, heightened fatigue [7], burnout [8], and stress among healthcare professionals [9], a byproduct of the ongoing crisis, may also have indirect implications for the duration of patient stays. These combined factors elucidate the multifaceted dynamics influencing hospital operations during these unprecedented times.

Numerous studies have explored predicting hospital resource needs for COVID-19 patients. Many of these investigations have leveraged machine learning (ML). ML has established itself as an invaluable tool in the medical realm, adept at sifting through and synthesizing vast amounts of data to discern intricate patterns. Most health-related challenges nowadays rely heavily on ML to disentangle the complexities inherent in large-scale data, facilitating informed healthcare decisions.

During outbreaks like COVID-19, forecasting the imminent demand for medical resources such as beds and nasal oxygen support becomes crucial. In this context, ML methodologies have proven invaluable [10,11]. For instance, researchers from London designed an ML algorithm that outperformed clinical experts in predicting COVID-19 patient mortality [12]. Another ML study successfully predicted which COVID-19 patients would transition into a severe respiratory phase with a 70–80% accuracy rate [13].

Furthermore, an AI-based tool named "ambient warning and response evaluation" has been employed to refine ICU clinical settings. This tool significantly enhanced timely and accurate decision-making, leading to a 37% reduction in LOS [14].

LOS estimation remains crucial for efficient healthcare management, offering insights into patient health trajectories, resource allocation, and the quality-of-care delivery. The state-of-the-art research listed encompasses a myriad of methodologies and priorities, thereby revealing both the advancements and the persisting gaps in LOS prediction.

Nemati et al. (2020) [15] utilized a global dataset and focused on a limited set of five variables, primarily age and sex, to estimate LOS. Their approach, which involved stagewise gradient boosting, did not venture into comprehensive features but mainly centered around symptoms onset date and symptoms. Given the minimalistic input feature set, this focus might limit its applicability in varied clinical settings.

Working in a tertiary care hospital in China, Hong et al. (2020) [16] used logistic regression with a set of 37 variables, including lymphocyte and neutrophil count, heart rate, and procalcitonin levels, D-dimer, and partial thrombin time. Their dataset was also relatively small, including 75 patients considering the number of predictors, and reached an AUC of 0.85 to classify prolonged (>14 days) versus normal ($\leq$14 days) hospital LOS. Their work lacks internal and external validation, indicating potential overfitting risks.

Ebinger et al. (2021) [17] embarked on an extensive exploration of 966 patients with 353 variables of electronic health records (EHRs) to classify patients based on extended stays (i.e., LOS > 8 vs. LOS $\leq$ 8 days) in the Cedars-Sinai Medical Center. Forty-two machine learning models were used as ensemble models of 12 base classifiers (including Elastic-net and random forest). Such models were trained using the first 1, 2, and 3 days of hospital admission. Advanced Average (AVG) Blender for the day 3 model outperformed the others. Age, Interleukin 6, blood urea nitrogen level, and oxygen flow rate were among the selected features. The best model had an area under the ROC curve (AUC) of 0.82 and a precision of 67%.

Usher et al. (2021) [18] analyzed data from 36 hospitals across Minnesota, Wisconsin, and the Dakotas. Using 20 variables, which included diverse features such as age, critical illness, mechanical ventilator (MV) application, and oxygen requirement, their approach adopted the random forest method, considering it as the best model. The classification output was the LOS $\leq$ 5 days (reference), LOS between 5 and 10 days, LOS between 10 and 15 days, and LOS > 15 days. With five-fold cross-validation, they achieved an AUC of 0.89, highlighting the potential of integrating diverse input features for LOS category prediction.

Mahboub et al. (2021) [19] at Rashid Hospital in Dubai took a distinct route by incorporating treatments as input features and variables such as urea, platelets, and D-dimer. Utilizing decision trees on a dataset of 2017 patients, they achieved a coefficient of determination (R^2) of 0.5, suggesting the relevance of treatment variables in predicting LOS.

Liuzzi et al. (2022) [20] from the Fondazione Don Carlo Gnocchi Living COVID-19 Registry in Italy incorporated a comprehensive set of 829 variables, with a focus on 55 primary variables spanning across admission clinical scales, symptoms, and therapies. Their method, employing sequential convolutional neural networks, was validated with repeated five-fold cross-validation, resulting in a median absolute deviation of 2.7 days.

Orooji et al. (2022) [21] in Iran, with data from 1225 patients, utilized 53 variables and emphasized 20 key features such as age, creatinine, and lymphocyte/neutrophil count. They applied statistical feature selection combined with multi-layer perceptron and 12 training algorithms, reaching a root-mean-square error (RMSE) of 1.6213 days.

In 2022, Alabbad et al. [22] from King Fahad University Hospital in Saudi Arabia classified ICU LOS into nine categories using 43 variables. The synthetic minority over-sampling technique (SMOTE) was used to balance the class distribution. Their best model employed random forest, and they also explored gradient boosting and extreme gradient boosting. With three-fold cross-validation, their model boasted a positive predictive value (PPV) of 94%, indicating high precision in prediction.

Alam et al. (2023) [23] from Prince Sultan Hospital in Riyadh incorporated 89 variables, including laboratory data, X-ray results, clinical data, and treatments, to classify LOS into seven categories. Their model utilized the Tab Transformer and achieved impressive results, with an F1 score of 93% for discharged patients. The SMOTE-N oversampling technique was also noted to balance the class distribution.

Zhang et al. (2023) [24] analyzed 83 variables, including immunotherapy and heparin, to predict LOS for 384 patients at Zhengzhou University Hospital. Using the least absolute shrinkage and selection operator (LASSO) and linear regression, they explained 30% of LOS variability ($R^2 = 0.30$). Missing data were managed with imputations, and results were verified via bootstrap validation.

Overall, while significant strides have been made in predicting LOS through diverse methodologies, ranging from classical regression models to neural networks, gaps in validation and comprehensive feature inclusion, conditioning on future events (e.g., therapies) resulting in selection bias, incorporating time-dependent predictors (e.g., treatments) as time-fixed, leading to immortal-time bias [25], and balancing the dataset, resulting in biased performance indices [26] remained a consistent challenge. Moreover, sample size insufficiency based on the number of input features [27] was the other problem of some methods proposed in the literature. Further research was required to enhance model generalizability across varied clinical settings.

Our research aimed to employ multivariable analysis and the Kolmogorov–Gabor polynomial to craft a predictive model. This model aimed to precisely forecast the LOS of COVID-19 patients in a nationally representative sample of the pediatric population in the Middle East and North Africa (MENA) based on their demographic and clinical data upon hospital admission.

Our primary model was designed to predict the continuous LOS. We evaluated and presented performance metrics for this continuous prediction. Additionally, we derived a binary representation of LOS from the predicted and actual data, categorizing it as either prolonged (LOS > 7 days) or normal (LOS ≤ 7 days). This binary classification's performance was also examined. We adopted this approach to accommodate the existing literature's categorical and continuous LOS representations. Our primary focus remained the continuous prediction model, which can seamlessly be converted to a binary prediction through straightforward post-processing.

2. Materials and Methods

2.1. Data Source

In this retrospective study, we examined the clinical records of N = 1600 confirmed COVID-19 cases with complete information from Isfahan, situated in the center of Iran, from 6 March to 7 May 2020. These patients were admitted to Khorshid Hospital, which caters to the vast metropolitan area of Isfahan, home to over 15 million residents. Given that this hospital functioned as the primary referral center for critical COVID-19 cases during this period, our study exclusively focused on the patients admitted to the hospital. Patients with a positive RT-PCR test confirming SARS-CoV-2 infection or confirmed chest computed tomography (CT) results were enrolled in this study.

All participants' LOS was calculated from their initial hospital ward or ICU admission until discharge. It is noteworthy to mention that this LOS represents the first recorded admission. Comprehensive information regarding the study design and the methods used to register variables can be found in our Khorshid COVID Cohort (KCC) study [28]. The data gathered included demographic details such as age and sex, pertinent dates including COVID-19 diagnosis and hospital or ICU admission, and the patient's most recent known clinical status.

2.2. Data Description and Pre-Processing

This study extracted and used patients' records, including non-clinical, clinical, and symptom data. Non-clinical data included sex, age, occupation, education, body mass index, family size, number of family members infected, house area, travel history, duration of symptoms before admission, and history of influenza vaccination. Clinical patient data included principal diagnosis, admission unit, medical history, and comorbidities. Laboratory data included the results of all blood tests performed at patient admission. The latest available laboratory tests included were CBC results, sodium (Na+), potassium (K+), urea, creatinine, alkaline phosphatase (ALP), aspartate transaminase (AST), alanine amino-transferase (ALT), bilirubin, international normalized ratio (INR), lactate dehydrogenase (LDH), C-reactive protein (CRP), ferritin, hemoglobin A1c (HbA1c), D-dimer, erythrocyte sedimentation rate (ESR), and vitamin D. To assess patient health status and identify the required level of care, parameters such as blood pressure, heart rate, and respiratory rate were recorded. Comorbidity categories were evaluated by the Charlson comorbidity index (CCI), which is one of the most commonly used methods to evaluate comorbid factors and predict mortality [29]. It was calculated based on age category, history of myocardial infarction (MI), congestive heart failure (CHF), peripheral vascular disease, history of a cerebrovascular accident or transient ischemic attacks, dementia, chronic obstructive pulmonary disease (COPD), connective tissue disease, peptic ulcer disease, liver disease, diabetes mellitus, hemiplegia, moderate to severe chronic kidney disease (CKD), presence of solid tumor, leukemia, lymphoma, and AIDS, ranging from 0 to 37. These medical conditions were classified by the International Classification of Diseases, Tenth Revision, Clinical Modification (ICD-10-CM) codes that are available in Appendix I Table SI-1 in [30]. The CCI was categorized into five groups: CCI score 0, CCI score 1–2, CCI score 3–4, CCI score 5–6, and CCI score ≥ 7 [31].

In addition to fever (body degree "up to 39.4 °C"), other symptoms, including fatigue, cough, sore throat, headache, nasal congestion, shortness of breath, severe chest pain, severe muscle pain, vomiting, dry cough, nausea, diarrhea, abdominal pain, muscle and joint pain, general weakness, smell-taste disorder, and dyspnea were identified by the medical interview [32]. Primary composite endpoints (PCEP) were defined as death, the use of mechanical ventilation, or admission to intensive care [33].

2.3. Statistical Data Analysis

Descriptive statistics, including means, frequencies, and proportions, are summarized for the collected data. The disease severity level stratifies summaries. Chi-squared and Fisher exact tests were used whenever appropriate to examine differences among categorical predictors. The endpoint of this study was LOS, which was calculated according to the number of days of hospitalization. The paired-sample t-test was used to identify if the LOS bias was statistically significant [34]. The Bland–Altman plot, (also known as the Tukey mean-difference plot) [35] was provided to analyze the LOS error. Patients were divided into two groups for descriptive analysis, according to the quartile LOS value: ≤ 7 days as normal and > 7 days as prolonged LOS [36]. Such a cutoff was used in terms of healthcare utilization. We considered $p < 0.05$ as statistically significant. Predictive modeling was performed offline using MATLAB version 9.6 R2019a (Natick, MA, USA: The MathWorks Inc.), while statistical analysis was performed using IBM SPSS Statistics for Windows, Version 29.0 (Armonk, NY, USA: IBM Corp).

2.4. Predictive Modeling

Volterra functional series, also known as Kolmogorov–Gabor polynomials [37], were used in our study for prediction. The level of interaction was limited to two to reduce the computational complexity and overfitting. The proposed model is provided in Equation (1).

$$y = a_0 + \sum_{i=1}^{m} a_i x_i + \sum_{i=1}^{m} \sum_{j=1}^{m} a_{ij} x_i x_j \tag{1}$$

where y is the output of the model (LOS), x_i is the ith input feature (i = 1, ..., m), m is the number of features, and the model parameters are a_0 (the offset) and a_{ij} (two-way interaction coefficients; i, j = 1, ..., m). Prior to estimation, the output variable was detrended by subtracting its average. After the model was constructed, this offset was subsequently added back. Since some input features were categorical, one-hot and ordinal encoding were used for nominal and ordinal features, respectively [38], allowing capturing the system's response for each of multiple generated binary features. Prior to estimating the coefficients, highly correlated (i.e., with an absolute correlation coefficient higher than or equal to 0.8) features and two-way interactions were identified, and some of those were selected to avoid collinearity and multicollinearity [39]. Note that multicollinearity was further reduced by dropping one of the one-hot encoded columns, also known as "dropping one level". Since Equation (1) contains all two combinations of the input coded features, we used regularized least squares (RLS) [40], with the Euclidean norm penalization, also known as the ridge regression, to estimate the coefficients in the under-determined system:

$$A_{RLS} = \left(X^T \times X + \lambda I \right)^{-1} \times X^T \times Y \tag{2}$$

where A_{RLS} are the estimated coefficients in Equation (1), Y is the target LOS vector, X is the data matrix for the selected input features of the training set, and T is the transpose operator. The regularization parameter (λ) was estimated during the cross-validation on the training set [40]. The ridge regression can help reduce the model's variance and improve its generalization to unseen data and more stable estimates, mitigate the risk of overfitting, manage model complexity, and provide some feature stability [41].

2.5. Model Validation

Five-fold cross-validation was used in our study, and the cross-validated results were provided. The goodness-of-fit of the LOS estimation algorithm was assessed using root-mean-squared-error (RMSE), mean and median absolute error, as well as the coefficient of determination (R^2) [42], adjusted R^2 (adj. R^2), and the concordance correlation coefficient (ρ_c) [43]. For the LoS_i and y_i pairs ($i = 1, \ldots, N$), such indices were calculated as follows:

$$R^2 = 1 - \frac{\sum_{i=1}^{N}(LoS_i - y_i)^2}{\sum_{i=1}^{N}(LoS_i - LoS_\mu)^2} \tag{3}$$

where,

$$\text{adj. } R^2 = 1 - \left(\frac{N-1}{N-p-1}\right) \times \left(1 - R^2\right) \tag{4}$$

where p is the number of selected input features of the model.

$$\rho_c = \frac{2 \times CoV(LoS, y)}{\sigma_{LoS}^2 + \sigma_y^2 + (\bar{y} - LoS_\mu)^2} \tag{5}$$

where CoV is the covariance, $\sigma_{LoS}^2 = \left(\frac{1}{N}\right) \times \sum_{i=1}^{N}(LoS_i - LoS_u)^2$ is the variance of the LOS, LOS_u is the mean of LOS, σ_y^2 is the variance of the predicted LOS, and $\bar{y}$ is the mean of the predicted LOS.

We further analyzed the binary outcome of the prediction system for the normal (LOS $\leq$ 7 days) and prolonged LOS (LOS > 7 days) [44]. The performance indices were calculated based on the cross-validated confusion matrix:

- TP (True Positives) = The number of accurately identified prolonged LOS
- TN (True Negatives) = The number of accurately identified normal LOS
- FP (True Positives) = The number of inaccurately identified prolonged LOS
- FN (True Positives) = The number of inaccurately identified normal LOS

The following performance indices were then calculated:

$$Se = \frac{TP}{TP + FN} \tag{6}$$

$$Sp = \frac{TN}{TN + FP} \tag{7}$$

$$PPV = \frac{TP}{TP + FP} \tag{8}$$

$$DOR = \frac{TP \times TN}{FP \times FN} \tag{9}$$

$$AUC = \frac{Se + Sp}{2} \tag{10}$$

$$F_1 = \frac{2 \times TP}{2 \times TP + FN + FP} \tag{11}$$

$$MCC = \frac{TP \times TN - FP \times FN}{\sqrt{(TP + FP) \times (TP + FN) \times (TN + FP) \times (TN + FN)}} \tag{12}$$

$$K(C) = \frac{2 \times (TP \times TN - FP \times FN)}{(TP + FP) \times (TP + FN) \times (TN + FP) \times (TN + FN)} \tag{13}$$

where Se is the sensitivity, Sp is the specificity, PPV is the positive predictive value, DOR is the diagnostic odds ratio, AUC is the balance diagnostic accuracy (area under the ROC curve), F_1 is the F_1 score, MCC is the Matthews's correlation coefficient [45,46], and K(C) is the Cohen's Kappa agreement rate.

Also, the unbiased PPV was calculated based on the sensitivity and specificity of the developed dichotomous LOS model using different prevalence (*P*) measures of the prolonged LOS in the hospital. PPV is the probability that a patient has prolonged LOS when the dichotomous LOS model results are positive. The related formula was presented in Equation (14). It was estimated using the Bayes' theorem [26]:

$$\text{unbiased PPV} = \frac{\text{Se} \times \text{P}}{\text{Se} \times \text{P} + (1 - \text{Sp}) \times (1 - \text{P})} \tag{14}$$

Following the Transparent reporting of a multivariable prediction model for individual prognosis or diagnosis (TRIPOD) guideline [47], a CI of 95% of the performance indices was reported.

2.6. Ethical Considerations

The study protocol was reviewed and approved by the Isfahan University of Medical Sciences Research Ethical Committee (IUMSREC), with the following approvals: Modeling of incidence and outcomes of COVID-19: IR.MUI.RESEARCH.REC.1399.479 and Longitudinal epidemiologic investigation of patients' characteristics with coronavirus infection referring to Isfahan Khorshid Hospital: IR.MUI.MED.REC.1399.029, conforming to the Declaration of Helsinki. Patient informed consent was obtained before admission to the current study. All data were kept confidential and had no personal identifiers. No minors participated in our study.

3. Results

Descriptive statistics were used to summarize the baseline characteristics of the study population. In our setting, 1600 COVID-19 patients were included in the study. Patients were categorized according to their LOS ($\leq$7 days (n = 1165) as "normal", >7 days (n = 435) as "prolonged") in univariate comparison analysis. The median length of stay during the study period was 7.2 (IQR 4–9) days. Tables 1 and 2 summarize the descriptive statistics and the characteristics and symptoms of the patients considered in the study according to the length of stay categories.

For an example, an 86-year-old male COVID-19 patient with fever, cough, myalgia, sore throat, dizziness, diarrhea, stomachache and weight loss symptoms, but without chest pain, headache, loss of smell, vomiting, nausea, and short breath with CCI of 4, maximum body temperature of 36°, heart rate of 84 (beats per minute), respiratory rate of 16 (breaths per minute), systolic blood pressure of 105 (mmHg), diastolic blood pressure of 66 (mmHg), %O_2 saturation minimum of 90, neutrophils of 715 ($\times 10^9$/L), lymphocytes of 264 ($\times 10^9$/L), hemoglobin of 12.10 (g/dL), platelet of 126.00 ($\times 10^9$/L), ferritin of 255.50 (ng/mL), CRP of 14.00 (mg/L), ESR of 26.00 (mm/h), LDH of 487.00 (U/L), D-dimer of 98.30 (mg/L), AST of 39.00 (IU/L), HCO3 of 31.00 (mEq/L), ALT of 16.00 (IU/L), creatinine of 0.73 (mg/dL), phosphorus of 2.56 (mg/dL), magnesium of 1.90 (mg/dL), sodium of 135.00 (mEq/L), potassium of 4.00 (mEq/L), BUN of 24.80 (mg/dL), and total bilirubin of 0.96 (mg/dL) had an LOS of 5 days in Khorshid Hospital.

Table 1. Characteristics of hospitalized patients with COVID-19 in the Khorshid Cohort Study.

Parameters	Total (n = 1600)	Length of Stay (LOS)		p-Value [b]
		≤7 Days "Normal" (n = 1165)	>7 Days "Prolonged" (n = 435)	
LOS, days [a]	6.01 (4.85)	3.76 (1.94)	12.11 (5.09)	<0.001
Age (>65 years)	562 (56.10%)	507 (43.50%)	55 (12.60%)	<0.001
Gender (% Female)	670 (48.80%)	464 (39.80%)	206 (47.30%)	0.001
Charlson Comorbidity Index (CCI) [a]	2.67 (2.13)	2.49 (2.11)	3.13 (2.13)	<0.001
Temperature maximum (≥38 degrees Celsius)	412 (25.75%)	322 (23.64%)	90 (20.68%)	0.745
Heart rate, beats per minute (<60 or >100)	478 (53.98%)	388 (33.3%)	90 (20.68%)	0.028
Respiratory rate, breaths per minute [a]	22.41 (5.67)	22.02 (5.27)	23.49 (6.56)	0.006
Systolic blood pressure (≥120 mmHg)	574 (35.80%)	247 (21.01%)	277 (63.70%)	<0.001
Diastolic blood pressure (≥90 mmHg)	218 (13.60%)	113 (9.60%)	105 (24.10%)	0.046
% O_2 saturation minimum (<90)	754 (47.10%)	606 (52.01%)	148 (34.02%)	0.001
Neutrophils (<4 × 10^9/L)	956 (59.75%)	620 (53.22%)	336 (77.20%)	0.028
Lymphocytes (<1 × 10^9/L)	900 (96.40%)	621 (53.30%)	279 (64.10%)	0.028
Hemoglobin (<12 g/dL)	356 (22.30%)	293 (20.50%)	63 (14.40%)	0.085
Platelets (<150 × 10^9/L)	678 (59.75%)	480 (41.20%)	198 (45.51%)	0.142
Ferritin (>500 ng/mL)	94 (5.80%)	72 (6.01%)	22 (5.05%)	0.298
CRP (>30 mg/L)	685 (42.80%)	542 (46.52%)	143 (32.87%)	0.017
ESR (>60 mm/h)	420 (26.30%)	245 (21.03%)	175 (40.20%)	0.027
LDH (>222 U/L)	672 (42.00%)	416 (35.70%)	256 (58.80%)	0.046
D-dimer (>0.5 mg/L)	381 (23.80%)	95 (8.20%)	286 (65.70%)	0.036
AST (>35 IU/L)	1156 (72.30%)	749 (64.30%)	407 (93.50%)	0.330
HCO3 (mEq/L)	23.65 (3.67)	17.25 (3.76)	20.45 (2.78)	0.0123
ALT (>45 IU/L)	401 (25.10%)	305 (26.18%)	96 (22.06%)	0.204
Creatinine (>1 mg/dL)	822 (51.40%)	591 (45.40%)	231 (53.10%)	<0.001
Phosphorus (mg/dL) [a]	3.06 (0.85)	2.97 (0.85)	3.24 (0.81)	<0.001
Magnesium (mg/dL) [a]	1.96 (0.51)	1.95 (0.27)	1.99 (0.74)	0.335
Sodium (mEq/L) [a]	136.30 (4.13)	136.42 (3.94)	136.09 (4.46)	0.054
Potassium (mEq/L) [a]	4.02 (0.56)	3.99 (0.54)	4.08 (0.60)	0.055
BUN (mg/dL) [a]	19.79 (13.47)	18.67 (12.37)	21.92 (15.13)	<0.001
Total bilirubin (mg/dL) [a]	1.03 (2.17)	1.06 (2.61)	0.98 (0.61)	0.361

[a] The percentage of the high-risk group (i.e., exposure) was provided in parentheses in total, "normal" or "prolonged" LOS subgroups when the high-risk cutoff was mentioned for parameters, and the standard deviation (SD) was provided (with "a" superscript) otherwise for the variables with an interval measurement scale. Such cutoffs were taken from the literature, and their citations were provided in the manuscript. For the predictor gender, the percentage of female subjects was provided in parentheses as the reference group. Statistical tests were selected based on the data's nature and the variables' distribution. [b] An independent-sample t-test was used for interval variables if the data were normally distributed; otherwise, the Mann–Whitney U test was employed. The Chi-square test was utilized for binary variables to compare proportions between the two independent groups. Note that the expected frequencies in any of the cells of the contingency table were more than five. ESR: erythrocyte sedimentation rate, LDH: lactate dehydrogenase, AST: aspartate transferase, HCO3: bicarbonate, ALT: alanine transaminase, BUN: blood urea nitrogen.

Figure 1 shows the frequency distribution of LOS, which was right-skewed. The median age of patients was 59 (IQR 47–79) years (range 5–91), and 58% were male. Comorbidities were present in more than half of the patients, with hypertension being the most common comorbidity, followed by diabetes. The Charlson comorbidity index is presented in Table 1 for patients admitted for ≤7 and >7 days in hospital. The comorbidities score was significantly higher in patients with longer LOS (p-value < 0.05, Table 1). The cutoffs used for high-risk exposures were provided by the following: age [48], body temperature [49], heart rate [50,51], blood pressure [52], oxygen saturation [53], neutrophils, lymphocytes, hemoglobin, platelets, D-dimer [54], ferritin [55], CRP [56], ESR [57], LDH [58], AST [59], ALT [60], and creatinine [61].

Table 2. Symptoms distribution between patients with normal and prolonged LOS.

Symptoms	Total	Length of Stay		*p*-Value [a]
		≤7 Days "Normal" (n = 1165)	>7 Days "Prolonged" (n = 435)	
Fever	1118 (69.9%)	721 (61.9%)	397 (91.3%)	< 0.001
Cough	1125 (70.3%)	990 (85.0%)	135 (31.0%)	< 0.001
Myalgia	838 (52.4%)	562 (48.2%)	276 (63.4%)	< 0.001
Throat pain	255 (15.9%)	168 (14.4%)	87 (20.0%)	0.058
Weight Loss	259 (16.2%)	164 (14.1%)	95 (21.8%)	0.018
Chest pain	394 (24.6%)	279 (23.9%)	115 (26.4%)	0.365
Dizziness	97 (6.1%)	64 (5.5%)	33 (7.6%)	0.540
Headache	515 (32.2%)	372 (31.9%)	143 (32.9%)	0.112
Loss of smell and taste	186 (11.6%)	134 (11.5%)	52 (12.0%)	0.260
Diarrhea	377 (23.6%)	247 (21.2%)	130 (29.9%)	0.113
Vomiting	352 (22.0%)	233 (20.0%)	119 (27.4%)	0.478
Nausea	543 (33.9%)	373 (32.0%)	170 (39.1%)	0.518
Shortness of breath	995 (62.2%)	646 (55.5%)	349 (80.2%)	0.032
Stomachache	243 (15.2%)	166 (14.2%)	77 (17.7%)	0.393

[a] The Chi-square test was utilized for binary variables to compare proportions between the two independent groups. Note that the expected frequencies in any of the cells of the contingency table were more than five.

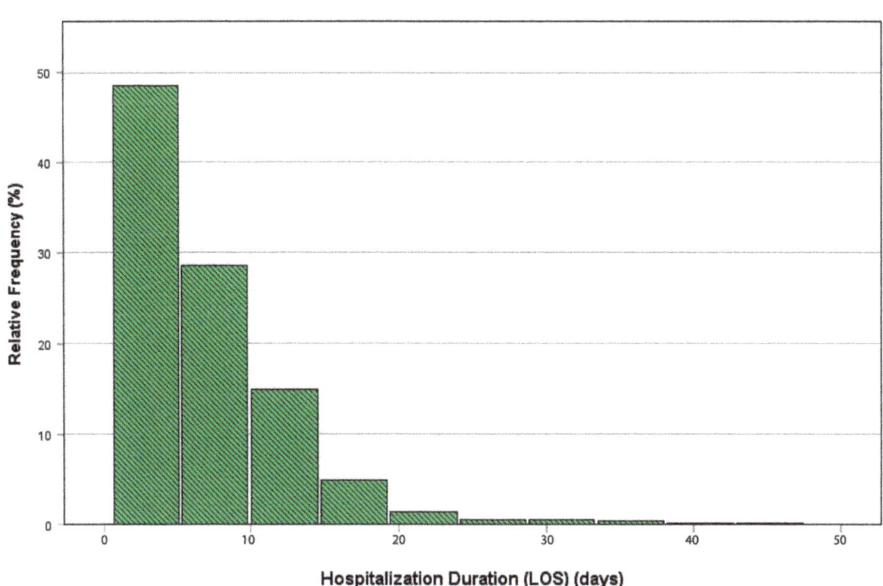

Figure 1. The length of stay (LOS) distribution.

Figure 2 shows the rate of different PCEP events among patients for both short and prolonged LOS. ICU admission is the most prevalent (55%) status among patients with prolonged LOS. There was a significant association between LOS and the PCEP binary variable (p-value < 0.001).

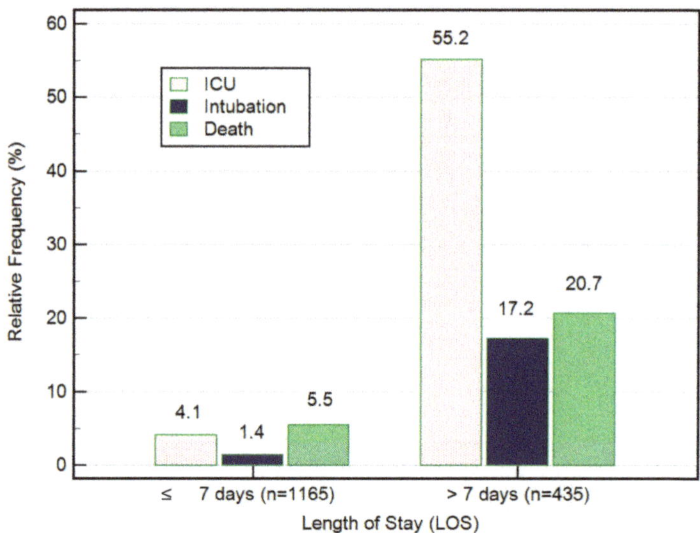

Figure 2. The distribution of different patient status PCEPs based on length of stay categories.

Fever (p-value < 0.001), cough (p-value < 0.001), myalgia (p-value < 0.001), weight loss (p-value 0.018), and shortness of breath (p-value 0.032) were significantly different in the LOS groups (Table 2). The cross-validated results of the proposed algorithm for the estimation of LOS as well as its "normal" and "prolonged" categories is provided in Table 3. The most important features of the LOS prediction system were inflammatory markers, HCO_3, and fever.

Table 3. The cross-validated results of the proposed prediction algorithm in percent.

Indices	RMSE	MAE_1	MAE_2	R^2	adj. R^2	ρ_c	Se	Sp	PPV	DOR	AUC	F_1	MCC	K(C)
Value	1.58	1.22	0.98	89	81	94	92	91	79	112	91	80	79	79
95% CI-Lower	1.51	1.16	0.92	88	79	93	89	89	75	71	89	76	77	75
95% CI-Upper	1.64	1.28	1.05	91	84	95	95	93	83	179	94	85	81	83

MAE_1: mean absolute deviation; MAE_2: median absolute deviation.

The Bland–Altman plot of residual analysis is provided in Figure 3. Although the bias was not statistically significant (p-value = 0.777; paired-sample t-test), the residual error was higher in higher target LOS than lower LOS values. The residual error was further analyzed in "normal" and "prolonged" LOS groups. The estimation error values of 3.9% and 3.2% subjects surpassed the lower or higher 1.96 limits in the "normal" and "prolonged" LOS groups, respectively.

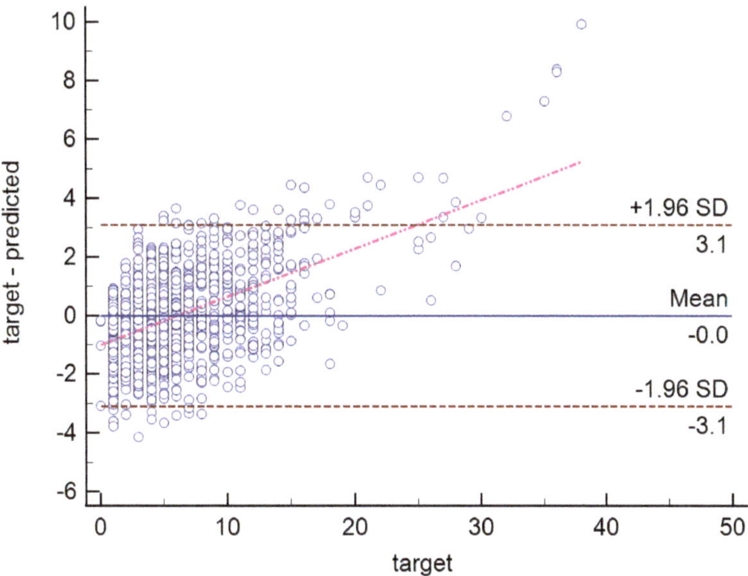

Figure 3. The Bland–Altman plot of the LOS prediction for cross-validated results. "Target" is the measured LOS, while "Predicted" is the estimated LOS. The regression line of the plot is provided in pink.

The ROC curve was then provided for the predicted LOS versus the binary ground truth (Figure 4). The best cutoff was calculated using the Youden index ($J = Se + Sp - 1$), estimated as >6.95, almost identical to our a-priori threshold.

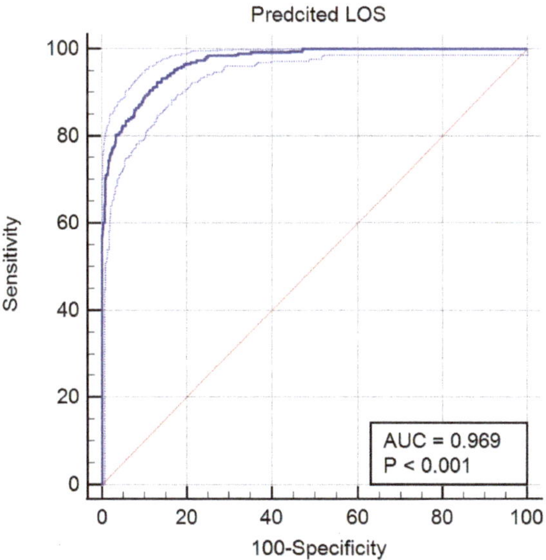

Figure 4. The receiver operating characteristic curve (ROC) (solid blue) for the predicted LOS versus the binary ground truth. The 95% confidence interval (CI) plots are shown in dotted blue. The reference line was also provided in pink.

We further predicted the importance of significant factors in the Kolmogorov–Gabor polynomials. The main predictors were only analyzed based on their normalized coefficients in the model. The seven most important factors are provided in Figure 5.

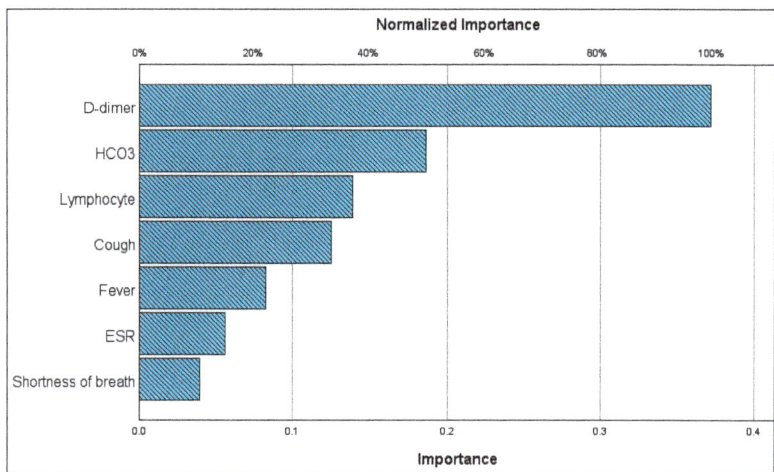

Figure 5. Seven most important main components in the Kolmogorov–Gabor polynomials model.

The unbiased PPV plot is provided in Figure 6 based on the prevalence of the prolonged LOS and Equation (14). The required parameters of Bayes' theorem were assessed from Table 3.

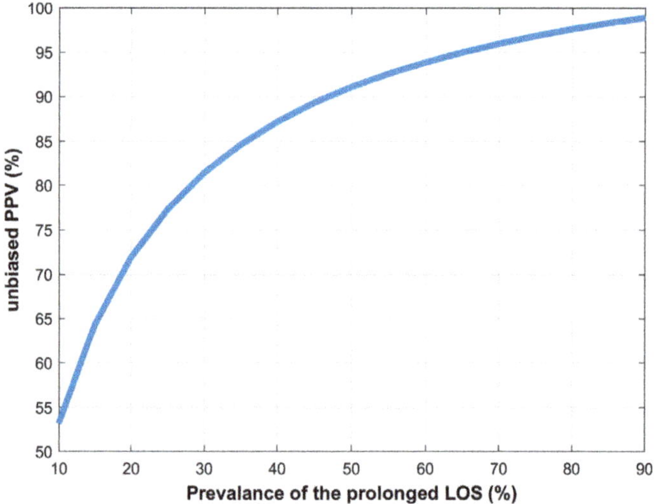

Figure 6. The unbiased PPV of the proposed binary LOS system based on the prevalence of prolonged LOS in the hospital.

4. Discussion

4.1. Implications

Medical researchers have recently been striving to enhance the quality and efficiency of healthcare systems and services. A significant aspect of this endeavor pertains to the LOS in the context of future outbreaks. Given the emergence of various variants of the virus responsible for COVID-19, accurately assessing or predicting LOS is becoming increasingly vital. An extended LOS not only impacts hospital capacity [62] but also escalates costs associated with outbreak management [63]. Hence, nations must plan for even the worst-case scenarios. This research delved into the risk factors of hospital admissions that influence the LOS among COVID-19 patients in Isfahan, Iran. We utilized a novel nonlinear artificial intelligence method for continuous data, focusing on comprehensive predictors. Our findings indicate that patients with prolonged hospital stays typically exhibited higher inflammatory markers, increased HCO_3, and more prevalent fever. These insights can guide clinicians in pinpointing specific risk factors linked to extended LOS. Moreover, our results serve as a benchmark for various models that could be applied in similar analyses, allowing healthcare professionals to narrow down critical variables for predicting LOS from the multitude recorded in hospital systems.

In our research, the median LOS was 7.2 days, with an interquartile range (IQR) of 4–9 days. It aligns closely with findings from a Chinese study [64], wherein the median hospital LOS fluctuated between 4 to 53 days over 45 domestic studies and 4 to 21 days across eight international studies. In contrast, a comprehensive report, drawing from data across 25 countries, recorded a median LOS of just 4 days and an IQR of 1–9 days [65]—substantially shorter than our recorded observations. Notably, our results surpassed the median LOS of 6 days documented in Saudi Arabia [66]. However, it fell short of the 16.4 days indicated in Indiana [67], aligning with the 8.5 days reported in the Mediterranean. These regional variances in LOS can be ascribed to many factors, including the infrastructure of healthcare facilities, the severity of treated cases, diverse admission and discharge protocols, and varying treatment approaches. Additionally, sociodemographic variables, especially age, are pivotal in influencing the observed international disparities in hospital stays.

4.2. Risk Factors

The consistency in hospital bed occupancy duration across various demographic groups in our research contrasts starkly with findings from a significant US study by Nguyen et al. [68]. Their research indicated that males typically had a more extended LOS than females. Due to the limited sample size in our cohort, we could not investigate the influence of gender on the likelihood of ICU admission. Furthermore, while our findings showed a consistent LOS regardless of demographic distinctions, European studies suggest a pronounced variability in hospital stays based on both gender and age [69]. In our data, the correlation between age and LOS was relatively weak (r = 0.134; *p*-value < 0.001).

In our research, the predominant symptoms upon admission, such as cough, fever, and shortness of breath, align with many earlier studies [70,71]. A systematic review and meta-analysis spanning 54 studies identified the most frequent symptoms in COVID-19 patients as follows: fever at 81.2% (95% CI: 77.9–84.4), cough at 58.5% (95% CI: 54.2–62.8), fatigue at 38.5% (95% CI: 30.6–45.3), dyspnea at 26.1% (95% CI: 20.4–31.8), and sputum production at 25.8% (95% CI: 21.1–30.4) [72]. Our findings concur with these percentages concerning fever and cough. However, the prevalence of shortness of breath in our study diverged.

Disturbances in total white blood cells, particularly lymphocytes, are often seen as the immune system's response to inflammation. There is growing evidence that lymphopenia, characterized by a reduced lymphocyte count, significantly influences the trajectory of COVID-19, right from its onset to the eventual development of viral sepsis. This decrease in lymphocytes has been identified as a symptom of acute COVID-19, potentially resulting from direct damage inflicted by the virus [73]. Our findings regarding lymphopenia echo those of previous studies. Earlier research has outlined prognostic models that gauge the severity of SARS-CoV-2 infection by monitoring the lymphocyte-to-leukocyte ratio [74,75].

Recent studies have illustrated that lymphocyte counts below 5% were predominantly observed in patients exhibiting severe symptoms upon follow-up. There also appears to be a trend wherein lymphopenia is more pronounced and persistent among the gravely affected patients [76,77]. These studies also highlighted that patients with extended hospital stays typically had increased circulating lymphocytes, whereas their neutrophil counts were marginally diminished. This surge in lymphocytes might be attributed to rejuvenated production, given their rise both as a percentage of total blood cells and in absolute terms. Notably, the lymphocyte count was elevated across all patient severity subgroups [78], suggesting its potential role in extended LOS or heightened mortality risk [79].

Our study underscores the substantial influence of D-dimer on the length of hospital stay, aligning with the conclusions of other meta-analyses. These analyses indicate that D-dimer correlates with factors such as comorbidities, demographics, specific laboratory tests, radiological findings, the duration of hospitalization, complications, and ultimate outcomes. Such findings propose that D-dimer is a distinct biomarker, interfacing with other inflammatory cytokine markers indicative of organ or tissue damage. Furthermore, the interaction of acute-phase proteins with D-dimer implies that infection-driven inflammation (comprising cytokines and chemokines) instigates a state of hyper-fibrinolysis, a notion reinforced by D-dimer's disconnect from the comprehensive coagulation panel [80,81].

Our study showed that patients with prolonged LOS among COVID-19 cases exhibited a significantly higher ESR. Many studies have assessed acute-phase responses to COVID-19 since the pandemic's onset, and these frequently included ESR data [82,83]. A meta-analysis [84] further highlighted that elevated ESR levels were particularly pronounced in severe and fatal cases of COVID-19. Another comprehensive meta-analysis by Zhang et al. [85], which analyzed 28 studies encompassing 4663 cases, discovered that 61.2% of cases with increased ESR had a longer length of stay and were at a heightened risk for severe disease. Notably, variations in sedimentation rates between the groups were not explored.

There is a noticeable gap in the literature regarding using HCO_3 values as predictors for LOS. It might be attributed to the understanding that abnormal HCO_3 levels already indicate extended hospitalization [86]. It is plausible that these levels act more as process variables than predictors, a sentiment echoed by our findings. The serum HCO_3 level indicates the acid–base balance within the human body and is commonly assessed in routine biochemical tests, particularly as renal diseases advance [87]. Certain clinical studies have posited a potential role for serum HCO_3 levels in forecasting mortality from ailments beyond progressive renal disease. For instance, diminished HCO_3 levels have been linked to mortality from malignancies, while elevated HCO_3 levels have been associated with cardiovascular disease complications and related mortalities [88].

Low HCO_3 serum levels upon ICU admission significantly predict both short-term and long-term mortality. Additionally, a reduced serum HCO_3 serves as an indicator of acidosis. Past research confirms that acidosis can diminish systemic vascular resistance, exacerbating conditions like circulatory shock, impaired myocardial contraction, and tissue malperfusion. This cascade of complications can ultimately precipitate end-organ failure, including acute kidney injury, which might primarily contribute to the grim prognosis observed in critically ill patients.

4.3. The Properties of Kolmogorov–Gabor Polynomials

Among the diverse techniques employed for continuous prediction, we utilized Kolmogorov–Gabor polynomials. These are more popularly recognized as the Volterra series. They serve as instrumental methodologies in identifying and modeling nonlinear systems. They can adeptly capture a broad spectrum of nonlinear behaviors by executing a series expansion based on system input. Within the context of a hospital setting, when compared with other prediction algorithms, the Volterra series boasts several advantages [89]:

Flexibility: The Volterra series can depict many nonlinear systems, endorsing its versatility across diverse modeling landscapes.

Interpretability: A standout feature of the Volterra series is its capacity to demystify the system's structure. It delineates the input–output relationship across linear, quadratic, and cubic terms. It facilitates a deeper comprehension of the system's nonlinearity and subsequent impact.

Theoretical Foundations: The mathematical underpinnings of Kolmogorov–Gabor polynomials are well-established and rigorously studied, ensuring a robust theoretical base for their application.

However, it is essential to note that while these advantages are compelling, particular challenges and potential drawbacks also emerge [37]:

Computational Complexity: The expansion order's escalation leads to an exponential growth in computational demands. It can hinder the feasibility of deploying high-order models, particularly when grappling with multitudinous inputs.

Overfitting: As with many adaptive models, there is an inherent overfitting risk when complexity overshadows the data's intricacy. Such scenarios necessitate a meticulous model selection process to safeguard against over-optimization and ensure genuine applicability to new datasets.

In our research, we have employed regularization techniques. Additionally, by capping the interaction level at two, we have strategically mitigated computational demands and curtailed the risk of overfitting.

4.4. Performance Indices

Guarding against testing hypotheses suggested by the data (Type III errors) was guaranteed by cross-validation. The LOS prediction method showed strong agreement with the measured LOS ($\rho_c = 0.94$), and strong goodness-of-fit ($R^2 = 0.8$), and did not show a significant bias (p-value = 0.777; paired-sample t-test). However, the Bland–Altman error regression showed higher errors for lower LOS values.

The binary classification algorithm, on the other hand, showed a statistical power of 92%, a Type I error of 0.09, and a precision of 79%. It also had an excellent balanced diagnosis accuracy (AUC = 0.91), a high correlation between predicted and observed class labels (MCC = 0.79), and an excellent class labeling agreement rate (K(C) = 0.79). However, it is not entirely clinically reliable, as Type I errors must be less than 0.05, and the precision must be higher than or equal to 95% [34].

4.5. Comparison with the State-of-the-Art

We searched "Embase" for journal papers with the key words "('length of stay'/exp OR 'length of stay') AND ('hospital'/exp OR 'hospital') AND ('prediction'/exp OR 'prediction') AND ('machine learning'/exp OR 'machine learning') AND ('COVID'/exp OR 'COVID'OR 'coronavirus'/exp OR 'coronavirus')" without publication date condition. Among 64 screened, 45 records were excluded after analyzing their abstracts since they did not predict LOS. Nineteen records were thus assessed for eligibility. Journal papers with at least one prediction performance index and a sound ML methodology were included in Table 4 (10 methods as the state-of-the-art, besides the proposed algorithm "this study").

Among the studies in Table 4, only Hong et al., 2020 [16] and our study followed the TRIPOD guideline [47] to report the 95% CI of the performance indices. In addition to transparency in reporting, it quantifies precision, uncertainty, reproducibility, and generalization. Only Alam et al. [23], Mahboub et al. [19], Liuzzi et al. [20], Usher et al. [18] and our study did not use missing imputations. The others used missing imputations. However, no analysis was performed to identify the reasons for missing data, i.e., missing completely at random (MCAR), missing at random (MAR), and missing not at random (MNAR), which is critical in missing data analysis [90]. In our study, we not only did not include ICU admission, mechanical ventilation, or treatments as the inputs, but we also only used baseline information at hospital admission, which was not the case for Ebinger et al. [17], Usher et al. [18], Liuzzi et al. [20], Mahboub et al. [19], and Alam et al. [23].

Table 4. The state-of-the-art to predict hospital LOS in COVID-19 patients.

Reference	Center/Region	Sample Size	Inputs	Important Features	Outputs	Models	Validation	Indices (the Best Method)	Important Characteristics
Ebinger et al., 2021 [17]	Cedars-Sinai Medical Center (Los Angeles), USA	966	353 variables	Age, respiratory rate, oxygen flow rate	LOS > 8 days vs. LOS $\leq$ 8 days	42 models	20% Hold-out	Se = 93% Sp = 63% F1 = 78% PPV = 67% AUC = 0.82	Missing imputation; cumulative day three information was used.
Hong et al., 2020 [16]	A tertiary care hospital in Zhejiang province, China	75	37 variables	Lymphocyte count, heart rate, cough, Epidermis, procalcitonin;	LOS > 14 days vs. LOS $\leq$ 14 days	Stepwise multivariable regression	No internal or external validation	AUC = 0.85 [CI 95: 0.75–0.94]	Missing imputation;
Orooji et al., 2022 [21]	Ayatollah Taleghani Hospital, Abadan, Iran	1225	53 variables	20 variables: Age, creatinine, WBC, lymphocyte/neutrophil count, BUN, ASP, ALT, LDH, activated PTT, coughing, hypertension, CVD, diabetes, dyspnea, oxygen therapy, pneumonia, GI complications, ESR, and CRP.	LOS	Statistical feature selection (correlation coefficient)+ MLP+ 12 training algorithms	10% Hold-out	RMSE = 1.6213 (days)	Patients who died within three days of admission were excluded (n = 128); selection bias. Missing data imputation.
Zhang et al., 2023 [24]	Zhengzhou University Hospital (Henan), China	384	83 variables	Immunotherapy, heparin, familial cluster, rhinorrhea (runny nose), and APTT	LOS	LASSO+ linear regression	Bootstrap validation (N = 2000)	R^2 = 0.30	Missing data imputation (10 imputations);
Alabbad et al., 2022 [22]	King Fahad University hospital, Saudi Arabia	895	43 variables	Age, C-reactive protein (CRP), nasal oxygen support days	9-class ICU LOS	Random forest (RF) (the best classifier), gradient boosting (GB), extreme gradient boosting (XGBoost), and ensemble models	3-fold cross-validation	PPV = 94% Se = 94% F_1 = 94%	Missing data imputation; SMOTE was used to balance nine classes to have 144 records each, biased performance indices. The original samples ranged from 12 to 144 for the classes; no admission date was provided.

Table 4. *Cont.*

Reference	Center/Region	Sample Size	Inputs	Important Features	Outputs	Models	Validation	Indices (the Best Method)	Important Characteristics
Nemati et al., 2020 [15]	Global dataset	1182	Five variables	Age, sex	LOS	Stagewise GB (the best method), IPCRidge, CoxPH, Coxnet, Component-wise GB, Fast SVM, Fast Kernel SVM	No internal or external validation	C-index = 0.71	No comprehensive features except symptoms onset date, symptoms, and chronic disease binary variable
Usher et al., 2021 [18]	36 hospitals (Minnesota, Wisconsin, and the Dakotas)	2665	20 variables	Various variables, including age, critical illness, oxygen requirement, weight loss, and nursing home admission	LOS at >5, >10 and >15 days	GLM, RF (the best model)	5-fold cross-validation	AUC = 0.89	ICU admission, mechanical ventilation, and mortality risk are among the input features; selection and immortal-time bias.
Liuzzi et al., 2022 [20]	28 centers (Fondazione Don Carlo Gnocchi (FDG) Living COVID-19 Registry), Italy	222	829	55 variables: anagraphical data, admission clinical scales, admission signs and symptoms, admission supports, COVID-19 therapy, therapy prior to COVID-19, hematochemics	LOS	Sequential convolutional neural network	Repeated (N = 10) 5-fold cross-validation	MAE_2 = 2.7 days (IQR = 3.0 days)	17 COVID-19 therapies were included in the input data; selection and immortal-time bias.
Mahboub et al., 2021 [19]	Rashid Hospital (Dubai), UAE	2017	22 variables	Urea, PLT, D-dimer, K+, anti-inflammatory medicine, antiviral medicine, mechanical ventilation, hemoglobin, azithromycin medicine, vitamin C medicine, painkiller medicine	LOS	Decision Tree	25% Hold-out	R^2 = 0.5	In addition to mechanical ventilation, treatments were used as input features; selection and immortal-time bias.
Alam et al., 2023 [23]	Prince Sultan Hospital (Riyadh), Saudi Arabia	308	89 variables	Laboratory, X-ray, clinical data, and treatments, including LDH and D-dimer levels, lymphocyte count, and comorbidities such as hypertension and diabetes	Seven-class LOS	Tab Transformer	30% stratified hold-out	Pr = 83%, Se = 93%, F_1 = 93% (discharged) Pr = 75%, Se = 98%, F_1 = 84% (dead)	SMOTE-N oversampling technique was used to balance the classes and biased performance indices. Treatments, including anticoagulants, antibiotics, antivirals, an immunomodulators, were used as the inputs; selection and immortal-time bias.

Table 4. *Cont.*

Reference	Center/Region	Sample Size	Inputs	Important Features	Outputs	Models	Validation	Indices (the Best Method)	Important Characteristics
Usher et al., 2021 [18]	36 hospitals (Minnesota, Wisconsin, and the Dakotas)	2665	20 variables	Various variables, including age, critical illness, oxygen requirement, weight loss, and nursing home admission	LOS at >5, >10 and >15 days	GLM, RF (the best model)	5-fold cross-validation	$AUC = 0.89$	ICU admission, mechanical ventilation, and mortality risk are among the input features; selection and immortal-time bias.
This study	Khorshid Hospital (Isfahan), Iran	1600	42	Inflammatory markers (ESR, D-dimer, lymphocyte counts), HCO_3, and fever	LOS and also $LOS \leq 7$ days vs. $LOS > 7$ days	The Kolmogorov–Gabor polynomial plus regularized least squares	Three-fold cross-validation	LOS: $R^2 = 0.89$ [0.88–0.91], $\rho_c = 0.94$ [0.93–0.95], $RMSE = 1.58$ [1.64–1.51] days $MAE_1 = 1.22$ [1.28–1.16] days, $MAE_2 = 0.98$ [0.92–1.05] days LOS categories: $Se = 92\%$ [89–95], $Sp = 91\%$ [89–93], $PPV = 79\%$ [75–83], $AUC = 0.87$ [84–89], $F_1 = 80\%$ [76–85]	No class balancing was used. ICU admission, mechanical ventilation, and treatments were not used as the input features.

MAE_1: mean absolute deviation; MAE_2: median absolute deviation.

Similarly to Alabbad et al. [22], Usher et al. [18], and Liuzzi et al. [20], we used cross-validation. Zhang et al. [24] used bootstrapped validation, though the 0.632+ bootstrap method is preferred in the literature [91]. Hold-out validation used by Ebinger et al. [17], Orooji et al. [21], Mahboub et al. [19], and Alam et al. [23] might introduce Type III error, and the repeated hold-out validation method is preferred. Also, Hong et al. [16] and Nemati et al. [15] did not use validation. Among the studies included in Table 4, only Nemati et al. [15], Usher et al. [18], and Liuzzi et al. [20] were multi-center. Orooji et al. [21] excluded subjects who died within 3 days of hospital admission, resulting in sampling bias. Our study is ranked in the top third based on the sample size. Moreover, Alabbad et al. [22] and Alam et al. [23] balanced the unbalanced training and test datasets, resulting in biased evaluation metrics, potential misleading improvement, and overfitting to the minority class. However, they had a better goodness-of-fit $R^2 = 0.80$ compared to other studies. Our study is the only one that reported the Bland–Altman plot critical to analyzing the residual error [35].

Like most studies in Table 4 [15–20], our study only focused on the first COVID-19 wave. However, Orooji et al. [21] considered the first, second, and third waves. Alam et al. [23] analyzed the first and second waves, and Zhang et al. [24] considered the Omicron variant. Thus, a direct comparison of the results of the proposed method and the other three methods [21,23,24] is not entirely rigorous.

4.6. Dichotomous LOS Definition

When the median LOS in our dataset was 7 days, then using a 7-day cutoff to dichotomize hospital LOS was statistically motivated: (1) Using the median as a cutoff point ensures that approximately half of the patients are categorized as "short stay" and the other half as "long stay"; (2) the median represents the robust central tendency of the data; patients with "prolonged" LOS are staying longer than the majority of patients, suggesting they might have different clinical characteristics, needs, or outcomes; (3) the median is robust to outliers and is not affected by very short or very long LOS values; and (4) the binary outcome can be directly tied to the dataset's inherent structure, making the results more interpretable in the context of the data. However, it might make direct comparisons between the binary LOS model and other datasets or studies more challenging unless they also use a median LOS of 7 days.

4.7. Limitations and Future Research

Our study has several limitations. Firstly, given its single-center design with 1600 COVID-19 patients at a major academic hospital following specific institutional treatment protocols, the findings might not directly apply to other hospitals throughout Asian countries. More samples are required to improve the statistical power of the proposed method. Secondly, while we tried to control for disease severity in our analysis, we could not account for more subjective factors, including the nuances of treatment that might influence endpoint decisions. To comprehensively evaluate the potential impact of treatment on LOS, a prospective randomized trial is imperative. Thirdly, the Bland–Altman analysis of residual error highlighted a non-uniform error across measured LOS. Integrating the Bland–Altman parameters into the cost function will be a focal point of our future endeavors. While our initial findings demonstrate promising results, expanding the validation scope will provide a more holistic understanding of the model's capabilities. Addressing these gaps in temporal and spatial validation will be instrumental in fostering confidence in our approach and ensuring its relevance across broader contexts, which is the focus of our future activity. Moreover, using multimodal image-processing prediction methods could, in principle, improve the reliability of the proposed algorithm [92,93], which is a focus of our future studies.

While the current model has been calibrated based on the original SARS-CoV-2 strain, the underlying framework holds potential for adaptation to newer strains. We can ensure its sustained relevance and accuracy in predicting LOS by continually updating and retraining the model with data from emerging variants. Integrating this dynamic model within the hospital health information system will facilitate real-time adaptability, making it a versatile tool for clinicians across different pandemic phases.

5. Conclusions

In this research, the utilization of machine learning models, notably the Volterra functional series, demonstrated a promising approach to predicting the length of stay (LOS) of COVID-19 patients. Validated on a significant dataset from Khorshid Hospital in Iran, the model showed strong performance metrics, including an R^2 of 0.8 and a concordance correlation coefficient of 0.94, indicating a good fit and a high agreement with the measured LOS. As noted in multiple studies, key features that played a vital role in LOS prediction were inflammatory markers, bicarbonate, and fever, aligning with the commonly observed symptoms in COVID-19 patients. The binary classification algorithm further provided insights into differentiating between "normal" and "prolonged" LOS groups. While the results present a substantial basis, there is room for improvement in the clinical reliability of the binary classification algorithm, especially concerning its Type I error and precision rate.

However, some limitations and considerations remain in the study. The Bland–Altman error regression indicated a higher error rate for patients with a lower LOS, suggesting potential areas for refinement in the model for this patient subgroup. Moreover, while our findings regarding the most prevalent symptoms upon admission were consistent with several other studies, there were notable discrepancies in the observed prevalence of shortness of breath. As healthcare providers and hospitals globally grapple with the challenges posed by the COVID-19 pandemic, findings from this research could pave the way for better resource management. Nonetheless, further temporal and spatial validation is imperative before generalized application. Future research endeavors could delve deeper into optimizing the model's clinical reliability and expanding the model's scope to other pertinent clinical outcomes. Further studies and medical regulations are essential to establish a dependable clinical prediction model suitable for smart hospitals.

Author Contributions: Conceptualization, H.M., M.Á.M., S.R.L. and M.M. (Marjan Mansourian); methodology, H.M., F.A., M.M. (Mohammadreza Mohebbian), K.T. and M.M. (Marjan Mansourian); software, M.N., F.A. and H.M.; validation, F.A., R.S., K.T. and M.M. (Marjan Mansourian); formal analysis, H.M., F.A., S.R.L. and M.M. (Marjan Mansourian); investigation, H.M., M.N., F.A. and M.M. (Marjan Mansourian); resources, R.S. and M.M. (Marjan Mansourian); data curation, F.A., M.N. and M.M. (Marjan Mansourian); writing—original draft preparation, H.M., M.N., F.A., K.T. and M.M. (Marjan Mansourian); writing—review and editing, M.M. (Mohammadreza Mohebbian), M.Á.M., S.R.L. and R.S.; visualization, H.M., F.A. and M.M. (Marjan Mansourian); supervision, M.Á.M., R.S. and M.M. (Marjan Mansourian); project administration, R.S. and M.M. (Marjan Mansourian); funding acquisition, H.M., M.N., M.Á.M., R.S. and M.M. (Marjan Mansourian). All authors have read and agreed to the published version of the manuscript.

Funding: This research was funded by the Beatriu de Pinós post-doctoral programme from the Office of the Secretary of Universities and Research from the Ministry of Business and Knowledge of the Government of Catalonia programme: 2020 BP 00261 (H.M.); National Funds through the FCT—Fundação para a Ciência e a Tecnologia, I.P., under the scope of the projects UIDB/00297/2020 and UIDP/00297/2020 (Center for Mathematics and Applications) (M.N.); the Ministry of Science and Innovation [Ministerio de Ciencia e Innovación (MICINN)], Spain, under contract PID2020-117751RB-I00 (M.A.M., S.R.L.). CIBER-BBN is an initiative of the Instituto de Salud Carlos III, Spain. The funders had no role in study design, data collection and analysis, decision to publish, or preparation of the manuscript.

Data Availability Statement: The data presented in this study are available on request from the corresponding author. The data are not publicly available due to Hospital privacy restrictions.

Acknowledgments: We sincerely thank the nurses and interns of Khorshid Hospital for their invaluable contribution to patient recruitment and follow-up data collection. Foremost, our appreciation goes to the patients who generously gave their consent to partake in this study.

Conflicts of Interest: The authors declare no conflict of interest. The funders had no role in the design of the study; in the collection, analyses, or interpretation of data; in the writing of the manuscript; or in the decision to publish the results.

References

1. Zhuang, Z.; Cao, P.; Zhao, S.; Han, L.; He, D.; Yang, L. The shortage of hospital beds for COVID-19 and non-COVID-19 patients during the lockdown of Wuhan, China. *Ann. Transl. Med.* **2021**, *9*, 200. [CrossRef] [PubMed]
2. Sen-Crowe, B.; Sutherland, M.; McKenney, M.; Elkbuli, A. A Closer Look into Global Hospital Beds Capacity and Resource Shortages During the COVID-19 Pandemic. *J. Surg. Res.* **2021**, *260*, 56–63. [CrossRef] [PubMed]
3. Jaziri, R.; Alnahdi, S. Choosing which COVID-19 patient to save? The ethical triage and rationing dilemma. *Ethics Med. Public Health* **2020**, *15*, 100570. [CrossRef] [PubMed]
4. Remuzzi, A.; Remuzzi, G. COVID-19 and Italy: What next? *Lancet* **2020**, *395*, 1225–1228. [CrossRef]
5. Deschepper, M.; Eeckloo, K.; Malfait, S.; Benoit, D.; Callens, S.; Vansteelandt, S. Prediction of hospital bed capacity during the COVID-19 pandemic. *BMC Health Serv. Res.* **2021**, *21*, 468. [CrossRef]
6. Pasquale, S.; Gregorio, G.L.; Caterina, A.; Francesco, C.; Beatrice, P.M.; Vincenzo, P.; Caterina, P.M. COVID-19 in Low- and Middle-Income Countries (LMICs): A Narrative Review from Prevention to Vaccination Strategy. *Vaccines* **2021**, *9*, 1477. [CrossRef]
7. Sasangohar, F.; Jones, S.L.; Masud, F.N.; Vahidy, F.S.; Kash, B.A. Provider Burnout and Fatigue During the COVID-19 Pandemic: Lessons Learned from a High-Volume Intensive Care Unit. *Anesth. Analg.* **2020**, *131*, 106–111. [CrossRef]
8. Sikaras, C.; Ilias, I.; Tselebis, A.; Pachi, A.; Zyga, S.; Tsironi, M.; Gil, A.P.R.; Panagiotou, A. Nursing staff fatigue and burnout during the COVID-19 pandemic in Greece. *AIMS Public Health* **2022**, *9*, 94–105. [CrossRef]
9. Sagherian, K.; Steege, L.M.; Cobb, S.J.; Cho, H. Insomnia, fatigue and psychosocial well-being during COVID-19 pandemic: A cross-sectional survey of hospital nursing staff in the United States. *J. Clin. Nurs.* **2023**, *32*, 5382–5395. [CrossRef]
10. Alsunaidi, S.J.; Almuhaideb, A.M.; Ibrahim, N.M.; Shaikh, F.S.; Alqudaihi, K.S.; Alhaidari, F.A.; Khan, I.U.; Aslam, N.; Alshahrani, M.S. Applications of Big Data Analytics to Control COVID-19 Pandemic. *Sensors* **2021**, *21*, 2282. [CrossRef]
11. Marateb, H.R.; Mohebbian, M.R.; Shirzadi, M.; Mirshamsi, A.; Zamani, S.; Abrisham chi, A.; Bafande, F.; Mañanas, M.Á. Reliability of machine learning methods for diagnosis and prognosis during the COVID-19 pandemic: A comprehensive critical review. In *High Performance Computing for Intelligent Medical Systems*; IOP Publishing: Bristol, UK, 2021; pp. 5-1–5-25. [CrossRef]
12. Steele, A.J.; Denaxas, S.C.; Shah, A.D.; Hemingway, H.; Luscombe, N.M. Machine learning models in electronic health records can outperform conventional survival models for predicting patient mortality in coronary artery disease. *PLoS ONE* **2018**, *13*, e0202344. [CrossRef] [PubMed]
13. Alimadadi, A.; Aryal, S.; Manandhar, I.; Munroe, P.B.; Joe, B.; Cheng, X. Artificial intelligence and machine learning to fight COVID-19. *Physiol. Genom.* **2020**, *52*, 200–202. [CrossRef] [PubMed]
14. Pickering, B.W.; Dong, Y.; Ahmed, A.; Giri, J.; Kilickaya, O.; Gupta, A.; Gajic, O.; Herasevich, V. The implementation of clinician designed, human-centered electronic medical record viewer in the intensive care unit: A pilot step-wedge cluster randomized trial. *Int. J. Med. Inf. Inform.* **2015**, *84*, 299–307. [CrossRef]
15. Nemati, M.; Ansary, J.; Nemati, N. Machine-Learning Approaches in COVID-19 Survival Analysis and Discharge-Time Likelihood Prediction Using Clinical Data. *Patterns* **2020**, *1*, 100074. [CrossRef] [PubMed]
16. Hong, Y.; Wu, X.; Qu, J.; Gao, Y.; Chen, H.; Zhang, Z. Clinical characteristics of Coronavirus Disease 2019 and development of a prediction model for prolonged hospital length of stay. *Ann. Transl. Med.* **2020**, *8*, 443. [CrossRef]
17. Ebinger, J.; Wells, M.; Ouyang, D.; Davis, T.; Kaufman, N.; Cheng, S.; Chugh, S. A Machine Learning Algorithm Predicts Duration of hospitalization in COVID-19 patients. *Intell. Based Med.* **2021**, *5*, 100035. [CrossRef]
18. Usher, M.G.; Tourani, R.; Simon, G.; Tignanelli, C.; Jarabek, B.; Strauss, C.E.; Waring, S.C.; Klyn, N.A.M.; Kealey, B.T.; Tambyraja, R.; et al. Overcoming gaps: Regional collaborative to optimize capacity management and predict length of stay of patients admitted with COVID-19. *JAMIA Open* **2021**, *4*, ooab055. [CrossRef]
19. Mahboub, B.; Bataineh, M.T.A.; Alshraideh, H.; Hamoudi, R.; Salameh, L.; Shamayleh, A. Prediction of COVID-19 Hospital Length of Stay and Risk of Death Using Artificial Intelligence-Based Modeling. *Front. Med.* **2021**, *8*, 592336. [CrossRef]
20. Liuzzi, P.; Campagnini, S.; Fanciullacci, C.; Arienti, C.; Patrini, M.; Carrozza, M.C.; Mannini, A. Predicting SARS-CoV-2 infection duration at hospital admission:a deep learning solution. *Med. Biol. Eng. Comput.* **2022**, *60*, 459–470. [CrossRef]
21. Orooji, A.; Shanbehzadeh, M.; Mirbagheri, E.; Kazemi-Arpanahi, H. Comparing artificial neural network training algorithms to predict length of stay in hospitalized patients with COVID-19. *BMC Infect. Dis.* **2022**, *22*, 923. [CrossRef]
22. Alabbad, D.A.; Almuhaideb, A.M.; Alsunaidi, S.J.; Alqudaihi, K.S.; Alamoudi, F.A.; Alhobaishi, M.K.; Alaqeel, N.A.; Alshahrani, M.S. Machine learning model for predicting the length of stay in the intensive care unit for COVID-19 patients in the eastern province of Saudi Arabia. *Inf. Inform. Med. Unlocked* **2022**, *30*, 100937. [CrossRef]
23. Alam, F.; Ananbeh, O.; Malik, K.M.; Odayani, A.A.; Hussain, I.B.; Kaabia, N.; Aidaroos, A.A.; Saudagar, A.K.J. Towards Predicting Length of Stay and Identification of Cohort Risk Factors Using Self-Attention-Based Transformers and Association Mining: COVID-19 as a Phenotype. *Diagnostics* **2023**, *13*, 1760. [CrossRef] [PubMed]
24. Zhang, J.; Li, L.; Hu, X.; Cui, G.; Sun, R.; Zhang, D.; Li, J.; Li, Y.; Shen, S.; He, P.; et al. Development of a model by LASSO to predict hospital length of stay (LOS) in patients with the SARS-CoV-2 omicron variant. *Virulence* **2023**, *14*, 2196177. [CrossRef] [PubMed]

25. Wolkewitz, M.; Allignol, A.; Harbarth, S.; de Angelis, G.; Schumacher, M.; Beyersmann, J. Time-dependent study entries and exposures in cohort studies can easily be sources of different and avoidable types of bias. *J. Clin. Epidemiol.* **2012**, *65*, 1171–1180. [CrossRef] [PubMed]
26. Pepe, M.S. *The Statistical Evaluation of Medical Tests for Classification and Prediction*; Oxford University Press: Oxford, UK; New York, NY, USA, 2003; 302p.
27. Steyerberg, E.W. *Clinical Prediction Models: A Practical Approach to Development, Validation, and Updating*; Springer: New York, NY, USA, 2019; p. 558.
28. Sami, R.; Soltaninejad, F.; Amra, B.; Naderi, Z.; Haghjooy Javanmard, S.; Iraj, B.; Haji Ahmadi, S.; Shayganfar, A.; Dehghan, M.; Khademi, N.; et al. A one-year hospital-based prospective COVID-19 open-cohort in the Eastern Mediterranean region: The Khorshid COVID Cohort (KCC) study. *PLoS ONE* **2020**, *15*, e0241537. [CrossRef]
29. Charlson, M.E.; Pompei, P.; Ales, K.L.; MacKenzie, C.R. A new method of classifying prognostic comorbidity in longitudinal studies: Development and validation. *J. Chronic Dis.* **1987**, *40*, 373–383. [CrossRef]
30. Glasheen, W.P.; Cordier, T.; Gumpina, R.; Haugh, G.; Davis, J.; Renda, A. Charlson Comorbidity Index: ICD-9 Update and ICD-10 Translation. *Am. Health Drug Benefits* **2019**, *12*, 188–197.
31. Comoglu, S.; Kant, A. Does the Charlson comorbidity index help predict the risk of death in COVID-19 patients? *North. Clin. Istanb.* **2022**, *9*, 117–121. [CrossRef]
32. Walker, H.; Hall, W.; Hurst, J. *Clinical Methods: The History, Physical, and Laboratory Examinations*, 3rd ed.; Butterworths: Boston, MA, USA, 1990.
33. Guan, W.J.; Ni, Z.Y.; Hu, Y.; Liang, W.H.; Ou, C.Q.; He, J.X.; Liu, L.; Shan, H.; Lei, C.L.; Hui, D.S.C.; et al. Clinical Characteristics of Coronavirus Disease 2019 in China. *N. Engl. J. Med.* **2020**, *382*, 1708–1720. [CrossRef]
34. Mansourian, M.; Marateb, H.R.; Mansourian, M.; Mohebbian, M.R.; Binder, H.; Mañanas, M.Á. Rigorous performance assessment of computer-aided medical diagnosis and prognosis systems: A biostatistical perspective on data mining. *Model. Anal. Act. Biopotential Signals Healthc.* **2020**, *2*, 17-11–17-24. [CrossRef]
35. Giavarina, D. Understanding Bland Altman analysis. *Biochem. Med.* **2015**, *25*, 141–151. [CrossRef] [PubMed]
36. Ofori-Asenso, R.; Liew, D.; Mårtensson, J.; Jones, D. The Frequency of, and Factors Associated with Prolonged Hospitalization: A Multicentre Study in Victoria, Australia. *J. Clin. Med.* **2020**, *9*, 3055. [CrossRef]
37. Madala, H.R.; Ivakhnenko, A.G.e. *Inductive Learning Algorithms for Complex Systems Modeling*; CRC Press: Boca Raton, FL, USA, 1994; p. 368.
38. Hancock, J.T.; Khoshgoftaar, T.M. Survey on categorical data for neural networks. *J. Big Data* **2020**, *7*, 28. [CrossRef]
39. Yoo, W.; Mayberry, R.; Bae, S.; Singh, K.; Peter He, Q.; Lillard, J.W., Jr. A Study of Effects of MultiCollinearity in the Multivariable Analysis. *Int. J. Appl. Sci. Technol.* **2014**, *4*, 9–19. [PubMed]
40. Beck, A. *Introduction to Nonlinear Optimization: Theory, Algorithms, and Applications with MATLAB*; Society for Industrial and Applied Mathematics, Mathematical Optimization Society: Philadelphia, PA, USA, 2014; p. 282.
41. Jain, R.K. Ridge regression and its application to medical data. *Comput. Biomed. Res.* **1985**, *18*, 363–368. [CrossRef] [PubMed]
42. Chicco, D.; Warrens, M.J.; Jurman, G. The coefficient of determination R-squared is more informative than SMAPE, MAE, MAPE, MSE and RMSE in regression analysis evaluation. *PeerJ Comput. Sci.* **2021**, *7*, e623. [CrossRef]
43. Lawrence, I.K.L. A Concordance Correlation Coefficient to Evaluate Reproducibility. *Biometrics* **1989**, *45*, 255–268. [CrossRef]
44. Rees, E.M.; Nightingale, E.S.; Jafari, Y.; Waterlow, N.R.; Clifford, S.; Pearson, C.A.B.; Group, C.W.; Jombart, T.; Procter, S.R.; Knight, G.M. COVID-19 length of hospital stay: A systematic review and data synthesis. *BMC Med.* **2020**, *18*, 270. [CrossRef]
45. Chicco, D.; Tötsch, N.; Jurman, G. The Matthews correlation coefficient (MCC) is more reliable than balanced accuracy, bookmaker informedness, and markedness in two-class confusion matrix evaluation. *BioData Min.* **2021**, *14*, 13. [CrossRef]
46. Chicco, D.; Jurman, G. The advantages of the Matthews correlation coefficient (MCC) over F1 score and accuracy in binary classification evaluation. *BMC Genom.* **2020**, *21*, 6. [CrossRef]
47. Collins, G.S.; Reitsma, J.B.; Altman, D.G.; Moons, K.G.M. Transparent reporting of a multivariable prediction model for individual prognosis or diagnosis (TRIPOD): The TRIPOD Statement. *BMC Med.* **2015**, *13*, 1. [CrossRef] [PubMed]
48. Yanez, N.D.; Weiss, N.S.; Romand, J.-A.; Treggiari, M.M. COVID-19 mortality risk for older men and women. *BMC Public Health* **2020**, *20*, 1742. [CrossRef]
49. Uchiyama, S.; Sakata, T.; Tharakan, S.; Ishikawa, K. Body temperature as a predictor of mortality in COVID-19. *Sci. Rep.* **2023**, *13*, 13354. [CrossRef] [PubMed]
50. Jin, H.; Yang, S.; Yang, F.; Zhang, L.; Weng, H.; Liu, S.; Fan, F.; Li, H.; Zheng, X.; Yang, H.; et al. Elevated resting heart rates are a risk factor for mortality among patients with coronavirus disease 2019 in Wuhan, China. *J. Transl. Int. Med.* **2021**, *9*, 285–293. [CrossRef] [PubMed]
51. Devgun, J.M.; Zhang, R.; Brent, J.; Wax, P.; Burkhart, K.; Meyn, A.; Campleman, S.; Abston, S.; Aldy, K.; Group, T.I.C.F.S. Identification of Bradycardia Following Remdesivir Administration Through the US Food and Drug Administration American College of Medical Toxicology COVID-19 Toxic Pharmacovigilance Project. *JAMA Netw. Open* **2023**, *6*, e2255815. [CrossRef]
52. Hopkins Tanne, J. US guidelines say blood pressure of 120/80 mm Hg is not "normal". *BMJ* **2003**, *326*, 1104. [CrossRef]
53. Mejía, F.; Medina, C.; Cornejo, E.; Morello, E.; Vásquez, S.; Alave, J.; Schwalb, A.; Málaga, G. Oxygen saturation as a predictor of mortality in hospitalized adult patients with COVID-19 in a public hospital in Lima, Peru. *PLoS ONE* **2020**, *15*, e0244171. [CrossRef]

54. Liu, X.; Zhang, R.; He, G. Hematological findings in coronavirus disease 2019: Indications of progression of disease. *Ann. Hematol.* **2020**, *99*, 1421–1428. [CrossRef]

55. Cheng, L.; Li, H.; Li, L.; Liu, C.; Yan, S.; Chen, H.; Li, Y. Ferritin in the coronavirus disease 2019 (COVID-19): A systematic review and meta-analysis. *J. Clin. Lab. Anal.* **2020**, *34*, e23618. [CrossRef]

56. Stringer, D.; Braude, P.; Myint, P.K.; Evans, L.; Collins, J.T.; Verduri, A.; Quinn, T.J.; Vilches-Moraga, A.; Stechman, M.J.; Pearce, L.; et al. The role of C-reactive protein as a prognostic marker in COVID-19. *Int. J. Epidemiol.* **2021**, *50*, 420–429. [CrossRef]

57. Maradit-Kremers, H.; Nicola, P.J.; Crowson, C.S.; Ballman, K.V.; Jacobsen, S.J.; Roger, V.L.; Gabriel, S.E. Raised erythrocyte sedimentation rate signals heart failure in patients with rheumatoid arthritis. *Ann. Rheum. Dis.* **2007**, *66*, 76–80. [CrossRef] [PubMed]

58. Nakakubo, S.; Unoki, Y.; Kitajima, K.; Terada, M.; Gatanaga, H.; Ohmagari, N.; Yokota, I.; Konno, S. Serum Lactate Dehydrogenase Level One Week after Admission Is the Strongest Predictor of Prognosis of COVID-19: A Large Observational Study Using the COVID-19 Registry Japan. *Viruses* **2023**, *15*, 671. [CrossRef] [PubMed]

59. Krishnasamy, N.; Rajendran, K.; Barua, P.; Ramachandran, A.; Panneerselvam, P.; Rajaram, M. Elevated Liver Enzymes along with Comorbidity Is a High Risk Factor for COVID-19 Mortality: A South Indian Study on 1512 Patients. *J. Clin. Transl. Hepatol.* **2022**, *10*, 120–127. [CrossRef] [PubMed]

60. Yin, L.K.; Tong, K.S. Elevated Alt and Ast in an Asymptomatic Person: What the primary care doctor should do? *Malays. Fam. Physician* **2009**, *4*, 98–99.

61. Hosten, A.O. BUN and Creatinine. In *Clinical Methods: The History, Physical, and Laboratory Examinations*; Walker, H.K., Hall, W.D., Hurst, J.W., Eds.; Butterworth Publishers: Boston, MA, USA, 1990.

62. Fine, M.J.; Pratt, H.M.; Obrosky, D.S.; Lave, J.R.; McIntosh, L.J.; Singer, D.E.; Coley, C.M.; Kapoor, W.N. Relation between length of hospital stay and costs of care for patients with community-acquired pneumonia. *Am. J. Med.* **2000**, *109*, 378–385. [CrossRef]

63. White, B.A.; Biddinger, P.D.; Chang, Y.; Grabowski, B.; Carignan, S.; Brown, D.F. Boarding inpatients in the emergency department increases discharged patient length of stay. *J. Emerg. Med.* **2013**, *44*, 230–235. [CrossRef]

64. Chang, R.; Elhusseiny, K.M.; Yeh, Y.-C.; Sun, W.-Z. COVID-19 ICU and mechanical ventilation patient characteristics and outcomes—A systematic review and meta-analysis. *PLoS ONE* **2021**, *16*, e0246318. [CrossRef]

65. Group, I.C.C.; Baillie, J.K.; Joaquin, B.; Abigail, B.; Lucille, B.; Fernando Augusto, B.; Tessa, B.; Aidan, B.; Gail, C.; Barbara Wanjiru, C.; et al. ISARIC COVID-19 Clinical Data Report issued: 27 March 2022. *medRxiv* **2022**. [CrossRef]

66. Alwafi, H.; Naser, A.Y.; Qanash, S.; Brinji, A.S.; Ghazawi, M.A.; Alotaibi, B.; Alghamdi, A.; Alrhmani, A.; Fatehaldin, R.; Alelyani, A.; et al. Predictors of Length of Hospital Stay, Mortality, and Outcomes Among Hospitalised COVID-19 Patients in Saudi Arabia: A Cross-Sectional Study. *J. Multidiscip. Healthc.* **2021**, *14*, 839–852. [CrossRef]

67. Garbacz, S. Average COVID-19 Hospital Stay Greater than Three Weeks. Available online: https://www.kpcnews.com/covid-19/article_8ab408ad-8fb0-5f74-8d57-11e586bd8a4f.html (accessed on 26 August 2023).

68. Nguyen, N.T.; Chinn, J.; De Ferrante, M.; Kirby, K.A.; Hohmann, S.F.; Amin, A. Male gender is a predictor of higher mortality in hospitalized adults with COVID-19. *PLoS ONE* **2021**, *16*, e0254066. [CrossRef]

69. Commission, E. Hospital Discharges and Length of Stay Statistics. Available online: https://ec.europa.eu/eurostat/statistics-explained/index.php?title=Hospital_discharges_and_length_of_stay_statistics&oldid=561104#Average_length_of_hospital_stay_for_in-patients (accessed on 26 August 2023).

70. Garibaldi, B.T.; Fiksel, J.; Muschelli, J.; Robinson, M.L.; Rouhizadeh, M.; Perin, J.; Schumock, G.; Nagy, P.; Gray, J.H.; Malapati, H.; et al. Patient Trajectories Among Persons Hospitalized for COVID-19: A Cohort Study. *Ann. Intern. Med.* **2021**, *174*, 33–41. [CrossRef]

71. Karagiannidis, C.; Mostert, C.; Hentschker, C.; Voshaar, T.; Malzahn, J.; Schillinger, G.; Klauber, J.; Janssens, U.; Marx, G.; Weber-Carstens, S.; et al. Case characteristics, resource use, and outcomes of 10 021 patients with COVID-19 admitted to 920 German hospitals: An observational study. *Lancet Respir. Med.* **2020**, *8*, 853–862. [CrossRef]

72. Alimohamadi, Y.; Sepandi, M.; Taghdir, M.; Hosamirudsari, H. Determine the most common clinical symptoms in COVID-19 patients: A systematic review and meta-analysis. *J. Prev. Med. Hyg.* **2020**, *61*, E304–E312. [CrossRef]

73. Huang, C.; Wang, Y.; Li, X.; Ren, L.; Zhao, J.; Hu, Y.; Zhang, L.; Fan, G.; Xu, J.; Gu, X.; et al. Clinical features of patients infected with 2019 novel coronavirus in Wuhan, China. *Lancet* **2020**, *395*, 497–506. [CrossRef]

74. Tan, L.; Wang, Q.; Zhang, D.; Ding, J.; Huang, Q.; Tang, Y.-Q.; Wang, Q.; Miao, H. Lymphopenia predicts disease severity of COVID-19: A descriptive and predictive study. *Signal Transduct. Target. Ther.* **2020**, *5*, 33. [CrossRef]

75. Henry, B.; Cheruiyot, I.; Vikse, J.; Mutua, V.; Kipkorir, V.; Benoit, J.; Plebani, M.; Bragazzi, N.; Lippi, G. Lymphopenia and neutrophilia at admission predicts severity and mortality in patients with COVID-19: A meta-analysis. *Acta Biomed.* **2020**, *91*, e2020008. [CrossRef]

76. Chen, R.; Sang, L.; Jiang, M.; Yang, Z.; Jia, N.; Fu, W.; Xie, J.; Guan, W.; Liang, W.; Ni, Z.; et al. Longitudinal hematologic and immunologic variations associated with the progression of COVID-19 patients in China. *J. Allergy Clin. Immunol.* **2020**, *146*, 89–100. [CrossRef]

77. Liang, J.; Nong, S.; Jiang, L.; Chi, X.; Bi, D.; Cao, J.; Mo, L.; Luo, X.; Huang, H. Correlations of disease severity and age with hematology parameter variations in patients with COVID-19 pre- and post-treatment. *J. Clin. Lab. Anal.* **2021**, *35*, e23609. [CrossRef]

78. Gelzo, M.; Cacciapuoti, S.; Pinchera, B.; De Rosa, A.; Cernera, G.; Scialò, F.; Mormile, M.; Fabbrocini, G.; Parrella, R.; Gentile, I.; et al. Prognostic Role of Neutrophil to Lymphocyte Ratio in COVID-19 Patients: Still Valid in Patients That Had Started Therapy? *Front. Public Health* **2021**, *9*, 664108. [CrossRef]
79. Rubio-Rivas, M.; Mora-Luján, J.M.; Formiga, F.; Corrales González, M.; García Andreu, M.D.M.; Moreno-Torres, V.; García García, G.M.; Alcalá Pedrajas, J.N.; Boixeda, R.; Pérez-Lluna, L.; et al. Clusters of inflammation in COVID-19: Descriptive analysis and prognosis on more than 15,000 patients from the Spanish SEMI-COVID-19 Registry. *Intern. Emerg. Med.* **2022**, *17*, 1115–1127. [CrossRef]
80. Coccheri, S. COVID-19: The crucial role of blood coagulation and fibrinolysis. *Intern. Emerg. Med.* **2020**, *15*, 1369–1373. [CrossRef]
81. Martín-Rojas, R.M.; Pérez-Rus, G.; Delgado-Pinos, V.E.; Domingo-González, A.; Regalado-Artamendi, I.; Alba-Urdiales, N.; Demelo-Rodríguez, P.; Monsalvo, S.; Rodríguez-Macías, G.; Ballesteros, M.; et al. COVID-19 coagulopathy: An in-depth analysis of the coagulation system. *Eur. J. Haematol.* **2020**, *105*, 741–750. [CrossRef]
82. Rodriguez-Morales, A.J.; Cardona-Ospina, J.A.; Gutiérrez-Ocampo, E.; Villamizar-Peña, R.; Holguin-Rivera, Y.; Escalera-Antezana, J.P.; Alvarado-Arnez, L.E.; Bonilla-Aldana, D.K.; Franco-Paredes, C.; Henao-Martinez, A.F.; et al. Clinical, laboratory and imaging features of COVID-19: A systematic review and meta-analysis. *Travel Med. Infect. Dis.* **2020**, *34*, 101623. [CrossRef]
83. Lu, R.; Qin, J.; Wu, Y.; Wang, J.; Huang, S.; Tian, L.; Zhang, T.; Wu, X.; Huang, S.; Jin, X.; et al. Epidemiological and clinical characteristics of COVID-19 patients in Nantong, China. *J. Infect. Dev. Ctries.* **2020**, *14*, 440–446. [CrossRef]
84. Henry, B.M.; de Oliveira, M.H.S.; Benoit, S.; Plebani, M.; Lippi, G. Hematologic, biochemical and immune biomarker abnormalities associated with severe illness and mortality in coronavirus disease 2019 (COVID-19): A meta-analysis. *Clin. Chem. Lab. Med.* **2020**, *58*, 1021–1028. [CrossRef]
85. Zhang, Z.L.; Hou, Y.L.; Li, D.T.; Li, F.Z. Laboratory findings of COVID-19: A systematic review and meta-analysis. *Scand. J. Clin. Lab. Investig.* **2020**, *80*, 441–447. [CrossRef]
86. Tan, L.; Xu, Q.; Li, C.; Chen, X.; Bai, H. Association Between the Admission Serum Bicarbonate and Short-Term and Long-Term Mortality in Acute Aortic Dissection Patients Admitted to the Intensive Care Unit. *Int. J. Gen. Med.* **2021**, *14*, 4183–4195. [CrossRef]
87. Erbel, R. Hypotensive Systolic Blood Pressure Predicts Severe Complications and In-Hospital Mortality in Acute Aortic Dissection. *J. Am. Coll. Cardiol.* **2018**, *71*, 1441–1443. [CrossRef]
88. Al-Kindi, S.G.; Sarode, A.; Zullo, M.; Rajagopalan, S.; Rahman, M.; Hostetter, T.; Dobre, M. Serum Bicarbonate Concentration and Cause-Specific Mortality: The National Health and Nutrition Examination Survey 1999-2010. *Mayo Clin. Proc.* **2020**, *95*, 113–123. [CrossRef]
89. *GMDH-Methodology and Implementation in MATLAB*; Imperial College Press: London, UK, 2014; p. 284.
90. Sterne, J.A.C.; White, I.R.; Carlin, J.B.; Spratt, M.; Royston, P.; Kenward, M.G.; Wood, A.M.; Carpenter, J.R. Multiple imputation for missing data in epidemiological and clinical research: Potential and pitfalls. *BMJ* **2009**, *338*, b2393. [CrossRef]
91. Efron, B.; Tibshirani, R. Improvements on Cross-Validation: The 632+ Bootstrap Method. *J. Am. Stat. Assoc.* **1997**, *92*, 548–560. [CrossRef]
92. Soda, P.; D'Amico, N.C.; Tessadori, J.; Valbusa, G.; Guarrasi, V.; Bortolotto, C.; Akbar, M.U.; Sicilia, R.; Cordelli, E.; Fazzini, D.; et al. AIforCOVID: Predicting the clinical outcomes in patients with COVID-19 applying AI to chest-X-rays. An Italian multicentre study. *Med. Image Anal.* **2021**, *74*, 102216. [CrossRef]
93. Rani, G.; Misra, A.; Dhaka, V.S.; Buddhi, D.; Sharma, R.K.; Zumpano, E.; Vocaturo, E. A multi-modal bone suppression, lung segmentation, and classification approach for accurate COVID-19 detection using chest radiographs. *Intell. Syst. Appl.* **2022**, *16*, 200148. [CrossRef]

 *information*

Article

A New Social Media Analytics Method for Identifying Factors Contributing to COVID-19 Discussion Topics

Fahim Sufi

School of Public Health and Preventive Medicine, Monash University, 553 St. Kilda Rd., Melbourne VIC 3004, Australia; fahim.sufi@monash.edu or research@fahimsufi.com

Abstract: Since the onset of the COVID-19 crisis, scholarly investigations and policy formulation have harnessed the potent capabilities of artificial intelligence (AI)-driven social media analytics. Evidence-driven policymaking has been facilitated through the proficient application of AI and natural language processing (NLP) methodologies to analyse the vast landscape of social media discussions. However, recent research works have failed to demonstrate a methodology to discern the underlying factors influencing COVID-19-related discussion topics. In this scholarly endeavour, an innovative AI- and NLP-based framework is deployed, incorporating translation, sentiment analysis, topic analysis, logistic regression, and clustering techniques to meticulously identify and elucidate the factors that are relevant to any discussion topics within the social media corpus. This pioneering methodology is rigorously tested and evaluated using a dataset comprising 152,070 COVID-19-related tweets, collected between 15th July 2021 and 20th April 2023, encompassing discourse in 58 distinct languages. The AI-driven regression analysis revealed 37 distinct observations, with 20 of them demonstrating a higher level of significance. In parallel, clustering analysis identified 15 observations, including nine of substantial relevance. These 52 AI-facilitated observations collectively unveil and delineate the factors that are intricately linked to five core discussion topics that are prevalent in the realm of COVID-19 discourse on Twitter. To the best of our knowledge, this research constitutes the inaugural effort in autonomously identifying factors associated with COVID-19 discussion topics, marking a pioneering application of AI algorithms in this domain. The implementation of this method holds the potential to significantly enhance the practice of evidence-based policymaking pertaining to matters concerning COVID-19.

Keywords: COVID-19 analytics; analysing COVID-19 discourse; social media analytics; regression; topic analysis

Citation: Sufi, F. A New Social Media Analytics Method for Identifying Factors Contributing to COVID-19 Discussion Topics. *Information* **2023**, *14*, 545. https://doi.org/10.3390/info14100545

Academic Editors: Arkaitz Zubiaga, Amar Ramdane-Cherif, Ravi Tomar and T.P. Singh

Received: 9 September 2023
Revised: 27 September 2023
Accepted: 4 October 2023
Published: 5 October 2023

1. Introduction

Social media analytics has been used in various ways to understand the impact of COVID-19. According to a study conducted by the World Health Organization (WHO), social media and other digital platforms created opportunities to keep people safe, informed, and connected during the pandemic [1].

By monitoring social media, analysts can also gauge public sentiment, track the spread of propaganda, and identify emerging narratives, thus offering insights into information operations and counter-messaging strategies. Furthermore, social media monitoring tools and algorithms (such as sentiment analysis, entity recognition, word frequency calculation, and topic analysis, as depicted in [2–4]) empower analysts to detect and analyse cyber threats in real-time, enabling proactive defence measures and the attribution of cyberattacks by identifying patterns, tracking malware propagation, and uncovering digital footprints left by threat actors. However, social media has also contributed to the spread of misinformation about COVID-19 [5–8]. In Ref. [5], a national survey by university researchers found that social media users are more likely to believe false claims about COVID-19, such as conspiracies, risk factors, and treatments. The survey also found that age, race, political party, and news source are some of the factors that influence the level of belief in COVID-19 misinformation. To eliminate these misconceptions and also to make strategic policy decisions on controlling COVID-19 crises, researchers and policymakers have been using Twitter analytics with artificial intelligence (AI) and natural language processing (NLP) (as shown in [9–25]). However, none of these tweet-mining technics in the area of COVID-19 (i.e., [9–25]) or other topics (e.g., [26–33]), have demonstrated a methodology to identify the factors correlated to Twitter discourse topics.

In this paper, an innovative methodology is proposed that uses AI-based services (Microsoft Cognitive Services [34]-based language detection, translation, and sentiment analysis) and algorithms (topic analysis, regression, and clustering) to autonomously identify the factors influencing COVID-19-related discussion topics, as shown in Figure 1. Moreover, the presented methodology was evaluated with 152,070 multilingual tweets, collected between 15th July 2021 and 20th April 2023. In summary, the following are the core contributions of this paper:

- An inventive framework, rooted in AI and NLP, is systematically employed. This framework integrates a spectrum of methodologies, including translation, sentiment analysis, topic analysis, regression, and clustering techniques, with the purpose of methodically discerning and expounding upon the factors that are pertinent to the diverse discourse topics encompassing COVID-19.
- This innovative approach underwent a rigorous examination and assessment, utilizing a dataset encompassing 152,070 tweets that were gathered within the temporal span from 15 July 2021 to 20 April 2023. Notably, this dataset encapsulates discourse in a wide array of 58 distinct languages.
- AI- and NLP-based regression identified and described 37 observations, of which 20 were found to be significant. Moreover, clustering techniques identified 15 observations, containing nine of significance.
- These 52 observations, generated through AI-driven methods, elucidated the relationships existing between topic confidences, encompassing Topic 1 confidence, Topic 2 confidence, Topic 3 confidence, Topic 4 confidence, and Topic 5 confidence, and an extensive array of factors. These factors included variables such as tweet time, followers, friends, retweets, language name, sentiment, positive sentiment confidence, neutral sentiment confidence, negative sentiment confidence, and predicted Topic.
- This methodology could be applied to identify factors related to any discussion topics within any micro-blogging social media platforms.

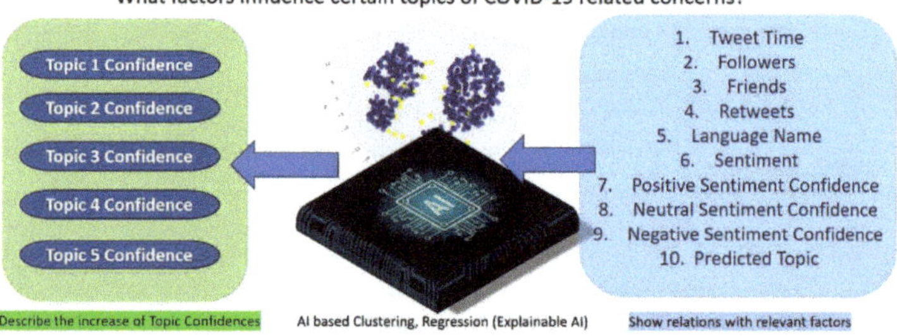

Figure 1. Conceptual diagram of the proposed system (factors 5 to 10 are NLP-based).

Within the rest of this paper, a background and literature review are provided (in Section 3), followed by the details of the proposed methodology (also in Section 3). Section 4 describes how the proposed methodology was evaluated with COVID-19-related tweets. Finally, Section 5 provides concluding remarks, limitations of this study, and future endeavours.

2. Background Context and Literature

In the realm of contemporary data analysis, the integration of multilingual, global sentiment analysis and topic analysis holds paramount significance when scrutinizing COVID-19-related tweets. This methodological approach encompasses a comprehensive investigation into the multifaceted linguistic expressions of a diverse global population during the pandemic. Multilingual sentiment analysis not only elucidates the emotional undercurrents within the discourse but also allows for the nuanced interpretation of sentiments across linguistic boundaries. Simultaneously, the employment of topic analysis facilitates the identification and categorization of emergent themes and topics within the vast corpus of COVID-19 tweets, ensuring a systematic exploration of the evolving narrative.

2.1. Global Perspective

A global perspective in COVID-19 tweet analysis is pivotal for recognizing international trends and disparities [9–21]. It enables us to identify common global concerns and regional variations, aiding policymakers in tailoring responses to specific contexts and populations.

2.2. Multilingual Analysis

The COVID-19 pandemic transcended linguistic barriers, impacting diverse populations worldwide. Multilingual analysis allows us to decipher sentiments and opinions expressed in various languages, providing a comprehensive view of global perceptions and concerns [13,14,21]. This inclusivity fosters a more accurate understanding of the pandemic's impact on different communities.

2.3. Sentiment Analysis

Sentiment analysis delves into the emotional undercurrents of COVID-19 tweets, shedding light on public sentiment towards the pandemic, government responses, and vaccination efforts. This knowledge is invaluable for gauging public support and addressing concerns, ultimately contributing to more effective public health communication [9–21,35–44].

2.4. Topic Analysis

COVID-19 tweet analysis through topic analysis identifies emerging themes and discussions within the vast tweet corpus [10,11,16,22]. This aids in tracking the evolution of public discourse, from early outbreak concerns to vaccine distribution and beyond. Understanding topics informs public health strategies and crisis communication [4]. Table 1 summarizes the existing research works on COVID-19 Twitter analytics that applied sentiment analysis and topic analysis on multilingual and global tweets.

Table 1. Literature review on COVID-19-based Twitter analytics.

Reference	Multilingual	Global	Sentiment Analysis	Topic Analysis	Identifying Factors of Topic
[9]	No	No	Yes	No	No
[23]	No	No	Yes	Yes	No
[10]	No	No	Yes	Yes	No
[11]	No	No	Yes	Yes	No
[22]	No	Yes	No	Yes	No
[12]	No	No	Yes	No	No
[13]	Yes	Yes	Yes	No	No
[24]	No	No	No	No	No
[25]	No	No	No	No	No
[14]	Yes	Yes	Yes	No	No
[15]	No	Yes	Yes	No	No
[16]	No	Yes	Yes	Yes	No
[17]	No	Yes	Yes	No	No
[18]	No	No	Yes	No	No
[19]	No	Yes	Yes	No	No
[20]	No	No	Yes	Yes	No
[21]	Yes	Yes	Yes	No	No
This Study	Yes	Yes	Yes	Yes	Yes

In summary, a comprehensive approach that integrates multilingual capabilities, global context, sentiment analysis, and topic analysis in COVID-19 tweet analysis is indispensable for capturing the nuanced dynamics of the pandemic's impact, sentiments, and evolving discourse on a global scale. This research-driven approach empowers decision-makers to make informed, data-driven choices in managing and mitigating the pandemic's effects. As seen in Table 1, none of the existing research work investigated the factors influencing COVID-19 discussion topics. This study reports the first academic work on identifying the factors behind COVID-19 discussion topics on Twitter by concurrently using sentiment analysis and topic analysis on multilingual and global tweets.

3. Materials and Methods

The proposed framework revolves around AI-driven processes of tweet acquisition, language detection, translation, sentiment analysis, topic analysis, and correlation analysis. The correlation analysis uses both regression and clustering techniques, and is demonstrated in Figure 2. Each of these steps are described within this section in detail.

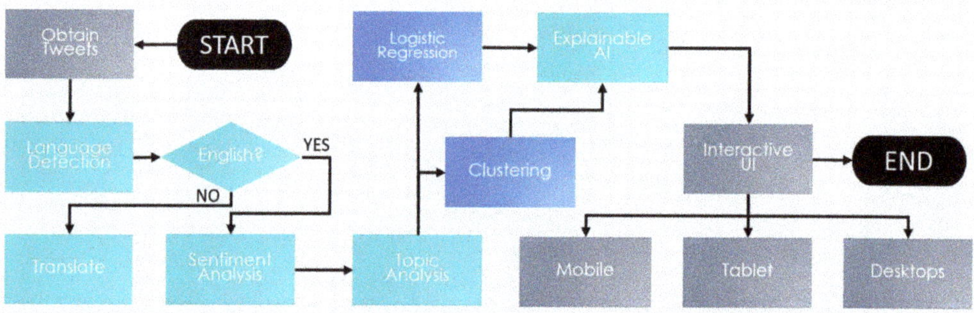

Figure 2. The process of analysing the correlated factors of COVID-19-related Twitter Topics.

3.1. Tweet Acquisition

At the inception of this analytical endeavour, we embark upon the acquisition of a corpus of multilingual tweets that are germane to the COVID-19 discourse. This foundational process entails the meticulous extraction of tweets that incorporate the keywords "COVID" or "CORONA". Notably, this endeavour is not confined to the mere capture of textual content but extends to the comprehensive cataloguing of contextual parameters that encapsulate the temporal, audience-related, and propagation-related dimensions of each tweet. These dimensions include the tweet text, tweet time, followers, friends, and retweets, among other pertinent attributes. This step orchestrates the crystallization of a heterogeneous dataset, the quintessence of the analytical journey that ensues.

3.2. Language Detection

Subsequently, a critical layer of linguistic scrutiny is introduced through the mechanism of language detection. The profusion of languages within the Twitterverse necessitates an astute differentiation, rendering this phase indispensable. Herein, we leverage cutting-edge APIs, notably those furnished by Microsoft Cognitive Services, to determine the linguistic origin of each tweet. This critical linguistic assignment is chronicled as the "Language Name". The veritable goal of this phase is the creation of a harmonious alignment of tweets with their respective linguistic affiliations, a foundational step for subsequent linguistic and sentiment analyses.

3.3. Translation (for Non-English Tweets)

In recognition of the global diversity that is inherent in Twitter discourse, where linguistic heterogeneity is the norm, an equilibrating mechanism is invoked for tweets that diverge from the English linguistic ambit. This mechanism, embodied in the translation process, endeavours to homogenize all tweets into the English language. Accordingly, those tweets that are identified as non-English in the preceding step undergo a transformational metamorphosis into English. This translation operation, facilitated by APIs such as those provided by Microsoft Cognitive Services, presents a unifying linguistic canvas, thereby fostering linguistic consistency for subsequent analytical endeavours.

3.4. Sentiment Analysis

The nuance of sentiment within the tweets, an elemental facet of the analysis, is meticulously unveiled through the prism of sentiment analysis. Each tweet within the standardized English dataset becomes a subject of scrutiny, wherein its emotional tenor in relation to the COVID-19 topic is artfully gauged. This nuanced analysis typically culminates in categorizations of tweets into one of three classes: positive, negative, or neutral. Notably, this classification is accompanied by quantified confidence scores, encapsulating the robustness of the categorization. The orchestration of this phase involves the utilization of sentiment analysis APIs, which, in the context herein, emanate from the domain of Microsoft Cognitive Services. Hence, the analytical outcome bestows upon each tweet a set of salient parameters: "Sentiment", "Positive Sentiment Confidence", "Neutral Sentiment Confidence", and "Negative Sentiment Confidence".

3.5. Topic Analysis (LDA-Based)

A pivotal stage in our analytical odyssey materializes with the advent of Latent Dirichlet Allocation (LDA)-based topic analysis. This modelling paradigm, founded upon probabilistic principles, aspires to uncover latent topics that are interwoven within the corpus of tweets. Each tweet assumes the role of a document, serving as a carrier of topic-related information. By engaging in the allocation of tweets to one or more topics, LDA bestows upon them topic affiliations, accompanied by associated confidence scores. This compositional orchestration of themes in the COVID-19 discourse begets a diverse set of parameters, most notably the "Predicted Topic" and the "Topic Confidence" scores for each tweet. This discourse-level dissection engenders insights into the salient themes permeating the Twitterverse in the context of COVID-19.

3.6. Correlation Analysis

At this juncture, the focus pivots toward the elucidation of associations, elucidating the intricate interplay between various parameters and COVID-19 discussion topics. Central to this endeavour is the endeavour to unearth correlations between the confidence levels assigned to each of the identified topics (e.g., Topic 1 confidence, Topic 2 confidence, and so forth) and a multifarious array of attributes. The palette of attributes encompasses diverse dimensions including temporal characteristics (e.g., tweet time), social dynamics (e.g., followers, friends, retweets), linguistic attributes (e.g., language name), sentiment attributes (e.g., sentiment, positive sentiment confidence, neutral sentiment confidence, negative sentiment confidence), and the very topics birthed from LDA-based topic analysis. This multifaceted inquiry invokes the services of AI-driven regression and clustering methods, eloquently weaving a tapestry of nuanced relationships, and revealing the underpinnings of the COVID-19 discourse.

Regression analysis automatically prioritizes and assesses the importance of factors for both categorical and numeric metrics. For numerical features, Microsoft's ML.NET SDCA regression [45] was employed, using linear regression, a fundamental supervised learning technique for solving regression problems. Linear regression predicts a continuous dependent variable based on independent variables, aiming to determine the best-fit line that accurately forecasts the continuous output, thereby establishing a linear relationship, represented by Equation (1).

$$y = b_0 + b_1 x_1 + \varepsilon \tag{1}$$

For categorical features, logistic regression was executed using L-BFGS logistic regression from ML.NET [46,47]. Logistic regression, a widely used supervised learning algorithm, serves purposes in both classification and regression problems. It predicts categorical dependent variables based on independent variables, employing Equation (2). Logistic regression outputs values between zero and one, making it suitable for tasks where probability estimates between two classes are needed, such as binary decisions like rainy or not rainy, 0 or 1, true or false, and so on.

$$\mathrm{Log}[y/y - 1] = b_0 + b_1 x_1 + b_2 x_2 + \dots b_n x_n \tag{2}$$

Initially, logistic regression operates as a regression model. However, when a threshold is introduced, it transforms into an effective classifier. The process begins with the utilization of the logistic or sigmoid function (the process described with Equations (3)–(9)).

$$\sigma(t) = \frac{1}{1 + e^{-t}} \tag{3}$$

The sigmoid function of Equation (3) maps real numbers to interval (0, 1). Then, a hypothesis function is defined with Equation (4).

$$h_\theta(x) = \sigma\left(\theta^T x\right) = \frac{1}{1 + e^{-\theta^T x}} \tag{4}$$

The classification decision is made on $y = 1$, when $h_\theta(x) \geq 0.5$ and $y = 0$ otherwise. The decision boundary is $\theta^T x = 0$. The cost function is shown with Equation (5).

$$j(\theta) = \sum_{i=1}^{m} H\left(y^{(i)}, h_\theta\left(x^{(i)}\right)\right) \tag{5}$$

where $H(p,q)$ is the cross-entropy of distribution q relative to distribution p and is shown with Equation (6).

$$H(p,q) = -\sum_i p_i \log q_i \tag{6}$$

In this case, $y^{(i)} \in \{0,1\}$ so $p_1 = 1$ and $p_2 = 0$. Therefore,

$$H\left(y^{(i)}, h_\theta\left(x^{(i)}\right)\right) = -y^{(i)} \log h_\theta\left(x^{(i)}\right) - \left(1 - y^{(i)}\right) \log\left(1 - h_\theta\left(x^{(i)}\right)\right) \tag{7}$$

Similar to the selection of the quadratic cost function in linear regression, the selection of this cost function is mainly driven by the fact that it is efficient, as shown in Equation (8).

$$\mathrm{grad}\, J(\theta) = \frac{\partial\, J(\theta)}{\partial\, \theta} = \begin{bmatrix} \frac{\partial}{\partial\, \theta_0} J(\theta) \\ \frac{\partial}{\partial\, \theta_1} J(\theta) \\ \cdot \\ \cdot \\ \cdot \\ \frac{\partial}{\partial\, \theta_n} J(\theta) \end{bmatrix} = X^T \left(h_\theta(X) - y\right) \tag{8}$$

Hence, the gradient descent for logistic regression could be reflected with Equation (9).

$$\theta(k + 1) = \theta(k) - s\mathrm{grad} J(\theta) \tag{9}$$

Both linear and logistic regressions were automatically applied utilizing NLP [48].

3.7. Explanatory Analysis (NLP-Based)

The analytical sojourn reaches its culmination with a synthesis that bridges the chasm between numerical correlations and human understanding. Enter the realm of natural language processing (NLP)-based explainable AI, an ingenious avenue wherein the multifarious correlations unearthed in the prior step are rendered intelligible through human-readable narratives. By employing sophisticated NLP algorithms, this phase aspires to provide lucid elucidations that elucidate not only the "what" but also the "why" behind the identified correlations. The resulting explanations serve as the lighthouse that guides scholars and practitioners through the labyrinth of interconnected parameters, thereby fostering an enriched comprehension of the COVID-19 discussion dynamics on Twitter.

In summary, the processes of tweet acquisition, language detection, translation, sentiment analysis, and topic analysis created various attributes or factors, as shown in Table 2. These attributes are used in the correlation process (i.e., clustering, logistic regression, and explainable AI) for identifying the factors that influence COVID-19-related discussion topics (as shown in Table 2). Figure 3 demonstrates how these attributes are created as well as how these attributes are used. Algorithm 1 demonstrates our implementation of this methodology. Various notations used within Algorithm 1 are portrayed in Table 3.

Table 2. Lifecycle of attributes/factors (processes that create or use the attributes).

Attribute Created by	Data Object/Attribute Name	Attribute Used by
Obtain Tweets	Multi-Lingual Tweets	Sentiment Analysis
Obtain Tweets	Tweet Time	Clustering, Logistic Regression, Explainable AI
Obtain Tweets	Followers	Clustering, Logistic Regression, Explainable AI
Obtain Tweets	Retweets	Clustering, Logistic Regression, Explainable AI
Translate	English Translated Tweets	Sentiment Analysis
Language Detection	Language Name	Clustering, Logistic Regression, Explainable AI
Sentiment Analysis	Sentiment	Clustering, Logistic Regression, Explainable AI
Sentiment Analysis	Positive Sentiment Confidence	Clustering, Logistic Regression, Explainable AI
Sentiment Analysis	Neutral Sentiment Confidence	Clustering, Logistic Regression, Explainable AI
Sentiment Analysis	Negative Sentiment Confidence	Clustering, Logistic Regression, Explainable AI
Topic Analysis	Predicted Topic	Clustering, Logistic Regression, Explainable AI
Topic Analysis	Topic 1 Confidence	Clustering, Logistic Regression, Explainable AI
Topic Analysis	Topic 2 Confidence	Clustering, Logistic Regression, Explainable AI
Topic Analysis	Topic 3 Confidence	Clustering, Logistic Regression, Explainable AI
Topic Analysis	Topic 4 Confidence	Clustering, Logistic Regression, Explainable AI
Topic Analysis	Topic 5 Confidence	Clustering, Logistic Regression, Explainable AI
Explainable AI	Explanations	Interactive UI

Table 3. Description of notations.

Notation	Description
T	Extracted tweets as the output of *ExtractTweetsContainingKeywords("COVID", "CORONA")*
m	Date and time of tweet as the output of *ExtractTweetsContainingKeywords("COVID", "CORONA")*
f	Follower count as the output of *ExtractTweetsContainingKeywords("COVID", "CORONA")*
d	Friend count as the output of *ExtractTweetsContainingKeywords("COVID", "CORONA")*
r	Retweet count as the output of *ExtractTweetsContainingKeywords("COVID", "CORONA")*
l	Tweet language as detected using *DetectLanguage(Tweet)*
s	Detected sentiment as the output of *SentimentAnalysis(tweet)*
p	Positive sentiment confidence as the output of *SentimentAnalysis(tweet)*
n	Negative sentiment confidence as the output of *SentimentAnalysis(tweet)*
u	Neutral sentiment confidence as the output of *SentimentAnalysis(tweet)*
Topic	Topic ID as the output of *PerformLDATopicAnalysis(T_EN)*
c_1	Topic 1 confidence as the output of *PerformLDATopicAnalysis(T_EN)*
c_2	Topic 2 confidence as the output of *PerformLDATopicAnalysis(T_EN)*
c_3	Topic 3 confidence as the output of *PerformLDATopicAnalysis(T_EN)*
c_4	Topic 4 confidence as the output of *PerformLDATopicAnalysis(T_EN)*
c_5	Topic 5 confidence as the output of *PerformLDATopicAnalysis(T_EN)*

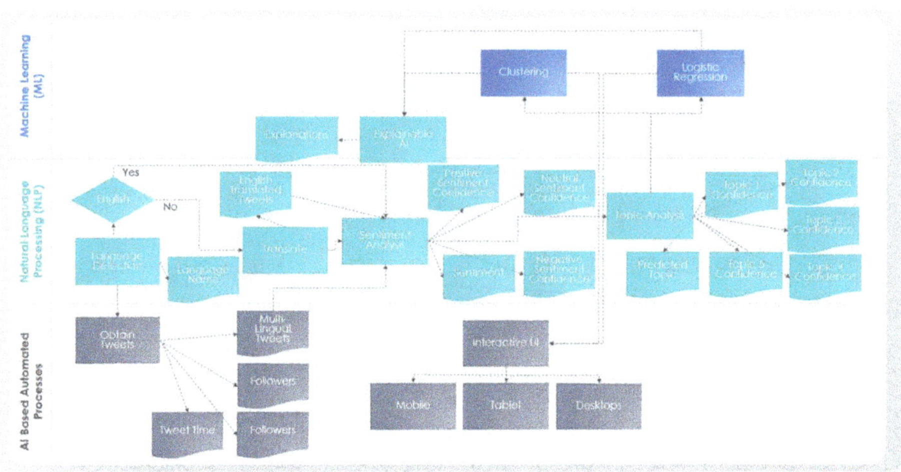

Figure 3. Detailed process map of the proposed system.

Algorithm 1: Analysing the correlated factors of COVID-19-related Twitter topics.

1. # Step 1: Tweet acquisition
   $T, m, f, d, r = ExtractTweetsContainingKeywords("COVID", "CORONA")$
2. # Step 2: Language detection
   for tweet in T:
3. $l = DetectLanguage(tweet)$
4. # Step 3: Translation (for non-English tweets)
   $T\_EN = []$
5. *for* tweet in T:
6.   *if* l is not "English":
7. $t\_EN = TranslateToEnglish(tweet)$
8. $T\_EN.append(t\_EN)$
9.   *else*:
10. $T\_EN.append(tweet)$
11. # Step 4: Sentiment analysis
    for tweet in T_EN:
12. $s, p, n, u = SentimentAnalysis(tweet)$
13. # Step 5: Topic analysis (LDA-based)
    $Topics, c_1, c_2, c_3, c_4, c_5 = PerformLDATopicAnalysis(T\_EN)$
14. # Step 6: Correlation analysis
    $Correlations = CorrelationAnalysis(\{c_1, c_2, c_3, c_4, c_5\} \rightarrow \{l, f, d, r, s, p, n, u\})$
15. # Step 7: Explanatory analysis (NLP-based)
    $Explanations = ExplainCorrelations(Correlations)$
16. # Display results or save to file
    $DisplayResults(Correlations, Explanations)$

In summation, this academic endeavour embodies a holistic and rigorous analytical framework for the in-depth examination of COVID-19 discourse within the Twitter ecosystem. This process, characterized by its methodical granularity, encompasses diverse facets of data acquisition, linguistic analysis, sentiment assessment, thematic exploration, correlation identification, and linguistic elucidation, thereby affording a comprehensive view of the intricate discourse surrounding the pandemic within the digital public sphere. Its integration of advanced AI and NLP techniques amplifies the depth and interpretability of the insights garnered, rendering it a valuable resource for scholars in the realms of data science, linguistics, and social sciences.

4. Results and Discussion

The methodology was tested and critically evaluated with 152,070 tweets from 15 July 2021 to 20 April 2023. During these 645 days, tweets in 58 distinct languages were analysed with AI-based language detection, translation, sentiment analysis, and LDA-based topic analysis. LDA-based topic analysis identified five topics on COVID-19-related discussion. Finally, AI- and NLP-based clustering and regression algorithms were used to identify and describe the correlations between the topic confidences against each of the related variables.

Table 4 provides the details of the five topics. These topics were (1) broad discussion on corona, (2) COVID statistics and vaccination, (3) wordplay on corona, (4) COVID experiences or updates, and finally, (5) likely context of COVID in India. As seen in Table 4, each of these discerned topics demonstrated distinct patterns of word occurrences and weights. For example, within Topic 3, the word "crown" and its variations appear prominently, along with "Corona". "Corona" in Latin means "crown", and the name of the virus is derived from this due to its appearance under the microscope. Moreover, the COVID virus appears as football (soccer) and hence the word "Corona_Futbol" appears with a weight of 582.

Table 4. Word weights across each of the five topics.

Topic 1: Broad Discussion on Corona		Topic 2: COVID Statistics and Vaccination		Topic 3: Wordplay on 'Corona'		Topic 4: COVID Experiences/Updates		Topic 5: Likely Context of COVID in India	
Word	Weight	Word	Weight	Word	Weight	Word	Weight	Word	Weight
Corona	19287	COVID	18257	crown	4871	COVID	9946	Corona	2560
corona	13595	COVID	15042	Corona	3743	COVID	6148	corona	2504
people	5770	vaccine	5295	Crown	1242	COVID	4899	COVID	932
vaccination	3255	COVID	4110	https://t.co	1161	get	3212	CORONA	710
also	3173	cases	3552	Corona_Futbol	582	people	3048	https://t.co	609
measures	2845	people	3413	first	517	corona	2811	India	589
would	2428	deaths	3404	crowned	495	days	2779	hai	533
like	2406	new	3379	City	490	like	2471	amp	446
one	2256	vaccines	2953	today	456	got	2250	exam	319
many	2241	https://t.co	2129	going	444	died	2134	narendramodi	290

As seen in Table 5, about 60,855 tweets were in English, 30,212 tweets were in German, followed by 22,226 tweets in Spanish, 7419 in Dutch, and 5748 in French. Most interestingly, as shown in Table 5, the language distribution against each of the topics has distinctive patterns, suggesting possible correlations between topics and languages. We can see in Table 5 that Topic 2, Topic 4, and Topic 5 contain mostly English tweets. However, Topic 1 and Topic 3 demonstrate a dominance of German and Spanish tweets, respectively. Figure 4 shows the word cloud for each of the five topics. Figure 4a is mostly in the German language. Figure 4c is mostly in Spanish. Figure 4b, Figure 4d, Figure 4e, and Figure 4f are predominantly in English. It should be mentioned that default stop-words like "am", "is", and "at" have been removed from Figure 4. Moreover, common terminologies like "COVID", "https", "rt", and "corona" have also been discarded from the word clouds shown in Figure 4. Finally, Table 6 depicts the distribution of sentiment confidences (i.e., the results of the sentiment analysis process), follower count, friend count, retweet count, and the number of distinct tweet languages against each of the topics.

Table 5. Most used Tweet languages for each of the topics.

Top 5 Ranks	All		Topic 1		Topic 2		Topic 3		Topic 4		Topic 5	
	Language	Tweets	Language	Tweets	Language	Tweets	Language	Tweets	Language	Tweets	Language	Tweets
1	English	60,855	German	25,477	English	27,050	Spanish	10,811	English	18,102	English	4717
2	German	30,212	English	8129	Spanish	3697	English	2857	Spanish	3863	Hindi	1212
3	Spanish	22,226	Dutch	5827	French	2713	Japanese	810	Portuguese	1806	Spanish	856
4	Dutch	7419	Spanish	2999	German	2147	German	523	German	1609	In	755
5	French	5748	French	1839	Portuguese	1613	Portuguese	418	French	860	Unidentified	647

Table 6. Details of NLP analysis for each of the predicted topics.

Prediction Topic	Count of TwitterID	Average Confidence-Negative Sentiment	Average Confidence-Neutral Sentiment	Average Confidence-Positive Sentiment	Average Follower Count	Average Friend Count	Average Retweet Count	Count of Tweet Language
Topic 1	50420	0.559371	0.293209	0.147265	5646.66	1154.27	350.71	51
Topic 2	43060	0.539859	0.369295	0.090684	20447.33	1653.85	961.3	54
Topic 3	17618	0.275259	0.485657	0.238882	17776.81	1265.74	314.4	43
Topic 4	30470	0.54395	0.252615	0.203355	3606.61	1346.51	1323.62	49
Topic 5	10502	0.318199	0.521049	0.160704	21259.78	1045.63	438.74	52

As seen in Table 6, each of these five topics appear to be in distinct patterns, and AI-based clustering and regression in subsequent processes would confirm all possible correlations against each of these topics.

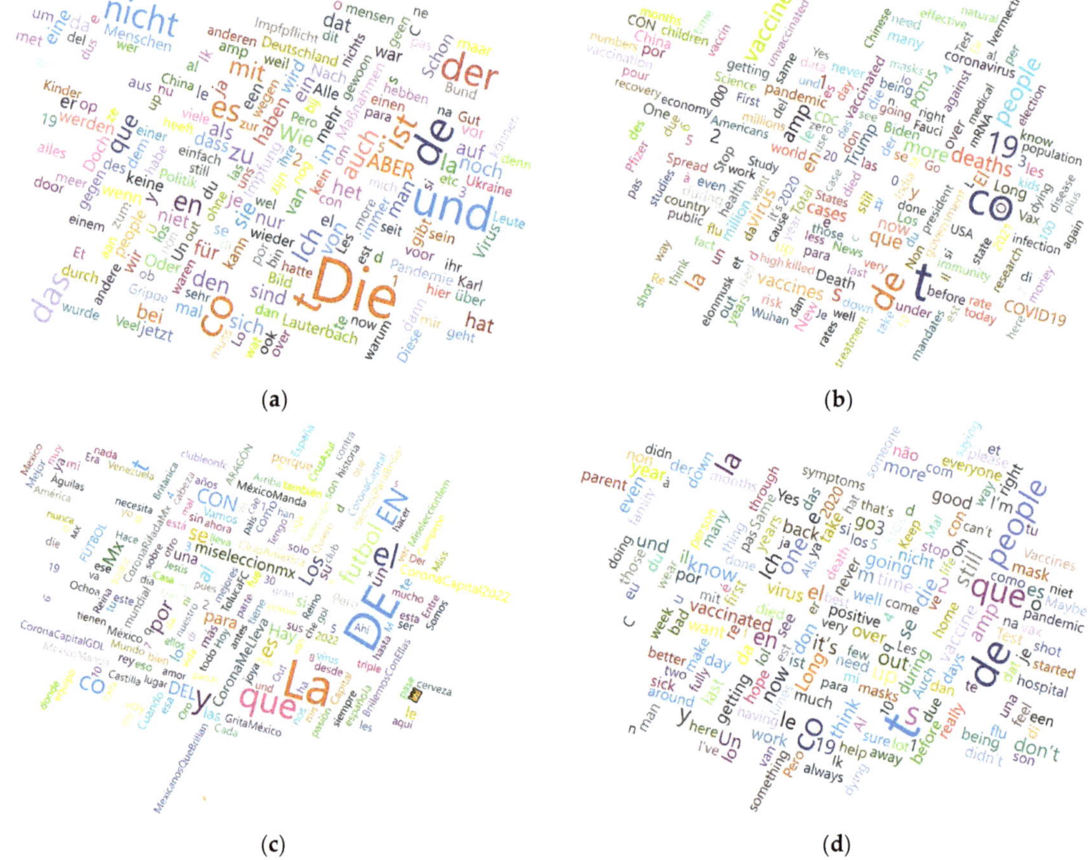

(**a**)　　　　　　　　(**b**)

(**c**)　　　　　　　　(**d**)

Figure 4. *Cont.*

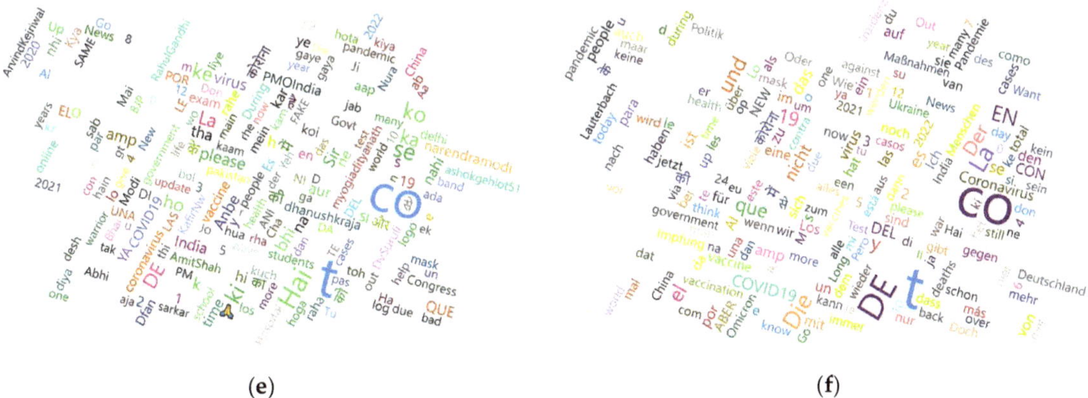

(e) **(f)**

Figure 4. Word Cloud for each of the analysed topics. (**a**) Topic 1 with 50,420 tweets (mostly German). (**b**) Topic 2 with 43,060 tweets (mostly English). (**c**) Topic 3 with 17,618 tweets (mostly Spanish). (**d**) Topic 4 with 30,470 tweets (mostly English). (**e**) Topic 5 with 10,502 tweets (mostly English). (**f**) All Topics with 152,070 tweets (mostly English).

4.1. Analysing the Correlated Factors for Topic 1

For Topic 1, six correlations were discovered using the AI-based regression method. Out of these six correlations, three of them are significant (as the correlation factor is greater than or equal to 0.1). This is observed from the result of the AI-based regression analysis as depicted in Figure 5a. The three significant factors that influence Topic 1 confidence (c_1) were identified to be language (l) and retweet count (r). The AI-based regression analysis uses NLP to describe these relationships. The following are three NLP-based descriptions of significant correlations:

- When the tweet language is 'de', the average Topic 1 confidence increases by 0.51;
- When the tweet language is 'nl', the average Topic 1 confidence increases by 0.38;
- When the average retweet count is 308 or less, the average Topic 1 confidence increases by 0.13.

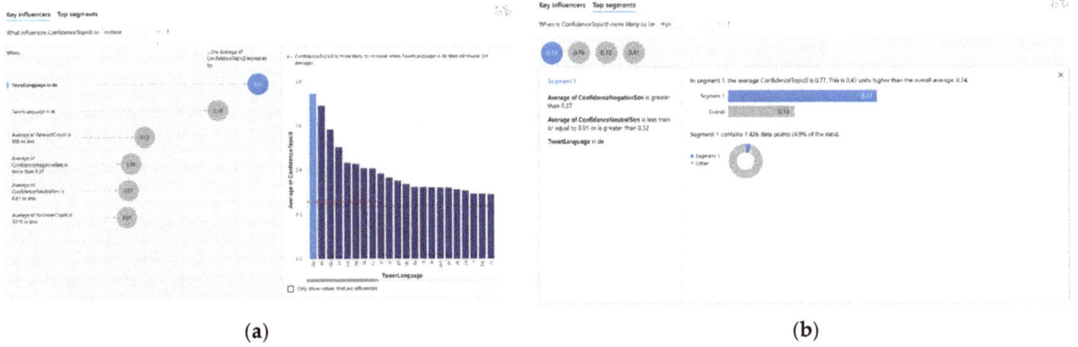

(a) **(b)**

Figure 5. Identifying the correlated factors for Topic 1—broad discussion on corona. (**a**) Identifying 6 correlations with regression. (**b**) Identifying 4 correlations with clustering.

These three significant correlations to Topic 1 confidence (c_1) are also portrayed in Equations (10)–(12). The insignificant correlations (i.e., a correlation factor less than 0.1) are portrayed in Equations (13)–(15).

$$c_1 \overset{0.51}{\leftarrow} \{l = {}'de'\} \tag{10}$$

$$c_1 \overset{0.38}{\leftarrow} \{l = {}'nl'\} \tag{11}$$

$$c_1 \overset{0.13}{\leftarrow} \{\bar{r} \leq 308\} \tag{12}$$

$$c_1 \overset{0.08}{\leftarrow} \{\bar{n} > 0.37\} \tag{13}$$

$$c_1 \overset{0.07}{\leftarrow} \{\bar{u} \leq 0.01\} \tag{14}$$

$$c_1 \overset{0.07}{\leftarrow} \left\{\bar{f} \leq 3319\right\} \tag{15}$$

The automated AI-based clustering technique also discovered four clusters, as shown in Figure 5b. All clusters were found to be significant, as the Topic 1 confidence (c_1) was more than or equal to 0.4.

Equations (16)–(19) depict the characteristics of these four significant clusters.

$$Cluster1 \overset{0.77}{\leftarrow} (\bar{n} > 0.37) \bigwedge (\bar{u} \leq 0.01 \vee \bar{u} > 0.32) \bigwedge (l = {}'de') \tag{16}$$

$$Cluster2 \overset{0.76}{\leftarrow} (\bar{n} > 0.37) \bigwedge (\bar{u} > 0.01 \vee \bar{u} \leq 0.032) \bigwedge (l = {}'de') \tag{17}$$

$$Cluster3 \overset{0.72}{\leftarrow} (\bar{n} \leq 0.37) \bigwedge (l = {}'de') \tag{18}$$

$$Cluster4 \overset{0.41}{\leftarrow} (\bar{n} > 0.37) \bigwedge (l \neq {}'de') \bigwedge (l \neq {}'en') \bigwedge (l \neq {}'es') \tag{19}$$

4.2. Analysing the Correlated Factors for Topic 2

For Topic 2, six correlations were discovered using the AI-based regression method. Out of these six correlations, four of them are significant (as the correlation factor is greater than or equal to 0.1). This is observed from the result of the AI-based regression analysis, as depicted in Figure 6a. The four significant factors that influence the Topic 2 confidence (c_2) were identified to be the language (l), retweet count (r), and positive sentiment confidence (p). The AI-based regression analysis uses NLP to describe these relationships. The following are four NLP-based descriptions of significant correlations for Topic 2 confidence (c_2):

- When the tweet language is 'en', the average Topic 2 confidence increases by 0.21;
- When the tweet language is 'fr', the average Topic 2 confidence increases by 0.17;
- When the average retweet count is more than 302, the average Topic 2 confidence increases by 0.14;
- When the average confidence-positive sentiment is 0.01 or less, the average Topic 2 confidence increases by 0.1.

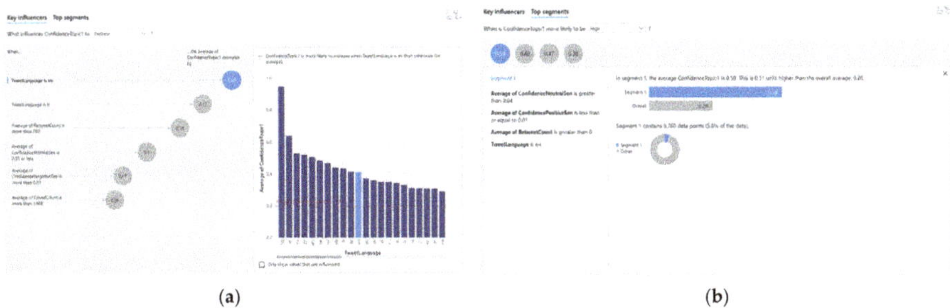

(a)   (b)

Figure 6. Identifying the correlated factors for Topic 2—COVID statistics and vaccination. (**a**) Identifying 6 correlations with regression. (**b**) Identifying 4 correlations with clustering.

These four significant correlations to the Topic 2 confidence (c_2) are also portrayed in Equations (20)–(23). The insignificant correlations (i.e., a correlation factor less than 0.1) are portrayed in Equations (24)–(25).

$$c_2 \overset{0,21}{\longleftarrow} \{l = 'en'\} \tag{20}$$

$$c_2 \overset{0,17}{\longleftarrow} \{l = 'fr'\} \tag{21}$$

$$c_2 \overset{0,14}{\longleftarrow} \{\bar{r} > 302\} \tag{22}$$

$$c_2 \overset{0.1}{\longleftarrow} \{\bar{p} \le 0.1\} \tag{23}$$

$$c_2 \overset{0,07}{\longleftarrow} \{\bar{n} > 0.7\} \tag{24}$$

$$c_2 \overset{0,06}{\longleftarrow} \{\bar{d} > 1448\} \tag{25}$$

The automated AI-based clustering technique also discovered four clusters, as shown in Figure 6b. Three out of the four clusters were found to be significant, as the Topic 2 confidence (c_2) was more than or equal to 0.4.

Equations (26)–(28) depict the characteristics of these three significant clusters. Equation (29) represents the insignificant cluster (i.e., Topic 2 confidence, $c_2 \le 0.4$).

$$Cluster1 \overset{0,59}{\longleftarrow} (\bar{u} > 0.04) \bigwedge \left(\bar{p} \le 0.01 \bigwedge (\bar{r} > 0) \right) \bigwedge (l = 'en') \tag{26}$$

$$Cluster2 \overset{0,48}{\longleftarrow} (\bar{u} \le 0.04) \bigwedge \left(\bar{p} \le 0.01 \bigwedge (\bar{r} > 0) \right) \bigwedge (l = 'en') \tag{27}$$

$$Cluster3 \overset{0,47}{\longleftarrow} \left(\bar{p} > 0.01 \vee \bar{p} \le 0.08 \bigwedge (\bar{r} > 0) \right) \bigwedge (l = 'en') \tag{28}$$

$$Cluster4 \overset{0,38}{\longleftarrow} \left(\bar{p} \le 0.01 \bigwedge (\bar{r} > 0) \right) \bigwedge (l \ne 'en') \bigwedge (l \ne 'de') \bigwedge (l \ne 'es') \tag{29}$$

4.3. Analysing the Correlated Factors for Topic 3

For Topic 3, eight correlations were discovered using the AI-based regression method. Out of these eight correlations, three of them are significant (as the correlation factor is greater than or equal to 0.1). This is observed from the result of the AI-based regression analysis, as depicted in Figure 7a. The four significant factors that influence the Topic 3 confidence (c_3) were identified to be the language (l), negative sentiment confidence

(n), and positive sentiment confidence (p). The AI-based regression analysis uses NLP to describe these relationships. The following are three NLP-based descriptions of significant correlations for Topic 3 confidence (c_3):

- When the tweet language is 'es,' the average Topic 3 confidence increases by 0.33;
- When the average confidence-negative sentiment is 0.01 or less, the average Topic 3 confidence increases by 0.17;
- When the average confidence-positive sentiment is more than 0.69, the average Topic 3 confidence increases by 0.12.

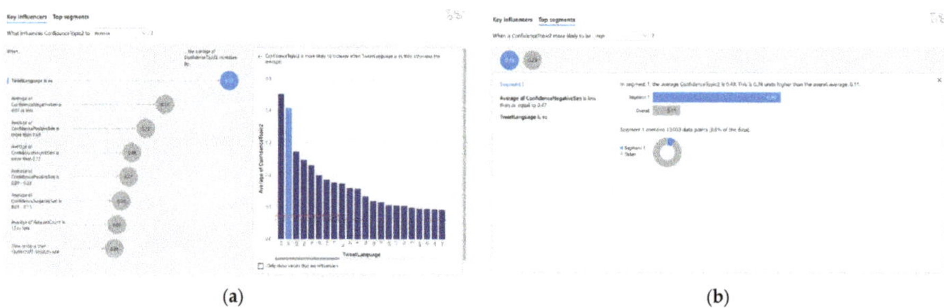

(a)   (b)

Figure 7. Identifying the correlated factors for Topic 3—wordplay on 'Corona'. (**a**) Identifying 8 correlations with regression. (**b**) Identifying 2 correlations with clustering.

These three significant correlations to Topic 3 confidence (c_3) are also portrayed in Equations (30)–(32). The insignificant correlations (i.e., a correlation factor less than 0.1) are portrayed in Equations (33)–(37).

$$c_3 \xleftarrow{0,33} \{l = {}'es'\} \tag{30}$$

$$c_3 \xleftarrow{0,17} \{\overline{n} \leq 0.01\} \tag{31}$$

$$c_3 \xleftarrow{0,12} \{\overline{p} > 0.69\} \tag{32}$$

$$c_3 \xleftarrow{0,08} \{\overline{u} > 0.73\} \tag{33}$$

$$c_3 \xleftarrow{0,07} \{0.09 \leq \overline{p} \leq 0.23\} \tag{34}$$

$$c_3 \xleftarrow{0,06} \{0.01 \leq \overline{n} \leq 0.11\} \tag{35}$$

$$c_3 \xleftarrow{0,05} \{\overline{r} \leq 17\} \tag{36}$$

$$c_3 \xleftarrow{0,04} \{m > \#19/04/2022\ 10:38:23\ AM\} \tag{37}$$

The automated AI-based clustering technique also discovered two clusters, as shown in Figure 7b. One out of the two clusters were found to be significant, as the Topic 3 confidence (c_3) was more than or equal to 0.4.

Equation (38) depicts the characteristics of the significant cluster. Equation (39) represents the insignificant cluster (i.e., Topic 3 confidence, $c_3 \leq 0.4$).

$$Cluster1 \xleftarrow{0,49} (\overline{n} \leq 0.47) \bigwedge (l = {}'es') \tag{38}$$

$$Cluster2 \overset{0,29}{\leftarrow} (\overline{n} > 0.47) \bigwedge (l = \,'es') \tag{39}$$

4.4. Analysing the Correlated Factors for Topic 4

For Topic 4, eight correlations were discovered using the AI-based regression method. Out of these eight correlations, four of them are significant (as the correlation factor is greater than or equal to 0.1). This is observed from the result of the AI-based regression analysis, as depicted in Figure 8a. The two significant factors that influence the Topic 4 confidence (c_4) were identified to be the retweet count (r) and language (l). The AI-based regression analysis uses NLP to describe these relationships. The following are four NLP-based descriptions of the significant correlations for Topic 4 confidence (c_4):

- When the average retweet count is more than 16,740, the average Topic 4 confidence increases by 0.31;
- When the tweet language is 'pt', the average Topic 4 confidence increases by 0.13;
- When the tweet language is 'en', the average Topic 4 confidence increases by 0.13;
- When the average retweet count is 1284–16740, the average Topic 4 confidence increases by 0.12.

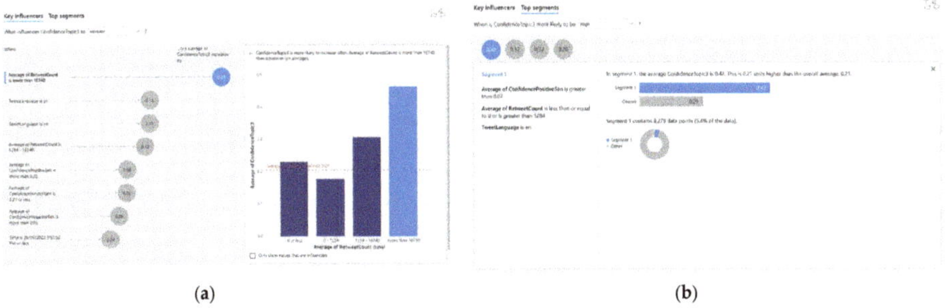

(a)   (b)

Figure 8. Identifying the correlated factors for Topic 4—COVID experiences or updates. (a) Identifying 8 correlations with regression. (b) Identifying 4 correlations with clustering.

These four significant correlations to Topic 4 confidence (c_4) are also portrayed in Equations (40)–(43). The insignificant correlations (i.e., a correlation factor less than 0.1) are portrayed in Equations (44)–(47).

$$c_4 \overset{0,31}{\leftarrow} \{\overline{r} > 16740\} \tag{40}$$

$$c_4 \overset{0,13}{\leftarrow} \{l = \,'pt'\} \tag{41}$$

$$c_4 \overset{0,13}{\leftarrow} \{l = \,'en'\} \tag{42}$$

$$c_4 \overset{0,12}{\leftarrow} \{1284 \leq \overline{r} \leq 16740\} \tag{43}$$

$$c_4 \overset{0,08}{\leftarrow} \{\overline{p} > 0.22\} \tag{44}$$

$$c_4 \overset{0,08}{\leftarrow} \{\overline{u} \leq 0.21\} \tag{45}$$

$$c_4 \overset{0,06}{\leftarrow} \{\overline{n} > 0.03\} \tag{46}$$

$$c_4 \xleftarrow{0.04} \{\overline{m} \leq \#26/09/2022 \ 6:53:52 \ PM\} \tag{47}$$

The automated AI-based clustering technique also discovered four clusters, as shown in Figure 8b. One out of the four clusters was found to be significant, as the Topic 4 confidence (c_4) was more than or equal to 0.4.

Equation (48) depicts the characteristics of the significant cluster. Equations (49)–(51) represent the insignificant clusters (i.e., Topic 4 confidence, $c_4 \leq 0.4$).

$$Cluster1 \xleftarrow{0.42} (\overline{p} > 0.07) \bigwedge (\overline{r} \leq 0 \vee \overline{r} > 1284) \bigwedge (l = {'}en{'}) \tag{48}$$

$$Cluster2 \xleftarrow{0.32} (\overline{u} \leq 0.21) \bigwedge (\overline{p} \leq 0.07) \bigwedge (\overline{r} \leq 0 \vee \overline{r} > 1284) \bigwedge (l = {'}en{'}) \tag{49}$$

$$Cluster3 \xleftarrow{0.32} (\overline{p} > 0.07) \bigwedge (\overline{r} > 0 \wedge \overline{r} \leq 1284) \bigwedge (l = {'}en{'}) \tag{50}$$

$$Cluster4 \xleftarrow{0.28} (\overline{u} \leq 0.21) \bigwedge (\overline{r} \leq 0 \vee \overline{r} > 1284) \bigwedge (l \neq {'}en{'}) \bigwedge (l \neq {'}de{'}) \tag{51}$$

4.5. Analysing the Correlated Factors for Topic 5

For Topic 5, nine correlations were discovered using the AI-based regression method. Out of these nine correlations, six of them are significant (as the correlation factor is greater than or equal to 0.1). This is observed from the result of the AI-based regression analysis, as depicted in Figure 9a. The three significant factors that influence the Topic 5 confidence (c_5) were identified to be the language (l), neutral sentiment confidence (u), follower count (f), and friend count (d). The AI-based regression analysis uses NLP to describe these relationships. The following are six NLP-based descriptions of the significant correlations:

- When the language is 'et', the average Topic 5 confidence increases by 0.8;
- When the language is 'hi', the average Topic 5 confidence increases by 0.43;
- When the language is 'und', the average Topic 5 confidence increases by 0.25;
- When the average confidence-neutral sentiment is more than 0.98, the average Topic 5 confidence increases by 0.15;
- When the average follower count is 2 or less, the average Topic 5 confidence increases by 0.14;
- When average friend count is 25 or less, the average Topic 5 confidence increases by 0.11.

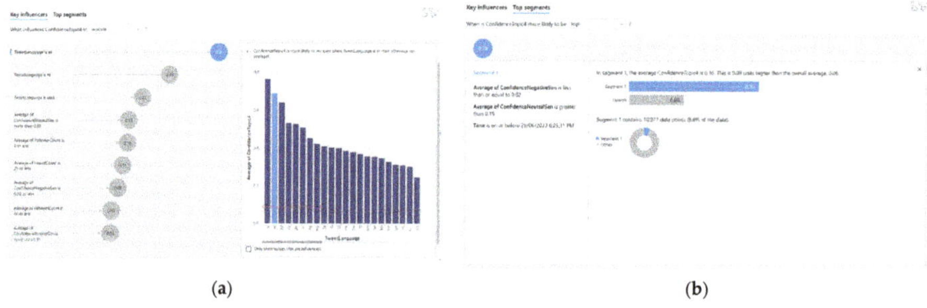

(a)   (b)

Figure 9. Identifying the correlated factors for Topic 5—Likely context of COVID in India. (a) Identifying 6 correlations with regression. (b) Identifying 1 correlation with clustering.

These six significant correlations to Topic 5 confidence (c_5) are also portrayed in Equations (52)–(57). The insignificant correlations (i.e., a correlation factor less than 0.1) are portrayed in Equations (58)–(60).

$$c_5 \xleftarrow{0,80} \{l = {'et'}\} \tag{52}$$

$$c_5 \xleftarrow{0,43} \{l = {'hi'}\} \tag{53}$$

$$c_5 \xleftarrow{0,25} \{l = {'und'}\} \tag{54}$$

$$c_5 \xleftarrow{0,15} \{\overline{u} > 0.98\} \tag{55}$$

$$c_5 \xleftarrow{0,14} \left\{\overline{f} \leq 2\right\} \tag{56}$$

$$c_5 \xleftarrow{0,11} \left\{\overline{d} \leq 25\right\} \tag{57}$$

$$c_5 \xleftarrow{0,08} \{\overline{n} \leq 0.02\} \tag{58}$$

$$c_5 \xleftarrow{0,03} \{\overline{r} \leq 80\} \tag{59}$$

$$c_5 \xleftarrow{0,02} \{\overline{p} > 0.01\} \tag{60}$$

The automated AI-based clustering technique also discovered one cluster, as shown in Figure 9b. This cluster was found to be significant, as the Topic 5 confidence (c_5) was more than or equal to 0.4.

Equation (61) depicts the insignificant characteristics of this cluster (i.e., Topic 5 confidence, $c_5 \leq 0.4$).

$$Cluster1 \xleftarrow{0,16} (\overline{n} \leq 0.02) \bigwedge (\overline{u} > 0.19) \bigwedge (m = \leq \#29/06/2022\ 6:25:31\ PM) \tag{61}$$

Finally, Table 7 summarizes the results of the cluster analysis for each of the topics (i.e., Topic 1 confidence, Topic 2 confidence, Topic 3 confidence, Topic 4 confidence, and Topic 5 confidence). Moreover, this table shows how many records (i.e., population count) were used to obtain the details of these clusters. As seen in Table 7, the significant clusters (i.e., a cluster confidence greater than or equal to 0.4) are highlighted in red.

Table 7. Fifteen observations found with AI-driven clustering (9 significant observations highlighted in red).

Cluster Characteristics	Cluster 1	Cluster 2	Cluster 3	Cluster 4
Avg. Topic 1 Confidence	0.77	0.76	0.72	0.41
Population Count	7426	10,678	12,108	20,351
Avg. Topic 2 Confidence	0.59	0.48	0.47	0.38
Population Count	8760	11,574	12,573	10,995
Avg. Topic 3 Confidence	0.49	0.29	-	-
Population Count	13,033	9193	-	-
Avg. Topic 4 Confidence	0.42	0.32	0.32	0.28
Population Count	8279	10,395	10,443	13,471
Avg. Topic 5 Confidence	0.16	-	-	-
Population Count	10,077	-	-	-

In essence, the methodology described within this paper autonomously generated 37 (six for Topic 1, six for Topic 2, eight for Topic 3, eight for Topic 4, and another nine for

Topic 9) with AI-driven regression. On the other hand, AI-driven clustering automatically generated 15 observations (four for Topic 1, four for Topic 2, two for Topic 3, four for Topic 4, and another one for Topic 5). These 52 (as represented with Equations (10)–(61)) AI-driven observations identified the factors that were deemed to be correlated with discussion topics found in COVID-19-related Twitter discourse. In Figure 10, the AI-driven observations (broken down into the total observation and significant observation) are portrayed with radar charts.

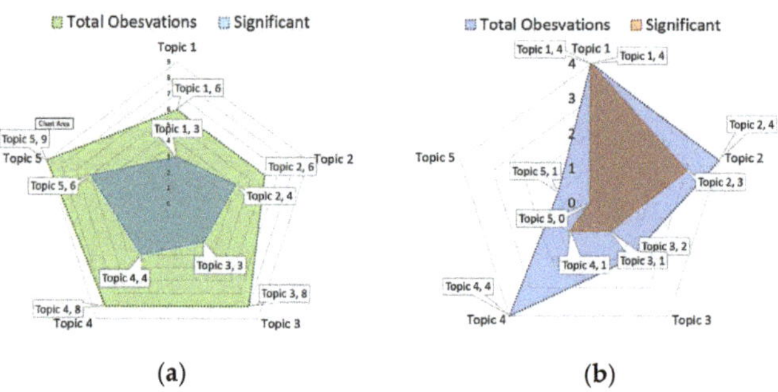

(a)   (b)

Figure 10. Total observations vs. significant observations for regression and cluster analysis. (**a**) Results of regression. (**b**) Results of clustering.

Since the proposed solution is designed to allow decision-makers to make evidence-based decisions on COVID-19-related issues based on Twitter analytics, this was deployed in mobile environments, both in iOS and Android. Figure 11 shows the deployed system in mobile environments, showing the correlation between retweets and the Topic 1 confidence (previously shown with Equation (12)).

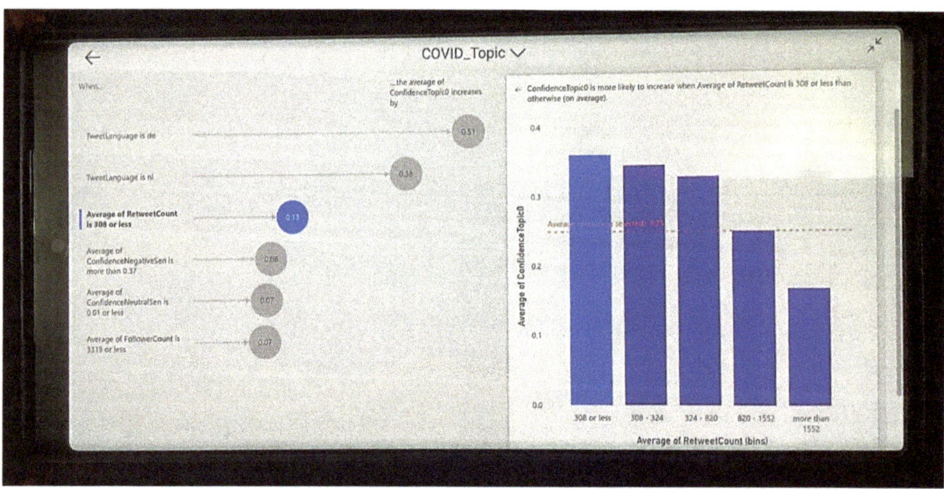

Figure 11. The proposed solution deployed on a Samsung Galaxy S23 Ultra Mobile running Android version 13.

With the deployed solution in mobile environments, a strategic decision-maker could be in a remote location, making evidence-based strategic decisions, whilst being completely mobile.

5. Conclusions

Since the emergence of the COVID-19 crisis, scholars and policymakers have adeptly harnessed Twitter as a principal reservoir for the meticulous scrutiny of public sentiments [9–25]. The perspicacious analysis of public sentiment engenders empirically grounded policymaking across a spectrum of COVID-19-related strategic imperatives, including, but not limited to, the imposition of lockdown measures, travel restrictions, vaccination campaigns, and the amelioration of misinformation dissemination. Consequently, the utilization of Twitter-based critical analysis has yielded substantive triumphs in the realm of COVID-19-driven decision making across multifarious dimensions.

However, none of these existing research works investigated the factors that drive COVID-19-based Twitter discourse. The present paper elucidates a systematic and methodological framework, employing artificial intelligence (AI) to autonomously unearth 52 distinct observations. This process, characterized by the utilization of both regression and clustering techniques, systematically unravels the intricate interplay between diverse factors and the topics encapsulating COVID-19 discussions on Twitter. Within this compendium of observations, 37 were ascertained through the AI-driven regression technique, while the AI-based clustering technique yielded an additional 15 observations. Furthermore, 29 of these observations bear considerable significance, denoting their pivotal role in shaping specific discourse themes.

These noteworthy observations discern an array of contributory variables encompassing tweet language, negative sentiment, positive sentiment, neutral sentiment, tweet timestamp, retweet count, friend count, and follower count, which exert discernible influences upon distinct discussion themes. Importantly, it merits emphasis that none of the extant studies on Twitter-based COVID-19 discourse, as indexed in [9–25], have proffered a methodology as innovative as the one advanced herein, integrating AI-powered regression and clustering techniques for the purpose of discerning the determinants of COVID-19-related discussion topics.

Furthermore, this research not only introduces an innovative methodological paradigm but also subjects this framework to rigorous evaluation, encompassing an extensive dataset spanning 645 days, commencing on 15 July 2021, and culminating on 20 April 2023. This dataset encompasses a multitude of multilingual tweets, spanning 58 distinct languages, thereby furnishing strategic decision-makers with a comprehensive toolkit for comprehending the manifold factors that govern the discourse surrounding COVID-19.

There are technical, qualitative, and ethical limitations of Twitter-based social media analytics, as apparent from [2,4,14]. Twitter has garnered acknowledgment as a fertile environment for the proliferation of disinformation and the dissemination of deceptive content, as noted in the scholarly discourse [49,50]. Within the confines of this specific investigation, a foundational proposition was laid out, positing the veracity of the entire corpus of 152,070 tweets with a cyber-related focus, subjected to scrutiny. Moreover, there are ethical issues pertaining to social media-based intelligence without the explicit permission of the social media users [51,52]. Research works in [51,52] portray ethical concerns in obtaining AI-driven intelligence from closed-network social media platforms like Facebook and LinkedIn. Users of Facebook, LinkedIn, and other closed-network platforms share their content only towards their closed group and do not consent to the intelligence acquisition of their data. In contrast, users of open platforms (like Twitter) are already aware that their contents are publicly available and could be subjected to intelligence acquisition. Consequently, an inherent constraint manifests itself in the shape of an absence of stringent validation protocols systematically applied to the open-source data sourced from the Twitter platform. As shown in Figure 12, the limitations of this work would shape the scope of future research in Twitter-based COVID-19 discourse.

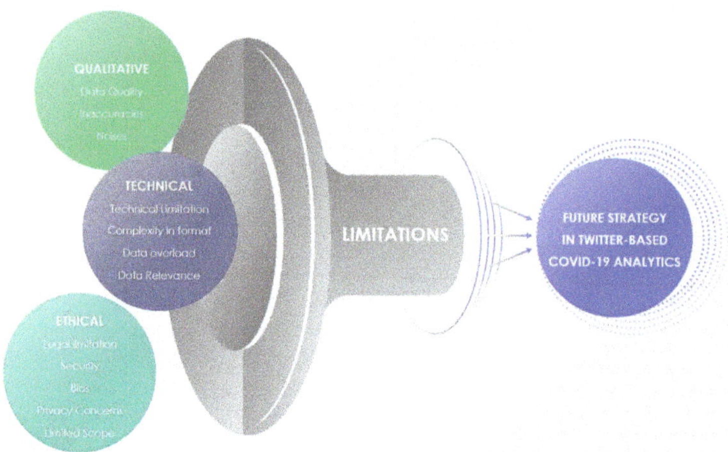

Figure 12. Limitations of the Twitter-based analysis of COVID-19 discourse.

Funding: This research received no external funding.

Data Availability Statement: Data would be provided upon request.

Acknowledgments: Special thanks to the COEUS Institute, Kennebunk, Maine, USA, where the author works as a Chief Technology Officer.

Conflicts of Interest: The author declares no conflict of interest.

References

1. World Health Organization. Social Media & COVID-19: A Global Study of Digital Crisis Interaction among Gen Z and Millennials. 2021. Available online: https://www.who.int/news-room/feature-stories/detail/social-media-covid-19-a-global-study-of-digital-crisis-interaction-among-gen-z-and-millennials (accessed on 1 September 2023).
2. Sufi, F. A New Social Media-Driven Cyber Threat Intelligence. *Electronics* **2023**, *12*, 1242. [CrossRef]
3. Sufi, F. Algorithms in Low-Code-No-Code for Research Applications: A Practical Review. *Algorithms* **2023**, *16*, 108. [CrossRef]
4. Sufi, F. A New AI-Based Semantic Cyber Intelligence Agent. *Future Internet* **2023**, *15*, 231. [CrossRef]
5. Northwestern. Social Media Contributes to Misinformation about COVID-19. 2020. Available online: https://news.northwestern.edu/stories/2020/09/social-media-contributes-to-misinformation-about-covid-19/ (accessed on 1 September 2023).
6. Hussain, A.; Ali, S.; Ahmed, M.; Hussain, S. The Anti-vaccination Movement: A Regression in Modern Medicine. *Cureus* **2018**, *10*, e2919. [CrossRef]
7. Johnson, N.F.; Velásquez, N.; Restrepo, N.J.; Leahy, R.; Gabriel, N.; Oud, S.E.; Zheng, M.; Manrique, P.; Wuchty, S.; Lupu, Y. The online competition between pro- and anti-vaccination views. *Nature* **2020**, *582*, 230–233. [CrossRef]
8. Benecke, O.; DeYoung, S.E. Anti-Vaccine Decision-Making and Measles Resurgence in the United States. *Glob. Pediatr. Health* **2019**, *6*, 2333794X19862949. [CrossRef]
9. Li, C.-Y.; Renda, M.; Yusuf, F.; Geller, J.; Chun, S.A. Public Health Policy Monitoring through Public Perceptions: A Case of COVID-19 Tweet Analysis. *Information* **2022**, *13*, 543. [CrossRef]
10. Gourisaria, M.K.; Chandra, S.; Das, H.; Patra, S.S.; Sahni, M.; Leon-Castro, E.; Singh, V.; Kumar, S. Semantic Analysis and Topic Modelling of Web-Scrapped COVID-19 Tweet Corpora through Data Mining Methodologies. *Healthcare* **2022**, *10*, 881. [CrossRef]
11. Kwok, S.W.H.; Vadde, S.K.; Wang, G. Tweet Topics and Sentiments Relating to COVID-19 Vaccination Among Australian Twitter Users: Machine Learning Analysis. *J. Med. Internet Res.* **2021**, *23*, e26953. [CrossRef]
12. Long, Z.; Alharthi, R.; Saddik, A.E. NeedFull—a Tweet Analysis Platform to Study Human Needs During the COVID-19 Pandemic in New York State. *IEEE Access* **2020**, *8*, 136046–136055. [CrossRef]
13. Sufi, F.K. Automatic identification and explanation of root causes on COVID-19 index anomalies. *MethodsX* **2023**, *10*, 101960. [CrossRef]
14. Sufi, F.K.; Razzak, I.; Khalil, I. Tracking Anti-Vax Social Movement Using AI-Based Social Media Monitoring. *IEEE Trans. Technol. Soc.* **2022**, *3*, 290–299. [CrossRef]
15. Narasamma, V.L.; Sreedevi, M.; Kumar, G.V. Tweet Data Analysis on COVID-19 Outbreak. In *Smart Technologies in Data Science and Communication*; Lecture Notes in Networks and Systems Book Series (LNNS); Springer: Berlin/Heidelberg, Germany, 2021; Volume 210.

16. Waheeb, S.A.; Khan, N.A.; Shang, X. Topic Modeling and Sentiment Analysis of Online Education in the COVID-19 Era Using Social Networks Based Datasets. *Electronics* **2022**, *11*, 715. [CrossRef]

17. Storey, V.; O'Leary, D. Text Analysis of Evolving Emotions and Sentiments in COVID-19 Twitter Communication. *Cognit. Comput.* **2022**. *epub ahead of print.* [CrossRef]

18. Kabakus, T. A novel COVID-19 sentiment analysis in Turkish based on the combination of convolutional neural network and bidirectional long–short term memory on Twitter. *Concurr. Comput.* **2022**, *34*, e6883. Available online: https://api.semanticscholar.org/CorpusID:246851122 (accessed on 3 September 2023). [CrossRef] [PubMed]

19. Joloudari, J.H.; Hussain, S.; Nematollahi, A.M.; Bagheri, R.; Fazl, F.; Alizadehsani, R.; Lashgari, R. BERT-deep CNN: State of the art for sentiment analysis of COVID-19 tweets. *Soc. Netw. Anal. Min.* **2022**, *13*, 99. [CrossRef]

20. Mir, A.A.; Sevukan, R. Sentiment analysis of Indian Tweets about Covid-19 vaccines. *J. Inf. Sci.* **2022**. *epub ahead of print.*

21. Sufi, F.; Alsulami, M. Identifying drivers of COVID-19 vaccine sentiments for effective vaccination policy. *Heliyon* **2023**, *9*, e19195. [CrossRef]

22. Lee, E.W.J.; Zheng, H.; Goh, D.H.-L.; Lee, C.S.; Theng, Y.L. Examining COVID-19 Tweet Diffusion Using an Integrated Social Amplification of Risk and Issue-Attention Cycle Framework. *Health Commun.* **2022**. *epub ahead of print.* [CrossRef]

23. Lanier, H.D.; Diaz, M.I.; Saleh, S.N.; Lehmann, C.U.; Medford, R.J. Analyzing COVID-19 disinformation on Twitter using the hashtags #scamdemic and #plandemic: Retrospective study. *PLoS ONE* **2022**, *17*, e0268409.

24. Slavik, C.E.; Buttle, C.; Sturrock, S.L.; Darlington, J.C.; Yiannakoulias, N. Examining Tweet Content and Engagement of Canadian Public Health Agencies and Decision Makers During COVID-19: Mixed Methods Analysis. *J. Med. Internet Res.* **2021**, *23*, e24883. [CrossRef] [PubMed]

25. Bijoy, B.S.; Saba, S.J.; Sarkar, S.; Islam, M.S.; Islam, S.R.; Amin, M.R.; Karmaker, S. COVID19α: Interactive Spatio-Temporal Visualization of COVID-19 Symptoms through Tweet Analysis. In Proceedings of the IUI '21 Companion: 26th International Conference on Intelligent User Interfaces—Companion, College Station, TX, USA, 14–17 April 2021.

26. Shin, H.-S.; Kwon, H.-Y.; Seung-Jin, R. A New Text Classification Model Based on Contrastive Word Embedding for Detecting Cybersecurity Intelligence in Twitter. *Electronics* **2020**, *9*, 1527. [CrossRef]

27. Zhao, J.; Yan, Q.; Li, J.; Shao, M.; He, Z.; Li, B. TIMiner: Automatically extracting and analyzing categorized cyber threat intelligence from social data. *Comput. Secur.* **2020**, *95*, 101867–101874. [CrossRef]

28. Schellekens, J. Release the bots of war: Social media and Artificial Intelligence as international cyber attack. *Przegląd Eur.* **2021**, *4*, 163–179. [CrossRef]

29. Sun, N.; Zhang, J.; Gao, S.; Zhang, L.Y.; Camtepe, S.; Xiang, Y. Data Analytics of Crowdsourced Resources for Cybersecurity Intelligence. In *Network and System Security, Proceedings of the 14th International Conference: NSS 2020, Melbourne, VIC, Australia, 25–27 November 2020*; Lecture Notes in Computer Science (including subseries Lecture Notes in Artificial Intelligence and Lecture Notes in Bioinformatics); Springer: Berlin/Heidelberg, Germany, 2020; Volume 12570, pp. 3–21.

30. Subroto, A.; Apriyana, A. Cyber risk prediction through social media big data analytics and statistical machine learning. *J. Big Data* **2019**, *6*, 1–19. [CrossRef]

31. Hee, V.; Jacobs, G.; Emmery, C.; Desmet, B.; Lefever, E.; Verhoeven, B.; De Pauw, G.; Daelemans, W.; Hoste, V. Automatic Detection of Cyberbullying in Social Media Text. *PLoS ONE* **2018**, *13*, e0203794.

32. Shu, K.; Sliva, A.; Sampson, J.; Liu, H. Understanding Cyber Attack Behaviors with Sentiment Information on Social Media. In *Social, Cultural, and Behavioral Modeling, Proceedings of the 11th International Conference: SBP-BRiMS 2018, Washington, DC, USA, 10–13 July 2018*; Lecture Notes in Computer Science (including subseries Lecture Notes in Artificial Intelligence and Lecture Notes in Bioinformatics); Springer: Berlin/Heidelberg, Germany, 2018; Volume 10899, pp. 377–388.

33. Alves, F.; Bettini, A.; Ferreira, P.M.; Bessani, A. Processing tweets for cybersecurity threat awareness. *Inf. Syst.* **2021**, *95*, 101586. [CrossRef]

34. Microsoft Documentation. Text Analytics: A Collection of Features from AI Language that Extract, Classify, and Understand Text within Documents. 2023. Available online: https://azure.microsoft.com/en-us/products/ai-services/text-analytics (accessed on 6 August 2023).

35. Pang, B.; Lee, L.; Vaithyanathan, S. Thumbs up? Sentiment classification using machine learning techniques. In Proceedings of the 2002 Conference on Empirical Methods in Natural Language Processing (EMNLP 2002), Philadelphia, PA, USA, 6–7 July 2002.

36. Turney, P.D. Thumbs up or thumbs down? Semantic orientation applied. In Proceedings of the 40th Annual Meeting of the Association for Computational Linguistics, Philadelphia, PA, USA, 6–12 July 2002.

37. Naseem, U.; Razzak, I.; Khushi, M.; Eklund, P.W.; Kim, J. COVIDSenti: A Large-Scale Benchmark Twitter. *IEEE Trans. Comput. Soc. Syst.* **2020**, *8*, 1003–1015. [CrossRef]

38. Li, L.; Zhang, Q.; Wang, X.; Zhang, J. Characterizing the Propagation of Situational Information in Social Media During COVID-19 Epidemic: A Case Study on Weibo. *IEEE Trans. Comput. Soc. Syst.* **2020**, *7*, 556–562. [CrossRef]

39. Cameron, D.; Smith, G.A.; Daniulaityte, R.; Sheth, A.P.; Dave, D.; Chen, L.; Anand, G.; Carlson, R.; Watkins, K.Z.; Falck, R. PREDOSE: A Semantic Web Platform for Drug Abuse Epidemiology using Social Media. *J. Biomed. Inform.* **2013**, *46*, 985–997. [CrossRef]

40. Chen, X.; Faviez, C.; Schuck, S.; Lillo-Le-Louët, A.; Texier, N.; Dahamna, B.; Huot, C.; Foulquié, P.; Pereira, S.; Leroux, V.; et al. Mining Patients' Narratives in Social Media for Pharmacovigilance: Adverse Effects and Misuse of Methylphenidate. *Front. Pharmacol.* **2018**, *9*, 541. [CrossRef] [PubMed]

41. McNaughton, E.C.; Black, R.A.; Zulueta, M.G.; Budman, S.H.; Butler, S.F. Measuring online endorsement of prescription opioids abuse: An integrative methodology. *Pharmacoepidemiol. Drug Saf.* **2012**, *21*, 1081–1092. [CrossRef]
42. Al-Twairesh, N.; Al-Negheimish, H. Surface and Deep Features Ensemble for Sentiment Analysis of Arabic Tweets. *IEEE Access* **2019**, *7*, 84122–84131. [CrossRef]
43. Vashisht, G.; Sinha, Y.N. Sentimental study of CAA by location-based tweets. *Int. J. Inf. Technol.* **2021**, *13*, 1555–1567. [CrossRef] [PubMed]
44. Ebrahimi; Yazdavar, H.; Sheth, A. Challenges of Sentiment Analysis for Dynamic Events. *IEEE Intell. Syst.* **2017**, *32*, 70–75. [CrossRef]
45. Yu, H.-F.; Hsieh, C.-J.; Chang, K.-W.; Lin, C.-J. Large Linear Classification When Data Cannot Fit in Memory. In Proceedings of the KDD '10: Proceedings of the 16th ACM SIGKDD international conference on Knowledge discovery and Data Mining, Washington, DC, USA, 25–28 July 2010.
46. Matthies, H.; Strang, G. The solution of non linear finite element equations. *Int. J. Numer. Methods Eng.* **1979**, *14*, 1613–1626. [CrossRef]
47. Nocedal, J. Updating Quasi-Newton Matrices with Limited Storage. *Math. Comput.* **1980**, *35*, 773–782. [CrossRef]
48. Microsoft Documentation. Choosing a Natural Language Processing Technology in Azure. 2020. Available online: https://docs.microsoft.com/en-us/azure/architecture/data-guide/technology-choices/natural-language-processing (accessed on 3 September 2023).
49. Gurajala, S.; White, J.S.; Hudson, B.; Voter, B.R.; Matthews, J.N. Profile characteristics of fake Twitter accounts. *Big Data Soc.* **2016**, *3*, 2053951716674236. [CrossRef]
50. Ajao, O.; Bhowmik, D.; Zargari, S. Fake News Identification on Twitter with Hybrid CNN and RNN Models. In Proceedings of the 9th International Conference on Social Media and Society, Copenhagen, Denmark, 18–20 July 2018.
51. Golder, S.; Ahmed, S.; Norman, G.; Booth, A. Attitudes Toward the Ethics of Research Using Social Media: A Systematic Review. *J. Med. Internet Res.* **2017**, *19*, e195. [CrossRef]
52. Mikal, J.; Hurst, S.; Conway, M. Ethical issues in using Twitter for population-level depression monitoring: A qualitative study. *BMC Med. Ethics* **2016**, *17*, 22. [CrossRef] [PubMed]

 *information*

Article

The Impact of Digital Business on Energy Efficiency in EU Countries

Aleksy Kwilinski [1,2,3,*], Oleksii Lyulyov [1,3] and Tetyana Pimonenko [1,3]

[1] Department of Management, Faculty of Applied Sciences, WSB University, 41-300 Dabrowa Gornicza, Poland; alex_lyulev@econ.sumdu.edu.ua (O.L.); tetyana_pimonenko@econ.sumdu.edu.ua (T.P.)
[2] The London Academy of Science and Business, 120 Baker St., London W1U 6TU, UK
[3] Department of Marketing, Sumy State University, 2, Rymsky-Korsakov St., 40007 Sumy, Ukraine
* Correspondence: a.kwilinski@london-asb.co.uk

Abstract: Digital business plays a crucial role in driving energy efficiency and sustainability by enabling innovative solutions such as smart grid technologies, data analytics for energy optimization, and remote monitoring and control systems. Through digitalization, businesses can streamline processes, minimize energy waste, and make informed decisions that lead to more efficient resource utilization and reduced environmental impact. This paper aims at analyzing the character of digital business' impact on energy efficiency to outline the relevant instruments to unleash EU countries' potential for attaining sustainable development. The study applies the panel-corrected standard errors technique to check the effect of digital business on energy efficiency for the EU countries in 2011–2020. The findings show that digital business has a significant negative effect on energy intensity, implying that increased digital business leads to decreased energy intensity. Additionally, digital business practices positively contribute to reducing CO_2 emissions and promoting renewable energy, although the impact on final energy consumption varies across different indicators. The findings underscore the significance of integrating digital business practices to improve energy efficiency, lower energy intensity, and advance the adoption of renewable energy sources within the EU. Policymakers and businesses should prioritize the adoption of digital technologies and e-commerce strategies to facilitate sustainable energy transitions and accomplish environmental objectives.

Keywords: e-commerce sales; e-commerce turnover; e-commerce web sales; digital economy; sustainable development

Citation: Kwilinski, A.; Lyulyov, O.; Pimonenko, T. The Impact of Digital Business on Energy Efficiency in EU Countries. *Information* **2023**, *14*, 480. https://doi.org/10.3390/info14090480

Academic Editor: Luis Borges Gouveia

Received: 5 July 2023
Revised: 1 August 2023
Accepted: 3 August 2023
Published: 29 August 2023

1. Introduction

The rapid advancement of digital technologies and digital business models has re-shaped various aspects of the world's development. Digital business continues to expand across industries, and it becomes crucial to examine its implications for energy efficiency and sustainability [1–6]. Digital business encompasses a wide range of activities, including e-commerce, cloud computing, data analytics, Internet of Things (IoT) applications, and smart city initiatives, among others [7–10]. These technologies and business models have the potential to transform traditional energy systems [11–13], providing new avenues for energy management, resource optimization [14–17], and environmental stewardship [18–22]. However, the full implications of digital business on energy efficiency are multifaceted and require a comprehensive analysis. One of the key areas where digital business intersects with energy efficiency is through its influence on energy demand and consumption patterns [23–25]. For instance, the growth of e-commerce has significantly changed the way products are distributed and delivered, affecting logistics and transportation systems. Digital business facilitates the integration of renewable energy sources into the power grid, enabling more efficient management of energy generation and consumption, declining carbon dioxide emissions, and improving the well-being of society [26–28]. Smart grid technologies, coupled with advanced data analytics, enable real-time monitoring and control, demand response mechanisms, and grid optimization, leading to more sustainable and efficient energy systems. However, digital business also poses challenges to energy efficiency and sustainability. The rapid proliferation of digital devices, data centers, and communication networks has led to increased energy consumption and associated environmental impacts [29–31]. Furthermore, issues such as electronic waste management, cybersecurity risks, and the ethical use of data in digital business need to be addressed to ensure a sustainable digital future [32–34]. In addition, digitalization requires sufficient green financial resources [35–38] and relevant digital knowledge and skills [39–50]. To fully harness the potential of digital business for energy efficiency and sustainability, it is essential to identify best practices, technological innovations, and policy frameworks that can promote energy-efficient digital transformation. This paper aims at analyzing the character of digital business' impact on energy efficiency to outline the relevant instruments to unleash EU countries' potential for attaining sustainable development. This study fills the scientific gaps on energy efficiency by developing approaches to explore the character of the digital business' impact on energy efficiency based on the panel-corrected standard errors technique (PSCE). The PCSE method is chosen for its appropriateness in analyzing small panel data while accounting for cross-sectional dependence. Furthermore, the approaches developed in this study provide accurate estimation of variability considering the panel error structures. This study makes an original contribution by using the panel-corrected standard errors technique (PCSE) to analyze the impact of digital business on energy efficiency. The accurate estimation of variability provided by the PCSE method enhances the robustness of the findings. Additionally, the research contributes to the existing literature by providing evidence of the positive relationship between digital business practices and energy intensity reduction, aligning with broader efforts towards a more sustainable and low-carbon economy in the EU. The research implications of this study underscore the potential of digital business in improving energy efficiency and reducing CO_2 emissions in the EU. Policymakers could utilize these findings to develop targeted policies that promote digitalization strategies among businesses to enhance sustainability efforts. However, the study acknowledges the complexity of the relationship between digitalization and energy intensity, highlighting the need for further research to understand the nonlinear dynamics and mediating factors involved.

The paper has the following structure: the literature review explores the theoretical background of energy efficiency, digital business, and links among them; materials and methods section describes the variables for analysis and sources for them and provides an explanation of the core stages of research and methodology to check the research hypothesis on digital business' effect on energy efficiency; the results section overviews empirical findings on testing the research hypothesis; and the discussion and conclusion sections explain the core results of the analysis, outlining the policy implication considering the findings, whilst identifying the limitations and further directions for research.

2. Literature Review

2.1. Energy Efficiency Assessment

Scholars [51] conclude that energy efficiency plays a crucial role in decarbonizing economic development. The findings of energy efficiency assessments could be used to identify opportunities for energy savings and propose cost-effective solutions to enhance energy efficiency, which is the primary goal of sustainable development. It should be noted that scholars developed a vast range of approaches for assessing energy efficiency. Scholars [52–55] outline that energy intensity refers to the amount of energy required to produce a unit of output or provide a specific service. It is a measure of the efficiency with which energy is utilized. Lower energy intensity indicates higher energy efficiency. Based on empirical findings, Dong et al. [52] outline that declining energy intensity allows increasing energy efficiency among Chinese provinces and promotes the energy capabilities of the country. Shahiduzzaman and Alam [53] empirically justify that energy intensity and carbon dioxide emissions are closely related to each other, which consequently affects the Australian energy efficiency. Applying the Granger causality test, scholars [54] conclude that energy efficiency depends on the energy consumption structure, economic structure, and energy intensity. Based on the results, scholars suggest decreasing coal energy consumption and boosting the development of green energy. Hosan et al. [55] show that energy intensity directly impacts sustainable economic growth within the energy efficiency of a country. Su et al. [56] developed the composite energy efficiency index to develop policy recommendations to improve energy efficiency in OECD countries. Scholars justify the crucial role of final energy consumption in energy independence. Furthermore, scholars [56,57] highlight that energy efficiency has a significantly positive effect on final energy consumption in a country. Paramati et al. [58] confirm that environmental technologies have a significant positive impact on energy efficiency by reducing energy consumption. These findings indicate that environmental technologies contribute to reducing overall energy consumption and improving energy efficiency in OECD countries. Studies [59–62] confirm that decreasing CO_2 emissions from fuel combustion improve the energy efficiency of a country. In addition, studies [63–67] show that renewable energy has the most significant impact on energy efficiency among all other dimensions. Thus, extending renewable energy allows for boosting the rapid growth of a country's energy efficiency [68,69].

2.2. Digital Business and Energy Efficiency

The analysis of the theoretical background on energy efficiency shows that digital business can have both linear and nonlinear effects on energy efficiency, depending on how it is implemented and utilized. Scholars [70–72] outline that digital technologies and solutions are used to optimize energy consumption and improve energy efficiency. Digital systems and smart meters enable real-time monitoring of energy consumption, allowing businesses to identify inefficiencies and implement corrective measures promptly. By analyzing large volumes of data generated by digital systems, businesses gain insights into energy usage patterns, identify areas of potential energy waste, and proactively schedule maintenance to prevent energy losses [73,74]. Digitalization facilitates the implementation of energy-efficient processes, such as automated controls, smart grid technologies, and demand response systems. These technologies allow optimization of energy use and reduction of waste. Pålsson et al. [75] found that e-commerce eliminates the need for physical retail spaces, and the increased reliance on transportation could result in energy consumption and associated environmental impacts. However, studies [76] outline that efficient logistics practices, such as route optimization, consolidation of shipments, and the use of electric vehicles or alternative fuels, help minimize the energy intensity of e-commerce delivery operations.

At the same time, studies [77–80] justify the nonlinear effect of digitalization on energy efficiency. Morley et al. [81] confirm that digitalization leads to increased energy consumption if the efficiency gains are offset by increased usage or new applications. For instance, the proliferation of digital devices and data centers contributes to higher overall energy consumption. Babu et al. [82] highlight that the production, use, and disposal of digital devices generates electronic waste. Improper handling and disposal of e-waste negatively impacts the environment, including energy-intensive recycling processes and resource depletion. The growth of digital business often requires expanding data centers and related infrastructure. Gunasekaran et al. [83] explain that e-commerce businesses often operate large-scale warehouses to manage inventory and fulfill online orders. These facilities require energy for lighting, heating, cooling, and operating material handling equipment. Optimizing warehouse design, implementing energy-efficient technologies, and adopting sustainable practices such as energy management systems and renewable energy integration enhance energy efficiency in these operations. Chen et al. [84] confirm the nonlinear effect of e-commerce on energy efficiency. This means that the initial effects of e-commerce on energy efficiency could be negative, but over time, as technology and practices evolve, they lead to positive energy efficiency outcomes. The U-shaped impact is represented by a curve that initially dips downward (indicating a negative impact) and then rises upward (indicating a positive impact). The U-shaped impact of e-commerce on energy efficiency underscores the importance of proactive measures, technological advancements, and sustainable practices to mitigate initial negative impacts and capitalize on the long-term potential for energy efficiency gains. It highlights the need for continuous improvement and collaboration among e-commerce businesses, policymakers, and consumers to ensure a sustainable and energy-efficient e-commerce ecosystem. Considering the abovementioned factors, this study tests the following hypotheses:

H1: *Digital business has a statistically significant impact on energy efficiency.*

H2: *The share of enterprises with e-commerce sales impacts energy efficiency.*

H3: *The share of enterprises with e-commerce sales of at least 1% turnover impacts energy efficiency.*

H4: *The share of enterprises with web sales impacts energy efficiency.*

The theoretical framework and core hypotheses are shown in Figure 1.

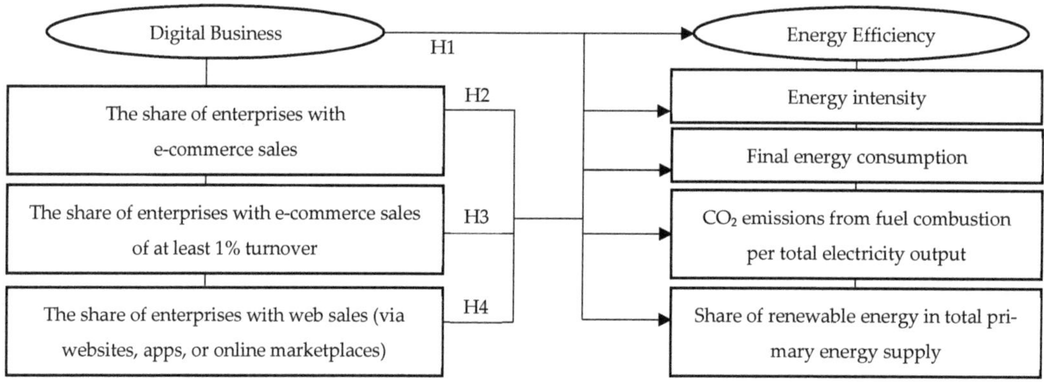

Figure 1. The theoretical framework of the research.

3. Materials and Methods

3.1. The Model and Estimation Procedure

The model specification for exploring the impact of digital business on energy efficiency and sustainability is built upon the foundational structure proposed by previous studies [6,13,17,22]:

$$Y_{i,t} = a_0 + b_1 X_{i,t} + b_2 Z_{i,t} + \varepsilon \tag{1}$$

where $Y_{i,t}$ is an explained variable; $X_{i,t}$ is a vector of explanatory variables; $Z_{i,t}$ is a vector of control variables; a_0 is constant; b_1, b_2 are search parameters of the model; and ε is an error term.

The estimation and verification process for the models involves the following steps: In the initial step, a series of diagnostic tests are conducted to ensure that the model is appropriately fitted. One crucial aspect is examining the correlation matrix, which provides an insight into the strength and direction of relationships among variables. Analyzing the correlation matrix can identify variables that exhibit strong associations or potential dependencies. Furthermore, the focus is on detecting multicollinearity, which occurs when independent variables are highly correlated with each other. Multicollinearity can lead to unreliable coefficient estimates and hinder the interpretability of the model. To assess the extent of multicollinearity, the variance inflation factor (VIF) is calculated for each variable. A high VIF suggests a high degree of multicollinearity, indicating the need for reconsideration or removal of variables from the model.

The second step involves conducting unit root tests to evaluate the stationarity of variables in the panel dataset. Stationarity refers to the property of a variable where its statistical properties, such as the mean and variance, remain constant over time. If variables are nonstationary, it can lead to spurious regression results and invalidate the model's conclusions. All data in the models are transformed into logarithmic form to address potential stationarity issues.

In the third step, panel data estimates are utilised, taking advantage of the larger availability of cross-sectional data units (27 EU countries) compared to the time series data (10 years). By including Croatia, which joined the EU in 2013, the study showed how adopting digital business practices and its impact on energy efficiency evolved during the new membership period. It offers valuable insights into how a relatively newer member state incorporates digital technologies into its economic practices and energy management strategies. This can be particularly enlightening for understanding how digital transformation and energy efficiency policies are adopted and adapted by countries that have recently joined the EU. On the other hand, excluding the United Kingdom, which exited the EU in 2020, allows the study to analyse the impact of digital business practices on energy efficiency without the influence of a country going through a significant political and economic separation from the Union. The withdrawal of the UK from the EU brought unique challenges and uncertainties, which overshadowed the focus on digital business practices' direct impact on energy efficiency during that period. Before proceeding with the estimation using fixed effects, random effects, or panel-corrected standard errors (PCSE), it is crucial to conduct various statistical tests to examine the characteristics of the data and carefully select the appropriate model specification and econometric method. These statistical tests serve the purpose of identifying specific characteristics within the data. First, the presence of heteroscedasticity is examined using the Wald test [85,86]. This test helps determine whether the variance of the errors varies across different observations. If heteroscedasticity is detected, it indicates that the assumption of constant error variance may be violated, and appropriate corrective measures need to be taken. Second, temporal autocorrelation is assessed using the Wooldridge test [87,88]. This test is employed to determine if there is a correlation between the error terms of the model at different time points. If temporal autocorrelation is detected, it implies that the assumption of independently distributed errors over time may not hold, necessitating appropriate adjustments in the model specification. Finally, the contemporaneous correlation among the cross-sectional units is examined using Pesaran's cross-section dependence test [89,90]. This test helps identify potential interdependence or spatial autocorrelation among observations across different countries. If cross-sectional dependence is found to exist, it indicates that the assumption of independently distributed errors across countries may not be valid, and alternative estimation methods or robust standard errors may be needed. To address potential issues of heteroscedasticity and the correlation of errors within the panel data, the panel-corrected standard errors (PCSE) technique was employed in the analysis. Heteroscedasticity refers to the situation where the variance of the error term varies across different levels of the independent variables. The correlation of errors within the panel arises due to unobserved factors affecting multiple observations within the panel. The PCSE technique adjusts the standard errors of the estimated coefficients to account for these issues, leading to robust statistical inference. By considering the heteroscedasticity and correlation structure of the data, the PCSE method yields more reliable coefficient estimates and valid hypothesis tests. Furthermore, applying this method helped minimize the impact of the disturbance at the end of the series, arising from a significant difference in the use of e-business during the COVID-19 pandemic. Additionally, robustness checks were conducted to assess the sensitivity of the results to the disturbance caused by the pandemic.

3.2. Data and Sources

Energy efficiency is a critical aspect to measure and evaluate progress towards sustainable energy practices and environmental goals. In the framework of this study, considering the studies [52–59], energy efficiency is explained via indicators (dependent variables): energy intensity; final energy consumption; CO_2 emissions from fuel combustion per total electricity output; and share of renewable energy in total primary energy supply.

- Energy intensity refers to the amount of energy used per unit of economic output, typically measured as energy consumed per unit of GDP. A decline in energy intensity over time indicates that an economy is becoming more energy efficient, producing more goods and services with less energy consumption. This indicator is essential for tracking the overall energy efficiency improvements of a country or region's economy.
- Final energy consumption represents the total energy utilized for end-use activities, such as transportation, residential, commercial, and industrial sectors. Monitoring final energy consumption helps identify trends in energy demand and highlights areas where energy-saving measures and technological advancements can lead to more efficient energy usage.
- CO_2 Emissions from Fuel Combustion per Total Electricity Output indicator relates to the amount of carbon dioxide (CO_2) emitted per unit of electricity generated from fuel combustion. A lower value indicates a cleaner and more energy-efficient electricity generation system, with reduced greenhouse gas emissions. By tracking this indicator, policymakers can gauge the environmental impact of electricity production methods and identify opportunities for transitioning to cleaner and more energy-efficient energy sources.
- The proportion of renewable energy sources in the total primary energy supply provides insights into the extent to which a country or region is transitioning towards more sustainable and environmentally friendly energy sources. A higher share of renewables suggests a reduced reliance on fossil fuels, contributing to energy diversification and improved energy efficiency.

These indicators offer a comprehensive assessment of energy efficiency from different angles, enabling policymakers and stakeholders to identify areas of improvement, set energy efficiency targets, and develop effective strategies to enhance energy performance and mitigate environmental impacts. By monitoring and analyzing these indicators over time, countries and regions can make informed decisions to foster a more sustainable energy landscape, promoting economic growth while reducing energy consumption and greenhouse gas emissions.

Based on prior studies [70–76], digital business is measured within the following independent variables: the share of enterprises with e-commerce sales, the share of enterprises with e-commerce sales of at least 1% turnover, and the share of enterprises with web sales (via websites, apps, or online marketplaces). Considering studies [70–76], digital business positively impacts energy efficiency by reducing the energy consumption associated with physical operations, optimizing supply chains, and enabling data-driven energy management strategies. Embracing digital business practices contributes to more sustainable and energy-efficient operations across various sectors. The first independent variable, the share of enterprises with e-commerce sales, is a crucial indicator of how businesses utilise digital platforms to conduct their transactions. By engaging in e-commerce, companies can reduce the need for physical operations, such as brick-and-mortar stores, potentially lowering their overall energy consumption associated with maintaining and running traditional retail outlets. The second variable, the share of enterprises with e-commerce sales of at least 1% turnover, further emphasises the significance of e-commerce in relation to energy efficiency. When businesses derive a substantial portion of their revenue from online sales, it indicates a higher reliance on digital technologies and a decreased reliance on energy-intensive physical infrastructure. The third variable, the share of enterprises with web sales through various digital channels, encompasses a broader spectrum of digital business practices, including sales via websites, apps, or online marketplaces. By leveraging these digital avenues, businesses can optimise their supply chains, streamline operations, and implement data-driven energy management strategies. This can lead to more efficient use of resources and reduced energy consumption in the production, distribution, and sales processes.

The study applies the following control variables:

1. GDP per capita: GDP per capita is an important indicator of economic development and can have a significant impact on energy intensity, final energy consumption, CO_2 emissions, and the adoption of digital business practices [5,91–93]. Higher GDP per capita generally indicates greater economic resources and technological advancements, which can influence energy efficiency and sustainability outcomes. Countries with higher GDP per capita may have more financial capacity to invest in energy-efficient technologies, implement sustainable practices, and adopt digital business strategies.

2. Trade openness [94–97]: Trade openness refers to the degree of integration of a country's economy with the global market. It can have implications for energy consumption and environmental impact. Countries with higher trade openness may have greater access to international markets, which can affect their energy consumption patterns, including imports and exports of energy-intensive goods.

3. Governance efficiency [98,99]: Governance efficiency reflects the effectiveness of a country's institutions and their ability to implement and enforce policies and regulations. Good governance can promote sustainable practices and enhance energy efficiency initiatives. Countries with more efficient governance systems may be better equipped to implement and enforce energy efficiency policies, incentivize digital business practices, and ensure compliance with sustainability standards.

4. Land Surface Temperature: The temperature conditions of the Earth's surface, as captured by Land Surface Temperature, have both direct and indirect effects on energy consumption and intensity [100,101]. Temperature variations directly impact heating and cooling demands, thereby influencing energy consumption patterns. Additionally, temperature conditions indirectly influence energy efficiency by affecting the need for energy-intensive processes such as air conditioning or heating systems.

5. Population density: Population density can affect energy consumption patterns [102–104]. Urban areas with higher population densities tend to have different energy requirements compared to rural areas. Factors such as the concentration of residential and commercial buildings, transportation needs, and infrastructure availability can all influence energy consumption. Furthermore, areas with higher population densities are more likely to have advanced infrastructure, including digital connectivity and technological innovations. This may facilitate the adoption of digital business practices, as well as the implementation of energy-efficient technologies.

The list of variables, their explanations, sources, and descriptive statistics is shown in Table 1.

Table 1. Descriptive statistics of the selected variables for analysis of the digital business effect on energy efficiency.

Variable	Explanation	Source	Mean	CV	Min	Max
EI	Energy intensity		5.076	0.090	3.798	6.201
FEC	Final energy consumption		2.763	0.495	−0.713	5.398
$SDG7_{co_2}$	CO_2 emissions from fuel combustion per total electricity output		0.215	3.057	−1.544	3.590
$SDG7_{ren}$	Share of renewable energy in total primary energy supply	Eurostat [105]	2.591	0.255	−0.010	3.783
E_s	The share of enterprises with e-commerce sales		2.853	0.166	1.335	3.846
E_t	The share of enterprises with e-commerce sales of at least 1% turnover		2.698	0.193	0.916	3.757
E_{ws}	The share of enterprises with web sales (via websites, apps, or online marketplaces)		2.299	0.226	0.875	3.329
GDP	GDP per capita		10.226	0.061	8.864	11.725
WGI	Governance efficiency	World Data Bank [106]	−0.141	−4.636	−2.442	0.627
TO	Trade openness		4.768	0.096	4.005	5.940
LST	Land Surface Temperature		2.676	0.186	0.272	3.416
PD	Population density		4.679	0.195	2.875	7.384

The correlation coefficients in the matrix range (Table 2) from -1 to 1, with values closer to -1 indicating a strong negative correlation, values closer to 1 indicating a strong positive correlation, and values close to 0 indicating a weak or no correlation. Positive correlations between digital business metrics (such as e-commerce sales, e-commerce turnover, or e-commerce web sales) and energy efficiency variables would suggest a positive relationship, indicating that an increase in digital business activities is associated with improved energy efficiency. All data are significant at 1% and 5%.

Table 2. The correlation matrix between the selected variables for analysis of the digital business' effect on energy efficiency.

Variables	EI	FEC	$SDG7_{co_2}$	$SDG7_{ren}$	E_s	E_t	E_{ws}	GDP	WGI	TO	LST	PD
EI	1.000											
FEC	-0.351	1.000										
$SDG7_{co_2}$	0.021	-0.313	1.000									
$SDG7_{ren}$	-0.126	0.158	-0.379	1.000								
E_s	-0.392	0.070	-0.174	-0.066	1.000							
E_t	-0.318	0.028	-0.212	0.007	0.966	1.000						
E_{ws}	-0.328	0.061	-0.036	-0.095	0.924	0.878	1.000					
GDP	-0.501	0.063	0.010	0.004	0.673	0.601	0.530	1.000				
WGI	0.155	-0.055	0.546	-0.534	0.174	0.143	0.185	0.239	1.000			
TO	0.041	-0.097	0.267	-0.498	-0.369	-0.383	-0.323	-0.434	0.069	1.000		
LST	-0.124	0.002	0.249	-0.719	-0.054	-0.160	0.003	0.105	0.355	0.424	1.000	
PD	-0.124	0.002	0.249	-0.719	-0.054	-0.160	0.003	-0.124	0.105	0.355	0.624	1.000

Based on the outputs of the correlation analysis, the regression model (1) was estimated separately for each explained variable (EI, FEC, $SDG7_{co_2}$, $SDG7_{ren}$). The results of the variance inflation factor (VIF) show that all values are less than threshold 5 (Table 3). This means that the variable is not highly correlated with other independent variables in the model [107]. The fundings allow rejection of the multicollinearity among selected variables.

Table 3. The results of VIF for the selected variables for analysis of the digital business' effect on energy efficiency.

Variables	EI			FEC			$SDG7_{co_2}$			$SDG7_{ren}$		
E_s	1.85	–	–	1.85	–	–	1.85	–	–	1.85	–	–
E_t	–	1.63	–	–	1.63	–	–	1.63	–	–	1.63	–
E_{ws}	–	–	1.45	–	–	1.63	–	–	1.45	–	–	1.45
GDP	3.21	3.21	3.21	3.21	3.21	3.21	3.21	3.21	3.21	3.21	3.21	3.21
WGI	4.01	3.97	3.57	4.01	3.97	3.97	4.01	3.97	3.57	4.01	3.97	3.57
TO	1.07	1.07	1.08	1.07	1.07	1.07	1.07	1.07	1.08	1.07	1.07	1.08
LST	3.29	3.28	3.38	3.29	3.28	3.28	3.29	3.28	3.38	3.29	3.28	3.38
PD	2.93	2.96	2.97	2.93	2.96	2.96	2.93	2.96	2.97	2.93	2.96	2.97

4. Results

The stationarity of the variables was examined using the Im-Pesaran-Shin unit root test [108] and the Fisher-based augmented Dickey–Fuller unit root test [109]. The results presented in Table 4 demonstrate that all variables become stationary after taking the first difference.

The results of the Hausman test presented in Table 5 indicate that all models exhibit statistically significant chi-square test statistics, with p values of 0.000. This implies a substantial disparity between the estimated coefficients of the fixed effects (FE) and random effects (RE) models for all variables. These findings suggest that the random effects model may not be suitable for the regression analysis, favoring the adoption of the fixed effects model instead. The significance of the results indicates the presence of unobserved time-invariant factors that impact the relationship between the explained and explanatory variables. Consequently, incorporating these fixed effects within the model enhances the reliability and accuracy of the coefficient estimates.

Table 4. The results of the unit root test.

| Variables | Im–Pesaran–Shin | | | | Augmented Dickey–Fuller | | | |
| | At Level | | At 1-st Difference | | At Level | | At 1-st Difference | |
	Statistic	p Value	Statistic	p Value	Statistic	p Value	Statistic	p Value
E_s	2.069	0.981	-10.884	0.000	1.422	0.078	32.594	0.000
E_t	1.303	0.904	-7.100	0.000	0.691	0.245	36.072	0.000
E_{ws}	1.804	0.964	-5.439	0.000	5.238	0.000	28.548	0.000
GDP	-2.776	0.003	-9.272	0.000	0.295	0.384	8.697	0.000
WGI	-3.856	0.000	-5.517	0.000	0.629	0.265	28.540	0.000
TO	-5.560	0.000	-4.068	0.000	-1.458	0.928	2.651	0.004
LST	-12.630	0.000	-14.190	0.000	9.914	0.000	35.966	0.000
PD	-5.804	0.000	-5.976	0.000	2.523	0.006	12.155	0.000

Table 5. Results of the Hausman test.

| Statistics | EI | | | FEC | | | $SDG7_{co_2}$ | | | $SDG7_{ren}$ | | |
	E_s	E_t	E_{ws}	E_s	E_t	E_{ws}	E_s	E_t	E_{ws}	E_s	E_t	E_{ws}
chi2	25.75	29.44	27.61	57.54	57.66	56.29	21.09	18.59	21.15	83.72	85.97	82.27
Prob > chi2	0.000	0.000	0.000	0.000	0.000	0.000	0.000	0.002	0.000	0.000	0.000	0.000

The statistically significant F-statistic values and low p values of the Wooldridge test (Table 6) suggest that heteroscedasticity is present in the regression models for all variables (EI, FEC, $SDG7_{co_2}$, $SDG7_{ren}$) and their respective explanatory variables (E_s, E_t, E_{ws}).

Table 6. Results of the Wooldridge test.

| Statistics | EI | | | FEC | | | $SDG7_{co_2}$ | | | $SDG7_{ren}$ | | |
	E_s	E_t	E_{ws}	E_s	E_t	E_{ws}	E_s	E_t	E_{ws}	E_s	E_t	E_{ws}
F(1,21)	41.045	40.989	45.229	82.008	82.164	83.051	50.510	72.701	55.248	30.908	30.547	31.383
Prob > F	0.000	0.000	0.000	0.000	0.000	0.000	0.000	0.000	0.000	0.000	0.000	0.000

The outputs presented in Table 7 provide the results of the modified Wald test, which aims to detect groupwise heteroscedasticity in the regression models. Groupwise heteroscedasticity occurs when the variances of the error terms in a regression model differ across distinct groups or subsets within the data.

Table 7. The results of the modified Wald test for groupwise heteroscedasticity.

| Statistics | EI | | | FEC | | | $SDG7_{co_2}$ | | | $SDG7_{ren}$ | | |
	E_s	E_t	E_{ws}	E_s	E_t	E_{ws}	E_s	E_t	E_{ws}	E_s	E_t	E_{ws}
chi2	469.19	257.77	1180.56	127.10	132.70	135.52	6939.78	4547.71	4467.72	391.29	250.14	198.59
Prob > chi2	0.000	0.000	0.000	0.000	0.000	0.000	0.000	0.000	0.000	0.000	0.000	0.000

Table 7 displays the Chi-square test statistics and their corresponding p values for each variable included in the regression models. The chi-square test statistic assesses the overall significance of the groupwise heteroscedasticity in the models. A statistically significant chi-square value, indicated by a very low p value ($p = 0.000$), suggests the presence of substantial groupwise heteroscedasticity. The presence of groupwise heteroscedasticity has implications for the interpretation of the regression results, similar to the implications of overall heteroscedasticity. When groupwise heteroscedasticity is present, the standard errors of the coefficient estimates could be biased, leading to incorrect inference and hypothesis testing.

The results presented in Table 8 showcase the outcomes of Pesaran's cross-section dependence test. This test is conducted to examine the hypothesis of cross-sectional independence in panel data models with N > T. It is specifically designed to detect contemporaneous correlation or cross-sectional dependence among the variables in the model. The decision to employ Pesaran's test over the Breusch–Pagan test is due to its suitability for detecting contemporaneous correlation in the model residuals. The Breusch–Pagan test is not well suited for panel data analysis with cross-sectional dependence [110]. Therefore, Pesaran's alternative approach is adopted to accurately assess the presence of cross-sectional dependence.

Table 8. The results of Pesaran's cross-section dependence test.

Statistics	EI			FEC				SDG7$_{co_2}$			SDG7$_{ren}$	
	E_s	E_t	E_{ws}	E_s	E_s	E_t	E_{ws}	E_s	E_s	E_t	E_{ws}	E_s
stat	20.519	20.748	23.443	8.603	7.732	9.275	2.033	2.098	3.555	14.976	14.952	17.449
Pr	0.000	0.000	0.000	0.000	0.000	0.000	0.042	0.036	0.000	0.000	0.000	0.000

The Pesaran's cross-section dependence test presented in Table 8 evaluates the null hypothesis of cross-sectional independence, which assumes no contemporaneous correlation among the variables. The test statistics (stat) and corresponding p values (Pr) for each dependent variable in the panel data models indicate that the null hypothesis of cross-sectional independence is rejected. This suggests the presence of cross-sectional dependence or contemporaneous correlation among the variables.

Based on the results of the Wooldridge test, the modified Wald test for groupwise heteroscedasticity, and Pesaran's cross-section dependence test, it is evident that the panel data models suffer from heteroscedasticity, contemporaneous correlation, and cross-sectional dependence. Considering these findings, employing panel-corrected standard error (PCSE) estimations would be a more appropriate approach. PCSE estimations effectively address the issues of heteroscedasticity and cross-sectional dependence in panel data models, providing robust standard errors that account for within-group correlation.

To validate the results obtained through PCSE estimations, the fixed effects model is employed. The results of the fixed effects model regression analysis (Table 9) indicate significant relationships between the independent variables E_s, E_t, and E_{ws} and the explained variables EI, SDG7$_{co_2}$, and SDG7$_{ren}$. However, for the variable FEC, the coefficient estimates for all indicators of digital business are not statistically significant ($p > 0.05$), suggesting a lack of significant relationship between these indicators and the explained variable. The F-statistic for all models is statistically significant with a p value of 0.000. This indicates that the regression models are a good fit for the data and that at least one independent variable has a significant relationship with the dependent variable. The high significance level suggests that the chosen independent variables collectively contribute to explaining the variation in the dependent variable. Furthermore, the correlation coefficient (rho) for all models in Table 9 is higher than 0.99, indicating a strong positive correlation between the dependent variable and the individual-specific effects. This suggests that a large proportion of the variation in the dependent variable is accounted for by the individual-specific effects. The high rho values reinforce the importance of considering the fixed effects in the model and suggest a substantial influence of the individual-specific effects on the dependent variable.

Table 9. The results of regression analysis: fixed effects model.

Variables	EI						FEC					
	E_s		E_t		E_{ws}		E_s		E_t		E_{ws}	
	Coef.	Prob.	Coef.	Prob.	Coef.	Prob.	Coef.	Prob.	Coef.	Prob.	Coef.	Prob.
E_i	−0.166	0.000	−0.166	0.000	−0.101	0.000	−0.016	0.241	−0.011	0.440	−0.006	0.548
GDP	−0.397	0.000	−0.410	0.000	−0.427	0.000	0.195	0.000	0.192	0.000	0.190	0.000
WGI	−0.068	0.111	−0.067	0.122	−0.092	0.036	−0.066	0.011	−0.067	0.010	−0.069	0.007
TO	−0.284	0.001	−0.264	0.002	−0.335	0.000	0.204	0.000	0.204	0.000	0.200	0.000
LST	−0.095	0.007	−0.100	0.005	−0.100	0.006	−0.030	0.157	−0.031	0.146	−0.031	0.145
PD	−1.390	0.000	−1.536	0.000	−1.411	0.000	−0.430	0.001	−0.442	0.001	−0.434	0.001
const	17.457	0.000	18.139	0.000	17.871	0.000	2.101	0.001	2.170	0.001	2.156	0.001
F	46.54	0.000	45.30	0.000	42.27	0.000	14.43	0.000	14.24	0.000	14.18	0.000
sigma_u	1.101		1.203		1.133		1.452		1.457		1.453	
sigma_e	0.059		0.060		0.061		0.036		0.036		0.036	
rho	0.997		0.998		0.997		0.999		0.999		0.999	
	$SDG7_{co_2}$						$SDG7_{ren}$					
E_i	−0.246	0.000	−0.280	0.000	−0.155	0.000	0.156	0.000	0.164	0.000	0.073	0.011
GDP	0.136	0.309	0.130	0.325	0.094	0.477	0.308	0.001	0.317	0.000	0.345	0.000
WGI	0.088	0.462	0.099	0.402	0.054	0.649	0.226	0.004	0.222	0.005	0.252	0.002
TO	−0.071	0.764	−0.030	0.898	−0.148	0.536	0.421	0.007	0.399	0.011	0.466	0.004
LST	0.067	0.492	0.064	0.512	0.061	0.536	0.126	0.049	0.129	0.043	0.133	0.041
PD	2.663	0.000	2.431	0.000	2.635	0.000	3.544	0.000	3.685	0.000	3.574	0.000
const	−12.368	0.000	−11.385	0.000	−11.779	0.000	−19.233	0.000	−19.866	0.000	−19.703	0.000
F	6.15	0.000	6.85	0.000	5.73	0.000	28.79	0.000	29.02	0.000	26.48	0.000
sigma_u	2.043		1.883		2.017		3.380		3.480		3.420	
sigma_e	0.165		0.164		0.166		0.108		0.108		0.110	
rho	0.993		0.992		0.993		0.999		0.999		0.999	

Table 10 presents the results of the regression analysis using PCSE estimations. For the dependent variable EI (Energy Intensity), all indicators of digital business show a statistically significant impact ($p < 0.05$). The negative coefficients for the indicators of digital business (E_s, E_t, E_{ws} are −0.112, −0.112, and −0.076, respectively) suggest that an increase in the levels of digital business is associated with a decrease in energy intensity in the EU. This implies that as the share of enterprises with e-commerce sales, the share of enterprises with e-commerce sales of at least 1% turnover, and the share of enterprises with web sales increase, the energy intensity in the EU is expected to decline. These findings have significant implications for the EU's sustainability goals and its efforts to reduce carbon emissions. The results indicate that embracing digitalization and promoting e-commerce practices contribute to improving energy efficiency and potentially lead to reduced energy consumption across various sectors. By actively supporting and encouraging digital business practices, the EU could enhance its energy efficiency measures, mitigate environmental impacts, and progress toward a more sustainable and environmentally friendly economy. The R^2 values range from 0.978 to 0.981, suggesting that the independent variables collectively explain a substantial portion of the variation in energy intensity. Additionally, the outputs for the variable $SDG7_{co_2}$ reveal that the adoption and expansion of digital business practices in the EU could also contribute to the reduction of CO_2 emissions. The coefficients for E_s, E_t, and E_{ws} are −0.179 ($p = 0.003$), −0.118 ($p = 0.002$), and −0.056 ($p = 0.033$), respectively. However, the R^2 values range from 0.215 to 0.245, indicating a moderate level of explanation.

In contrast, the components of digital business have a positive impact on the share of renewable energy in the total primary energy supply (E_s, E_t, E_{ws} are 0.033 ($p = 0.046$), −0.011 ($p = 0.018$), and 0.020 ($p = 0.046$), respectively). This implies that when digitization is implemented in the business sector, it leads to an increase in the adoption of renewable energy sources. By embracing digitalization and integrating modern technologies (big data and cloud computing) into the business sector, the EU countries could enhance their efforts to promote renewable energy. These technologies enable businesses to optimize their energy consumption, improve resource efficiency, and facilitate the integration of renewable energy sources. Conversely, the results of Table 10 suggest that the variables E_s, E_t, and E_{ws} have varying impacts on final energy consumption. While the share of enterprises with e-commerce sales of at least 1% turnover shows a statistically significant positive impact ($E_t = 0.152$, $p = 0.018$), the relationship for the share of enterprises with e-commerce sales ($E_s = 0.130$, $p = 0.089$) and the share of enterprises with web sales ($E_{ws} = 0.021$, $p = 0.668$) is not statistically significant. This finding underscores the importance of considering the characteristics and scale of enterprises when assessing the energy implications of e-commerce activities and highlights the need for targeted policies and measures to encourage energy-efficient practices among businesses with significant e-commerce operations. The R^2 values for both models range from 0.837 to 0.853, suggesting that the independent variables collectively explain a substantial portion of the variation in the explanatory variables. The Wald chi2 statistics for all models demonstrate the overall significance of the regression models, with high chi2 values indicating a significant relationship between the independent variables E_s, E_t, E_{ws} and EI, FEC, $SDG7_{co_2}$, and $SDG7_{ren}$. The intraclass correlation (rho) values for all models in the analysis range from 0.663 to 0.904. These values indicate that a substantial proportion of the variance in the output is attributable to the differences across entities, accounting for approximately 66.3% to 90.4% of the total variance.

Table 10. The results of regression analysis: PCSE estimations.

| | EI | | | | | | FEC | | | | | |
| Variables | E_s | | E_t | | E_{ws} | | E_s | | E_t | | E_{ws} | |
	Coef.	Prob.	Coef.	Prob.	Coef.	Prob.	Coef.	Prob.	Coef.	Prob.	Coef.	Prob.
E_i	−0.112	0.001	−0.112	0.000	−0.076	0.002	0.130	0.089	0.152	0.032	0.021	0.668
GDP	0.628	0.000	0.622	0.000	0.639	0.000	0.284	0.003	0.291	0.002	0.311	0.001
WGI	−0.170	0.000	−0.166	0.000	−0.153	0.001	−0.011	0.916	−0.027	0.797	0.020	0.810
TO	−0.176	0.000	−0.178	0.000	−0.175	0.000	1.958	0.000	1.968	0.000	1.849	0.000
LST	−0.096	0.023	−0.085	0.036	−0.103	0.018	−0.280	0.000	−0.286	0.000	−0.215	0.001
PD	−0.055	0.091	−0.068	0.034	−0.046	0.163	0.848	0.000	0.861	0.000	0.799	0.000
const	11.461	0.000	11.413	0.000	11.412	0.000	5.817	0.000	5.715	0.000	5.399	0.000
Wald chi2	527.73	0.000	507.37	0.000	547.35	0.000	699.06	0.000	737.31	0.000	533.59	0.000
R^2	0.979		0.981		0.978		0.837		0.836		0.853	
rho	0.663		0.682		0.656		0.736		0.728		0.851	
	$SDG7_{co_2}$						$SDG7_{ren}$					
E_i	−0.179	0.003	−0.118	0.002	−0.056	0.033	0.033	0.046	0.011	0.018	0.020	0.046
GDP	−0.044	0.731	−0.053	0.681	−0.064	0.620	0.022	0.048	0.019	0.083	0.018	0.073
WGI	−0.044	0.020	−0.058	0.044	−0.075	0.027	0.032	0.087	0.038	0.052	0.036	0.053
TO	0.816	0.563	0.852	0.493	0.806	0.542	0.462	0.000	0.458	0.000	0.452	0.000
LST	0.100	0.057	0.106	0.051	0.106	0.049	−0.082	0.028	−0.084	0.024	−0.085	0.018
PD	0.197	0.008	0.189	0.007	0.206	0.008	0.416	0.000	0.416	0.000	0.417	0.000
const	−3.804	0.071	−4.060	0.053	−3.999	0.063	7.128	0.000	7.153	0.000	7.096	0.000
Wald chi2	47.06	0.000	47.90	0.000	41.62	0.000	196.83	0.000	190.72	0.000	178.35	0.000
R^2	0.232		0.245		0.215		0.844		0.844		0.846	
rho	0.904		0.889		0.903		0.836		0.840		0.852	

Based on the provided results, governance efficiency (WGI) has a significant influence on the dependent variables, excluding final energy consumption. This indicates that the effectiveness of governance in terms of efficiency and regulatory frameworks directly affects energy intensity, CO_2 emissions, or the adoption of renewable energy sources in the analyzed context. At the same time, WGI is conducive to extending renewable energy, and the growth of WGI provokes a decline in carbon dioxide emissions and energy intensity. GDP has a significant positive impact on final energy consumption, renewable energy, and energy intensity, but it does not significantly affect CO_2 emissions. The results show that TO has different effects on the dependent variables. It is positively associated with final energy consumption and the share of renewable energy, suggesting that countries with more open trade policies tend to prioritize the use of renewable energy sources. However, it has a negative effect on energy intensity. This means that countries with more open trade policies tend to have less energy intensity. Growth of TO leads to declining EI on average for all models with variables E_s, E_t, and E_{ws} by 0.17.

Land surface temperature (LST) is negatively associated with all dependent variables, excluding CO_2 emissions. It allows the conclusion that countries experiencing higher temperatures tend to have lower energy intensity, renewable energy, and energy consumption. This shows that countries with higher temperatures could face obstacles in increasing the adoption of renewable energy sources. Thus, LST growth led to declining energy intensity on average by 0.09, final energy consumption by 0.25, and renewable energy consumption by 0.08. However, LST is positively associated with CO_2 emissions, indicating that higher land surface temperatures may contribute to higher carbon emissions. Increasing LST provokes an increase in CO_2 emissions by 0.1. Population density has significant positive effects on all variables, excluding energy intensity. This means that higher population density is associated with less energy intensity and with higher final energy consumption, CO_2 emissions, and renewable energy. The significant positive effect between population density and the share of renewable energy indicates that areas with higher population density have a greater potential for adopting and integrating renewable energy sources. This could be driven by factors such as policy support, economies of scale, and the availability of infrastructure for renewable energy installations.

5. Discussion

The empirical findings from this investigation support our four hypotheses, indicating that digital business, the share of enterprises with e-commerce sales, enterprises with e-commerce sales of at least 1% turnover, and enterprises with web sales have a statistically significant impact on energy efficiency.

The substantial impact of all digital business indicators on energy intensity reinforces the potential of digitalization to enhance energy efficiency and reduce energy consumption in the EU. This aligns with the conclusions drawn by previous scholars [73–75], providing further evidence of the positive relationship between digital business practices and energy intensity reduction. These consistent findings highlight the importance of embracing digital technologies and strategies to promote energy efficiency across various sectors.

However, it is essential to acknowledge the presence of contrasting results in some studies [77–80], suggesting a nonlinear effect of digitalization on energy intensity. These divergent outcomes emphasize the complexity of the relationship, indicating that the impact of digital business practices on energy efficiency may vary based on specific contextual factors or interactions with other influencing variables. Such complexities warrant further investigation and in-depth analysis to better understand the nuanced nature of this relationship.

Additionally, the statistically significant positive impact of the share of enterprises with e-commerce sales of at least 1% turnover on final energy consumption echoes findings from another study [76]. This finding indicates that the energy demands associated with e-commerce activities, including order fulfillment, warehousing, and transportation, contribute to increased energy consumption. It underscores the importance of considering energy-efficient measures in e-commerce operations and implementing strategies to mitigate their energy-intensive processes.

6. Conclusions

This study aims at analyzing the impact of the share of enterprises with e-commerce sales, enterprises with e-commerce sales of at least 1% turnover, and enterprises with web sales (via websites, apps, or online marketplaces) on energy efficiency. This study arises from the increasing significance of digital business practices in the EU and their potential impact on energy efficiency and sustainability. As digitalization and e-commerce continue to expand, there is a growing need to understand how these trends influence energy consumption, CO_2 emissions, and the adoption of renewable energy sources. By exploring the associations between digital business indicators and energy efficiency, this research aims to provide valuable insights for policymakers and businesses to enhance energy-saving strategies, promote renewable energy integration, and advance the transition to a more sustainable and low-carbon economy. Understanding the implications of digital business on energy efficiency is essential in shaping effective policies and practices to tackle environmental challenges and meet energy sustainability targets in the EU.

The empirical results demonstrate the significant impact of digital business on energy intensity, CO_2 emissions, and the share of renewable energy in the EU. All indicators of digital business exhibit a statistically significant negative association with energy intensity, indicating that increased levels of digital business contribute to improved energy efficiency and reduced energy consumption. Moreover, the findings highlight the positive role of digital business practices in reducing CO_2 emissions. By embracing digitalization and integrating modern technologies like big data and cloud computing, EU countries can enhance their efforts to promote renewable energy implementation and mitigate carbon emissions. The share of enterprises with e-commerce sales of at least 1% turnover has a statistically significant positive impact on final energy consumption, emphasizing the need for energy-efficient measures in e-commerce operations. However, the relationships for the share of enterprises with e-commerce sales and the share of enterprises with web sales are not statistically significant.

7. Policy Implications

Considering the findings, the following policy implications for improving energy efficiency within digital business could be outlined:

1. It is necessary to establish energy efficiency standards specifically tailored for digital businesses, including e-commerce platforms. These standards should cover areas such as data centers, server utilization, packaging practices, and logistics operations. Implementing mandatory energy efficiency standards will promote sustainable practices and optimize energy consumption in the digital business sector.
2. The government should incentivize and facilitate the adoption of renewable energy sources by digital businesses, including e-commerce platforms. This could be achieved by providing financial incentives, tax breaks, or grants for investing in renewable energy infrastructure, such as on-site solar panels or wind turbines.
3. Policymakers should facilitate collaboration between digital businesses and energy providers to optimize energy consumption and promote energy efficiency. This should involve partnerships that offer preferential energy pricing for e-commerce platforms based on their commitment to energy efficiency and demand response programs. By collaborating with energy providers, digital businesses access expertise and technologies to improve their energy efficiency performance.
4. The EU countries should intensify investment in smart grid infrastructure to support the energy needs of an expanding digital business sector. Smart grids enable real-time monitoring, efficient energy distribution, and demand-side management. This facilitates the integration of renewable energy sources, enhances energy efficiency, and enables load balancing to optimize energy consumption. In addition, it is crucial to allocate funding for research and innovation projects that address energy efficiency challenges in the digital business sector. Encouraging collaboration between academia, industry, and research institutions will drive innovation and the development of energy-efficient practices and technologies.
5. Governments should mandate energy audits and reporting for digital businesses, particularly those engaged in e-commerce activities. Energy audits help identify energy-saving opportunities and enable businesses to track their energy consumption over time. Requiring regular reports of energy performance will drive transparency and accountability, encouraging businesses to improve their energy efficiency practices.
6. It is necessary to promote skills development and training programs focused on energy efficiency for digital businesses. This can include initiatives to upskill and reskill professionals in energy management, data analytics, and sustainable practices. The EU countries should launch public awareness campaigns to educate consumers about the energy implications of e-commerce and the importance of sustainable

purchasing behaviors. This could include raising awareness about energy-efficient delivery options, encouraging consolidated shipments, and promoting responsible consumption. Informed consumer choices could drive market demand for energy-efficient digital services and products.

8. Research Limitations and Further Investigations

Despite the reliable and valuable findings, this study has limitations that could be investigated in future analyses. The results highlight the importance of digital business practices in enhancing energy efficiency, reducing energy intensity, and promoting using renewable energy sources in the EU. Policymakers and businesses should consider the adoption of digital technologies and e-commerce practices to drive sustainable energy transitions and achieve environmental goals. However, further research is needed to explore the mechanisms through which digital business practices impact energy consumption and better understand these relationships' nuances. Furthermore, the chosen data series of analysis include the period of the COVID-19 pandemic. The pandemic's unprecedented disruptions to daily life and business operations may have significantly influenced energy consumption patterns, digital business practices, and overall economic activity. This aspect should be thoroughly investigated to understand the impact of pandemic-induced changes in consumer behavior, work-from-home practices, and supply chain disruptions on energy efficiency and adopting renewable energy sources. In addition, this study focuses on analyzing the EU countries, limiting the findings' implications for other countries. Future studies should extend the list of variables that could boost energy efficiency improvement and digital business development, such as investment, e-governance, and internet penetration. The study's limitation lies in the neglect of technology-related cause-effect relationships and the absence of considering the time-factor influence (time lag), suggesting that future investigations should incorporate these factors to gain a more comprehensive understanding of the digital business–energy efficiency relationship. Additionally, future investigations should incorporate different stakeholders (energy companies; digital enterprises; etc.) and develop the recommendations for them for improving energy efficiency. Future investigations should consider the influence of global indicators, such as GDP per capita, on both e-commerce and traditional trade. Accounting for these broader economic factors, the study will consider significant drivers that affect energy efficiency in both spheres and do not miss the opportunity for a comprehensive comparison between digital and traditional business models.

Author Contributions: Conceptualization, A.K., O.L. and T.P.; methodology, A.K., O.L. and T.P.; software, A.K., O.L. and T.P.; validation, A.K., O.L. and T.P.; formal analysis, A.K., O.L. and T.P.; investigation, A.K., O.L. and T.P.; resources, A.K., O.L. and T.P.; data curation, A.K., O.L. and T.P.; writing—original draft preparation, A.K., O.L. and T.P.; writing—review and editing, A.K., O.L. and T.P.; visualization, A.K., O.L. and T.P. All authors have read and agreed to the published version of the manuscript.

Funding: This research was funded by the Ministry of Education and Science of Ukraine, grant number 0121U100468.

Data Availability Statement: Not applicable.

Conflicts of Interest: The authors declare no conflict of interest.

References

1. Pietrzak, P.; Takala, J. Digital trust—Asystematic literature review. *Forum Sci. Oeconomia* **2021**, *9*, 59–71. [CrossRef]
2. Skvarciany, V.; Jurevičienė, D. An approach to the measurement of the digital economy. *Forum Sci. Oeconomia* **2021**, *9*, 89–102. [CrossRef]
3. Pudryk, D.; Kwilinski, A.; Lyulyov, O.; Pimonenko, T. Towards Achieving Sustainable Development: Interactions between Migration and Education. *Forum Sci. Oeconomia* **2023**, *11*, 113–132. [CrossRef]
4. Chen, Y.; Lyulyov, O.; Pimonenko, T.; Kwilinski, A. Green development of the country: Role of macroeconomic stability. *Energy Environ.* **2023**. [CrossRef]

5. Hussain, H.I.; Haseeb, M.; Kamarudin, F.; Dacko-Pikiewicz, Z.; Szczepańska-Woszczyna, K. The role of globalization, economic growth and natural resources on the ecological footprint in Thailand: Evidence from nonlinear causal estimations. *Processes* **2021**, *9*, 1103. [CrossRef]

6. Kuzior, A.; Sira, M.; Brozek, P. Using blockchain and artificial intelligence in energy management as a tool to achieve energy efficiency. *Virtual Econ.* **2022**, *5*, 69–90. [CrossRef]

7. Zhanibek, A.; Abazov, R.; Khazbulatov, A. Digital transformation of a country's image: The case of the Astana international finance center in Kazakhstan. *Virtual Econ.* **2022**, *5*, 71–94. [CrossRef]

8. Chygryn, O.; Kuzior, A.; Olefirenko, O.; Uzik, J. Green Brand as a New Pattern of Energy-Efficient Consumption. *Mark. Manag. Innov.* **2022**, *3*, 78–87. [CrossRef]

9. Chygryn, O.; Bektas, C.; Havrylenko, O. Innovation and Management of Smart Transformation Global Energy Sector: Systematic Literature Review. *Bus. Ethics Leadersh.* **2023**, *7*, 105–112. [CrossRef]

10. Szczepańska-Woszczyna, K.; Gatnar, S. Key Competences of Research and Development Project Managers in High Technology Sector. *Forum Sci. Oeconomia* **2022**, *10*, 107–130. [CrossRef]

11. Bag, S.; Omrane, A. The relationship between the personality traits of entrepreneurs and their decision-making process: The role of manufacturing SMEs' institutional environment in India. *Forum Sci. Oeconomia* **2021**, *9*, 103–122. [CrossRef]

12. Wyrwa, J.; Zaraś, M.; Wolak, K. Smart solutions in cities during the COVID-19 pandemic. *Virtual Econ.* **2021**, *4*, 88–103. [CrossRef]

13. Ziabina, Y.; Navickas, V. Innovations in Energy Efficiency Management: Role of Public Governance. *Mark. Manag. Innov.* **2022**, *4*, 218–227. [CrossRef]

14. Saługa, P.W.; Zamasz, K.; Dacko-Pikiewicz, Z.; Szczepańska-Woszczyna, K.; Malec, M. Risk-adjusted discount rate and its components for onshore wind farms at the feasibility stage. *Energies* **2021**, *14*, 6840. [CrossRef]

15. Trushkina, N.; Abazov, R.; Rynkevych, N.; Bakhautdinova, G. Digital Transformation of Organizational Culture under Conditions of the Information Economy. *Virtual Econ.* **2020**, *3*, 7–38. [CrossRef]

16. Dzwigol, H.; Aleinikova, O.; Umanska, Y.; Shmygol, N.; Pushak, Y. An Entrepreneurship Model for Assessing the Investment Attractiveness of Regions. *J. Entrep. Educ.* **2019**, *22* (Suppl. S1), 1–7.

17. Vaníčková, R.; Szczepańska-Woszczyna, K. Innovation of business and marketing plan of growth strategy and competitive advantage in exhibition industry. *Pol. J. Manag. Stud.* **2020**, *21*, 425–445. [CrossRef]

18. Kolosok, S.; Saher, L.; Kovalenko, Y.; Delibasic, M. Renewable Energy and Energy Innovations: Examining Relationships Using Markov Switching Regression Model. *Mark. Manag. Innov.* **2022**, *2*, 151–160. [CrossRef]

19. Karnowski, J.; Miśkiewicz, R. Climate Challenges and Financial Institutions: An Overview of the Polish Banking Sector's Practices. *Eur. Res. Stud. J.* **2021**, *XXIV*, 120–139. [CrossRef]

20. Arefieva, O.; Polous, O.; Arefiev, S.; Tytykalo, V.; Kwilinski, A. Managing sustainable development by human capital reproduction in the system of company's organizational behavior. *IOP Conf. Ser. Earth Environ. Sci.* **2021**, *628*, 012039. [CrossRef]

21. Kharazishvili, Y.; Kwilinski, A.; Sukhodolia, O.; Dzwigol, H.; Bobro, D.; Kotowicz, J. The systemic approach for estimating and strategizing energy security: The case of Ukraine. *Energies* **2021**, *14*, 2126. [CrossRef]

22. Miśkiewicz, R.; Rzepka, A.; Borowiecki, R.; Olesińki, Z. Energy Efficiency in the Industry 4.0 Era: Attributes of Teal Organizations. *Energies* **2021**, *14*, 6776. [CrossRef]

23. Dźwigoł, H. The uncertainty factor in the market economic system: The microeconomic aspect of sustainable development. *Virtual Econ.* **2021**, *4*, 98–117. [CrossRef]

24. Miśkiewicz, R.; Matan, K.; Karnowski, J. The Role of Crypto Trading in the Economy, Renewable Energy Consumption and Ecological Degradation. *Energies* **2022**, *15*, 3805. [CrossRef]

25. Miśkiewicz, R. The Impact of Innovation and Information Technology on Greenhouse Gas Emissions: A Case of the Visegrád Countries. *J. Risk Financ. Manag.* **2021**, *14*, 59. [CrossRef]

26. Matvieieva, Y.; Hamida, H.B. Modeling and Forecasting Energy Efficiency Impact on the Human Health. *Health Econ. Manag. Rev.* **2022**, *3*, 78–85. [CrossRef]

27. Vakulenko, I.; Lieonov, H. Renewable Energy and Health: Bibliometric Review of Non-Medical Research. *Health Econ. Manag. Rev.* **2022**, *3*, 44–53. [CrossRef]

28. Letunovska, N.; Saher, L.; Vasylieva, T.; Lieonov, S. Dependence of public health on energy consumption: A cross-regional analysis. In Proceedings of the E3S Web of Conferences, Odesa, Ukraine, 16 April 2021; p. 250. [CrossRef]

29. Zhou, X.; Zhou, D.; Zhao, Z.; Wang, Q. A framework to analyze carbon impacts of digital economy: The case of China. *Sustain. Prod. Consum.* **2022**, *31*, 357–369. [CrossRef]

30. Sovacool, B.K.; Upham, P.; Monyei, C.G. The "whole systems" energy sustainability of digitalization: Humanizing the community risks and benefits of Nordic datacenter development. *Energy Res. Soc. Sci.* **2022**, *88*, 102493. [CrossRef]

31. Katal, A.; Dahiya, S.; Choudhury, T. Energy efficiency in cloud computing data centers: A survey on software technologies. *Clust. Comput.* **2022**, *26*, 1845–1875. [CrossRef] [PubMed]

32. Dzwigol, H.; Trushkina, N.; Kwilinski, A. The organizational and economic mechanism of implementing the concept of green logistics. *Virtual Econ.* **2021**, *4*, 41–75. [CrossRef]

33. Ziabina, Y.; Kovalenko, Y. Regularities in the Development of The Theory of Energy Efficiency Management. *SocioEconomic Chall.* **2021**, *5*, 117–132. [CrossRef]

34. Drożdż, W.; Kinelski, G.; Czarnecka, M.; Wójcik-Jurkiewicz, M.; Maroušková, A.; Zych, G. Determinants of Decarbonization—How to Realize Sustainable and Low Carbon Cities? *Energies* **2021**, *14*, 2640. [CrossRef]
35. Chygryn, O.; Miskiewicz, R. New trends and patterns in green competitiveness: A bibliometric analysis of evolution. *Virtual Econ.* **2022**, *5*, 24–41. [CrossRef] [PubMed]
36. Gao, X.; Huang, W.; Wang, H. Financial twitter sentiment on bitcoin return and high-frequency volatility. *Virtual Econ.* **2021**, *4*, 7–18. [CrossRef]
37. Pavlyk, V. Assessment of green investment impact on the energy efficiency gap of the national economy. *Financ. Mark. Inst. Risks* **2020**, *4*, 117–123. [CrossRef]
38. Prokopenko, O.; Miśkiewicz, R. Perception of "green shipping" in the contemporary conditions. *Entrep. Sustain. Issues* **2020**, *8*, 269–284. [CrossRef]
39. Hmoud, B. The adoption of artificial intelligence in human resource management. *Forum Sci. Oeconomia* **2021**, *9*, 105–118. [CrossRef]
40. Dacko-Pikiewicz, Z. Building a family business brand in the context of the concept of stakeholder-oriented value. *Forum Sci. Oeconomia* **2019**, *7*, 37–51. [CrossRef]
41. Miśkiewicz, R. Challenges facing management practice in the light of Industry 4.0: The example of Poland. *Virtual Econ.* **2019**, *2*, 37–47. [CrossRef]
42. Dzwigol, H. Research Methodology in Management Science: Triangulation. *Virtual Econ.* **2022**, *5*, 78–93. [CrossRef]
43. Stępień, S.; Smędzik-Ambroży, K.; Polcyn, J.; Kwiliński, A.; Maican, I. Are small farms sustainable and technologically smart? Evidence from Poland, Romania, and Lithuania. *Cent. Eur. Econ. J.* **2023**, *10*, 116–132. [CrossRef]
44. Yamoah, F.A.; Ul Haque, A. Strategic management through digital platforms for remote working in the higher education industry during and after the COVID-19 pandemic. *Forum Sci. Oeconomia* **2022**, *10*, 111–128. [CrossRef]
45. Ramadania, R.; Ratnawati, R.; Juniwati, J.; Afifah, N.; Heriyadi, H.; Darma, D.C. Impulse buying and hedonic behavior: A mediation effect of positive emotions. *Virtual Econ.* **2022**, *5*, 43–64. [CrossRef] [PubMed]
46. Novikov, V.V. Digitalization of Economy and Education: Path to Business Leadership and National Security. *Bus. Ethics Leadersh.* **2021**, *5*, 147–155. [CrossRef]
47. Miśkiewicz, R. The importance of knowledge transfer on the energy market. *Polityka Energetyczna* **2018**, *21*, 49–62. [CrossRef]
48. Trzeciak, M.; Jonek-Kowalska, I. Monitoring and Control in Program Management as Effectiveness Drivers in Polish Energy Sector. Diagnosis and Directions of Improvement. *Energies* **2021**, *14*, 4661. [CrossRef]
49. Trzeciak, M.; Kopec, T.P.; Kwilinski, A. Constructs of Project Programme Management Supporting Open Innovation at the Strategic Level of the Organization. *J. Open Innov. Technol. Mark. Complex.* **2022**, *8*, 58. [CrossRef]
50. Vorontsova, A.; Vasylieva, T.; Lyeonov, S.; Artyukhov, A.; Mayboroda, T. Education expenditures as a factor in bridging the gap at the level of digitalization. In Proceedings of the 2021 11th International Conference on Advanced Computer Information Technologies, ACIT 2021—Proceedings 2021, Deggendorf, Germany, 15–17 September 2021; pp. 242–245. [CrossRef]
51. Hassan, T.; Song, H.; Khan, Y.; Kirikkaleli, D. Energy efficiency a source of low carbon energy sources? Evidence from 16 high-income OECD economies. *Energy* **2022**, *243*, 123063. [CrossRef]
52. Dong, K.; Sun, R.; Hochman, G.; Li, H. Energy intensity and energy conservation potential in China: A regional comparison perspective. *Energy* **2018**, *155*, 782–795. [CrossRef]
53. Shahiduzzaman, M.; Alam, K. Changes in energy efficiency in Australia: A decomposition of aggregate energy intensity using logarithmic mean Divisia approach. *Energy Policy* **2013**, *56*, 341–351. [CrossRef]
54. Feng, T.; Sun, L.; Zhang, Y. The relationship between energy consumption structure, economic structure and energy intensity in China. *Energy Policy* **2009**, *37*, 5475–5483. [CrossRef]
55. Hosan, S.; Karmaker, S.C.; Rahman, M.M.; Chapman, A.J.; Saha, B.B. Dynamic links among the demographic dividend, digitalization, energy intensity and sustainable economic growth: Empirical evidence from emerging economies. *J. Clean. Prod.* **2022**, *330*, 129858. [CrossRef]
56. Su, B.; Goh, T.; Ang, B.W.; Ng, T.S. Energy consumption and energy efficiency trends in Singapore: The case of a meticulously planned city. *Energy Policy* **2022**, *161*, 112732. [CrossRef]
57. Taylor, P.G.; d'Ortigue, O.L.; Francoeur, M.; Trudeau, N. Final energy use in IEA countries: The role of energy efficiency. *Energy Policy* **2010**, *38*, 6463–6474. [CrossRef]
58. Paramati, S.R.; Shahzad, U.; Doğan, B. The role of environmental technology for energy demand and energy efficiency: Evidence from OECD countries. *Renew. Sustain. Energy Rev.* **2022**, *153*, 111735. [CrossRef]
59. Khan, S.; Murshed, M.; Ozturk, I.; Khudoykulov, K. The roles of energy efficiency improvement, renewable electricity production, and financial inclusion in stimulating environmental sustainability in the Next Eleven countries. *Renew. Energy* **2022**, *193*, 1164–1176. [CrossRef]
60. McLaughlin, E.; Choi, J.K.; Kissock, K. Techno-economic impact assessments of energy efficiency improvements in the industrial combustion systems. *J. Energy Resour. Technol.* **2022**, *144*, 082109. [CrossRef]
61. Lipiäinen, S.; Kuparinen, K.; Sermyagina, E.; Vakkilainen, E. Pulp and paper industry in energy transition: Toward energy-efficient and low carbon operation in Finland and Sweden. *Sustain. Prod. Consum.* **2022**, *29*, 421–431. [CrossRef]
62. Alajmi, R.G. Carbon emissions and electricity generation modeling in Saudi Arabia. *Environ. Sci. Pollut. Res.* **2022**, *29*, 23169–23179. [CrossRef]

63. Zhao, J.; Dong, K.; Dong, X.; Shahbaz, M. How renewable energy alleviate energy poverty? A global analysis. *Renew. Energy* **2022**, *186*, 299–311. [CrossRef]
64. Wen, J.; Okolo, C.V.; Ugwuoke, I.C.; Kolani, K. Research on influencing factors of renewable energy, energy efficiency, on technological innovation. Does trade, investment and human capital development matter? *Energy Policy* **2022**, *160*, 112718. [CrossRef]
65. Akadiri, S.S.; Adebayo, T.S. Asymmetric nexus among financial globalization, nonrenewable energy, renewable energy use, economic growth, and carbon emissions: Impact on environmental sustainability targets in India. *Environ. Sci. Pollut. Res.* **2022**, *29*, 16311–16323. [CrossRef] [PubMed]
66. Jahanger, A.; Ozturk, I.; Onwe, J.C.; Joseph, T.E.; Hossain, M.R. Do technology and renewable energy contribute to energy efficiency and carbon neutrality? Evidence from top ten manufacturing countries. *Sustain. Energy Technol. Assess.* **2023**, *56*, 103084. [CrossRef]
67. Chen, H.; Shi, Y.; Zhao, X. Investment in renewable energy resources, sustainable financial inclusion, and energy efficiency: A case of US economy. *Resour. Policy* **2022**, *77*, 102680. [CrossRef]
68. Miskiewicz, R. Efficiency of electricity production technology from postprocess gas heat: Ecological, economic and social benefits. *Energies* **2020**, *13*, 6106. [CrossRef]
69. Saługa, P.W.; Szczepańska-Woszczyna, K.; Miśkiewicz, R.; Chład, M. Cost of equity of coal-fired power generation projects in Poland: Its importance for the management of decision-making process. *Energies* **2020**, *13*, 4833. [CrossRef]
70. Iqbal, N.; Kim, D.H. An Iot task management mechanism based on predictive optimization for efficient energy consumption in smart residential buildings. *Energy Build.* **2022**, *257*, 111762.
71. Rusch, M.; Schöggl, J.P.; Baumgartner, R.J. Application of digital technologies for sustainable product management in a circular economy: A review. *Bus. Strategy Environ.* **2023**, *32*, 1159–1174. [CrossRef]
72. Xue, Y.; Tang, C.; Wu, H.; Liu, J.; Hao, Y. The emerging driving force of energy consumption in China: Does digital economy development matter? *Energy Policy* **2022**, *165*, 112997. [CrossRef]
73. Ghaffarianhoseini, A.; Ghaffarianhoseini, A.; Tookey, J.; Omrany, H.; Fleury, A.; Naismith, N.; Ghaffarianhoseini, M. The essence of smart homes: Application of intelligent technologies toward smarter urban future. In *Artificial Intelligence: Concepts, Methodologies, Tools, and Applications*; IGI Global: Hershey, PA, USA, 2017; pp. 79–121.
74. Yang, A.; Han, M.; Zeng, Q.; Sun, Y. Adopting building information modeling (BIM) for the development of smart buildings: A review of enabling applications and challenges. *Adv. Civ. Eng.* **2021**, *2021*, 1–26. [CrossRef]
75. Pålsson, H.; Pettersson, F.; Hiselius, L.W. Energy consumption in e-commerce versus conventional trade channels-Insights into packaging, the last mile, unsold products and product returns. *J. Clean. Prod.* **2017**, *164*, 765–778. [CrossRef]
76. Reijnders, L.; Hoogeveen, M.J. Energy effects associated with e-commerce: A case-study concerning online sales of personal computers in The Netherlands. *J. Environ. Manag.* **2001**, *62*, 317–321. [CrossRef]
77. Lei, X.; Ma, Y.; Ke, J.; Zhang, C. The Non-Linear Impact of the Digital Economy on Carbon Emissions Based on a Mediated Effects Model. *Sustainability* **2023**, *15*, 7438. [CrossRef]
78. Xu, Q.; Zhong, M.; Li, X. How does digitalization affect energy? International evidence. *Energy Econ.* **2022**, *107*, 105879. [CrossRef]
79. Cardinali, P.G.; De Giovanni, P. Responsible digitalization through digital technologies and green practices. *Corp. Soc. Responsib. Environ. Manag.* **2022**, *29*, 984–995. [CrossRef]
80. Wang, W.; Yang, X.; Cao, J.; Bu, W.; Adebayo, T.S.; Dilanchiev, A.; Ren, S. Energy internet, digital economy, and green economic growth: Evidence from China. *Innov. Green Dev.* **2022**, *1*, 100011. [CrossRef]
81. Morley, J.; Widdicks, K.; Hazas, M. Digitalization, energy and data demand: The impact of Internet traffic on overall and peak electricity consumption. *Energy Res. Soc. Sci.* **2018**, *38*, 128–137. [CrossRef]
82. Babu, B.R.; Parande, A.K.; Basha, C.A. Electrical and electronic waste: A global environmental problem. *Waste Manag. Res.* **2007**, *25*, 307–318.
83. Gunasekaran, A.; Marri, H.B.; McGaughey, R.E.; Nebhwani, M.D. E-commerce and its impact on operations management. *Int. J. Prod. Econ.* **2002**, *75*, 185–197. [CrossRef]
84. Chen, X.; Mao, S.; Lv, S.; Fang, Z. A Study on the Non-Linear Impact of Digital Technology Innovation on Carbon Emissions in the Transportation Industry. *Int. J. Environ. Res. Public Health* **2022**, *19*, 12432. [CrossRef]
85. Dogan, E.; Majeed, M.T.; Luni, T. Are clean energy and carbon emission allowances caused by bitcoin? A novel time-varying method. *J. Clean. Prod.* **2022**, *347*, 131089. [CrossRef]
86. Huang, F.L.; Wiedermann, W.; Zhang, B. Accounting for heteroskedasticity resulting from between-group differences in multilevel models. *Multivar. Behav. Res.* **2022**, *58*, 637–657. [CrossRef]
87. Orji, E.N.; John-Akamelu, C.R. Book Tax Differences and Financial Distress of Public Listed Consumer Goods Firms in Nigeria. *J. Glob. Account.* **2023**, *9*, 346–382.
88. Bhimavarapu, V.M.; Rastogi, S.; Gupte, R.; Pinto, G.; Shingade, S. Does the Impact of Transparency and Disclosure on the Firm's Valuation Depend on the ESG? *J. Risk Financ. Manag.* **2022**, *15*, 410. [CrossRef]
89. Pesaran, M.H. A simple panel unit root test in the presence of cross-section dependence. *J. Appl. Econom.* **2007**, *22*, 265–312.
90. Baltagi, B.H.; Hashem Pesaran, M. Heterogeneity and cross section dependence in panel data models: Theory and applications introduction. *J. Appl. Econom.* **2007**, *22*, 229–232. [CrossRef]

91. Pao, H.T.; Tsai, C.M. Multivariate Granger causality between CO_2 emissions, energy consumption, FDI (foreign direct investment) and GDP (gross domestic product): Evidence from a panel of BRIC (Brazil, Russian Federation, India, and China) countries. *Energy* **2011**, *36*, 685–693. [CrossRef]
92. Bakhsh, K.; Rose, S.; Ali, M.F.; Ahmad, N.; Shahbaz, M. Economic growth, CO_2 emissions, renewable waste and FDI relation in Pakistan: New evidence from 3SLS. *J. Environ. Manag.* **2017**, *196*, 627–632. [CrossRef]
93. Alqaralleh, H. On the nexus of CO_2 emissions and renewable and nonrenewable energy consumption in Europe: A new insight from panel smooth transition. *Energy Environ.* **2021**, *32*, 443–457. [CrossRef]
94. Liu, Y.; Sadiq, F.; Ali, W.; Kumail, T. Does tourism development, energy consumption, trade openness and economic growth matters for ecological footprint: Testing the Environmental Kuznets Curve and pollution haven hypothesis for Pakistan. *Energy* **2022**, *245*, 123208. [CrossRef]
95. Szczepańska-Woszczyna, K.; Gedvilaitė, D.; Nazarko, J.; Stasiukynas, A.; Rubina, A. Assessment of Economic Convergence among Countries in the European Union. *Technological and Economic Development of Economy* **2022**, *28*, 1572–1588. [CrossRef]
96. Karnowski, J.; Rzońca, A. Should Poland join the euro area? The challenge of the boom-bust cycle. *Argum. Oeconomica* **2023**, *1*, 227–262. [CrossRef]
97. Shahbaz, M.; Nasreen, S.; Ahmed, K.; Hammoudeh, S. Trade openness–carbon emissions nexus: The importance of turning points of trade openness for country panels. *Energy Econ.* **2017**, *61*, 221–232. [CrossRef]
98. Ziabina, Y.; Pimonenko, T.; Starchenko, L. Energy Efficiency of National Economy: Social, Economic and Ecological Indicators. *SocioEconomic Chall.* **2020**, *4*, 160–174. [CrossRef]
99. Wołowiec, T.; Kolosok, S.; Vasylieva, T.; Artyukhov, A.; Skowron, Ł.; Dluhopolskyi, O.; Sergiienko, L. Sustainable governance, energy security, and energy losses of Europe in turbulent times. *Energies* **2022**, *15*, 8857. [CrossRef]
100. Kafy, A.A.; Al Rakib, A.; Fattah, M.A.; Rahaman, Z.A.; Sattar, G.S. Impact of vegetation cover loss on surface temperature and carbon emission in a fastest-growing city, Cumilla, Bangladesh. *Build. Environ.* **2022**, *208*, 108573. [CrossRef]
101. Jamali, A.A.; Kalkhajeh, R.G.; Randhir, T.O.; He, S. Modeling relationship between land surface temperature anomaly and environmental factors using GEE and Giovanni. *J. Environ. Manag.* **2022**, *302*, 113970. [CrossRef]
102. Luan, F.; Yang, X.; Chen, Y.; Regis, P.J. Industrial robots and air environment: A moderated mediation model of population density and energy consumption. *Sustain. Prod. Consum.* **2022**, *30*, 870–888. [CrossRef]
103. Mahumane, G.; Mulder, P. Urbanization of energy poverty? The case of Mozambique. *Renew. Sustain. Energy Rev.* **2022**, *159*, 112089. [CrossRef]
104. Jiang, L.; Shi, X.; Wu, S.; Ding, B.; Chen, Y. What factors affect household energy consumption in mega-cities? A case study of Guangzhou, China. *J. Clean. Prod.* **2022**, *363*, 132388. [CrossRef]
105. Eurostat. Available online: https://ec.europa.eu/eurostat (accessed on 10 April 2023).
106. World Data Bank. Available online: https://databank.worldbank.org (accessed on 10 April 2023).
107. Anderson, B.J.; Slater, L.J.; Dadson, S.J.; Blum, A.G.; Prosdocimi, I. Statistical attribution of the influence of urban and tree cover change on streamflow: A comparison of large sample statistical approaches. *Water Resour. Res.* **2022**, *58*, e2021WR030742. [CrossRef]
108. Im, K.S.; Pesaran, M.H.; Shin, Y. Testing for unit roots in heterogeneous panels. *J. Econom.* **2003**, *115*, 53–74. [CrossRef]
109. Choi, I. Unit root tests for panel data. *J. Int. Money Financ.* **2001**, *20*, 249–272. [CrossRef]
110. Marques, A.C.; Junqueira, T.M. European energy transition: Decomposing the performance of nuclear power. *Energy* **2022**, *245*, 123244. [CrossRef]

 *information*

Article

Health Monitoring Apps: An Evaluation of the Persuasive System Design Model for Human Wellbeing

Asif Hussian [1], Abdul Mateen [1], Farhan Amin [2,*], Muhammad Ali Abid [3] and Saeed Ullah [1]

[1] Department of Computer Science, Federal Urdu University of Arts, Science and Technology, Islamabad 44000, Pakistan; asif.baig@kiu.edu.pk (A.H.); abdulmateen@fuuastisb.edu.pk (A.M.); saeedullah@gmail.com (S.U.)

[2] Department of Information and Communication Engineering, Yeungnam University, Gyeongsan 38541, Republic of Korea

[3] Faculty of Smart Engineering, The University of Agriculture, Dera Ismail Khan 29220, Pakistan; draliabidawan@gmail.com

* Correspondence: farhan@ynu.ac.kr

Abstract: In the current era of ubiquitous computing and mobile technology, almost all human beings use various self-monitoring applications. Mobile applications could be the best health assistant for safety and adopting a healthy lifestyle. Therefore, persuasive designing is a compulsory element for designing such apps. A popular model for persuasive design named the Persuasive System Design (PSD) model is a generalized model for whole persuasive technologies. Any type of persuasive application could be designed using this model. Designing any special type of application using the PSD model could be difficult because of its generalized behavior which fails to provide moral support for users of health applications. There is a strong need to propose a customized and improved persuasive system design model for each category to overcome the issue. This study evaluates the PSD model and finds persuasive gaps in users of the Mobile Health Monitoring application, developed by following the PSD model. Furthermore, this study finds that users misunderstand health-related problems when using such apps. A misunderstanding of this nature can have serious consequences for the user's life in some cases.

Keywords: big data; human-computer interaction; persuasive technologies; mobile health monitoring apps; persuasive system design model

Citation: Hussian, A.; Mateen, A.; Amin, F.; Abid, M.A.; Ullah, S. Health Monitoring Apps: An Evaluation of the Persuasive System Design Model for Human Wellbeing. *Information* **2023**, *14*, 412. https://doi.org/10.3390/info14070412

Academic Editor: Willy Susilo

Received: 13 June 2023
Revised: 11 July 2023
Accepted: 14 July 2023
Published: 16 July 2023

1. Introduction

Modern technologies that are designed to change the attitudes or behaviors of the users through persuasion and social influence and not through force are considered persuasive technologies. Such technologies are heavily used in trade, negotiation, politics, religion, military training, public health, and management, as well as in various areas of human-to-human or human-to-computer communication. Mobile applications are very effective tools that promote health, better attitudes, and good behavior in their users. As the number of smartphone users is rapidly increasing day by day, the number of applications is also increasing in smartphone app stores. Therefore, it is very important to build bridges between mobile Human-Computer Interactions (HCIs) and persuasive technologies as well as health psychology [1]. To influence the attitudes or behavior of users, persuasive technologies are used in different mobile applications. A framework introduced by researchers, known as the PSD Model [2], strongly influences the behaviors and attitudes of the user. Evaluation of the current state of mobile applications through persuasion to promote physical activity by changing the behavior of the user is very necessary. The PSD Model also helps to evaluate persuasive features in mobile applications [3]. The increasing influence of mobile health applications in the current technological era cannot be underestimated because they inform, educate, and persuade consumers. The current era of mobile applications enables a user to access suitable health education from different reliable sources. It is the will of researchers to ensure that this expanding field will reach its greatest potential, so it is necessary to understand the current mobile technology resources which can be used to improve the wellbeing of people. It is also very important for governments to introduce design principles that influence the usefulness of persuasive technologies to be understood [3]. This research work endorses all basic categories for persuasive system design principles, i.e., Primary Task Support, Dialogue Support, System Credibility Support, and Social Support of the PSD. This study emphasizes that Health Monitoring Apps (HMA) must provide results and recommendations for the user by keeping emotional and mental values in view. As we all know, HMAs are directly linked with a user's life. Any misunderstanding or mistake can even be a threat to the user's life. This research work recommends that there is a strong need to develop a customized Persuasive System Design model for HMAs.

The rest of the article is organized as follows: Section 2 elaborates on the literature survey. Section 3 contains a description of the Persuasive System Design (PSD) Model. Section 4 describes a summary of the PSD model. Section 5 elaborates on some current research work in the field of study. Section 6 describes the user studies, results, and the actual problem statement and research objectives. Finally, Section 7 concludes the article with research contributions and future work.

2. Literature Review

Mobile applications are very effective tools that promote health, better attitudes, and behavior in their users. According to [4], 12 out of 57 studies show that self-monitoring was the most common behavior change technique. As the number of smartphone users is rapidly increasing day by day, the number of applications is also increasing in the app stores of smartphones. Therefore, it is very important to bridge the gap between mobile HCI and persuasive technologies as well as health psychology. Principles and theories must be considered in designing persuasive mobile apps for health and safety promotion, and on how to rigorously extend Mobile HCI evaluation methods to measure the effectiveness of such apps [1]. The lack of appropriate human behavior causes the degradation of the environment, but social societies try their best to overcome the issues for the sake of human wellbeing. They develop and maintain the world we live in. Ref. [5] describes good behavior, named pro-environmental behavior (PEB), which leads to benefits for the environment. The popular thesis on human behavior, known as the theory of planned behavior (TPB), describes that "an individual's intention towards behavior, subjective norms, and perceived control over his/her behavior together leads to intentions and behavior". In the presence of the generalized PSD Model, for designing and analyzing persuasive technologies, Ref. [5] suggested and proposed a specialized model for developing pro-environmental behavior. Ref. [6] conducted a review of current persuasive technology design strategies and gathered the frequency of each strategy being studied by the researchers. However, there is a strong need for user studies to be performed with the actual end-users of HMA, so that we can observe the persuasive gaps and analyze the results in light of existing persuasive principles. What are the troubles and hardships facing the end-users of mobile HMAs? In light of the above facts, there is a need for a specialized persuasive design model for health monitoring apps. Because mobile HMAs are directly linked with the user's life, these apps need special attention during the design process. Mobile health apps named Sehha and Mawid were developed by following the basic principles of the PSD model by the health ministry of the Saudi Arabia Government [7]. The Sehha and Mawid apps were found to lack social support. The study guesses about doubtful results of application social support, either due to a lack of developers' knowledge or the nature of the application. However, the study [7] endorses that the mobile health apps (Sehha and Mawid) had been developed by following the PSD model. The above facts also lead to work on the PSD model and need to be examined.

2.1. Persuasive Technologies

All the interactive information technologies which are designed for changing users' attitudes or behaviors are known as persuasive technologies [2]. Ref. [8] Persuasive technologies are defined as "an attempt to shape, reinforce, or change behaviors, feelings, or thoughts about an issue, object, or action". To determine and identify the behavior change applications, a study was conducted in 2018 [9] by researchers which found 212,352 apps. A total of 5018 apps remained after applying the filter criteria. Out of the total, only 344 applications were found to be persuasive (behavior change) applications [9]. That study [9] recommended improvements in the designing of apps to help users adopt a justifiable and substantial lifestyle.

2.2. Persuasion in Mobile Health Apps

All mobile applications are designed to influence the attitudes and behaviors of the human being. The main objective behind the scene is to convince them to buy the product. Ref. [10] suggests that "smartphone applications have shown promise in supporting people to adopt healthy lifestyles". Various human health-related mobile apps are easily available to every smartphone user. Ref. [11] A usable application can be designed and developed with the help of HCI modeling. Mental health care and suicide prevention inequities may also be effectively recognized, acknowledged, and addressed. A digital therapeutic alliance might also benefit from it.

2.3. Health Monitoring Apps

Applications that are dedicated to monitoring health problems and self-assistance to adopt a healthy lifestyle are considered health monitoring apps. The following are examples of health monitoring apps.

- iCare Health Monitor (Figure 1)
- Wii Fit
- Wii Zumba Fitness or Wii Sports Resort

Figure 1. iCare Health Monitor (HMA).

Runtastic—Laufen and Fitness Fit4Life [12] In the literature, there are various theories and frameworks [13] centered around designing a persuasive system or technology, such as:

- Theory of Reasoned Action/Theory of Planned Behavior
- Reasoned Action Approach
- Technology Acceptance Model
- Information Processing Model
- Captology [6]
- Principles of Influence
- Persuasive Systems Design (PSD) Model
- Virtual Narrator [14].

Mobile health apps that track the symptoms of the patient were examined in a systematic review [15]. They have used the words cancer, oncology, and symptom tracker to search relevant apps in the iOS App Store and Android Google Play. Patients with cancer could record their symptoms and PROs by utilizing apps that included a symptom-tracking feature. A mobile app rating scale was used to assess each app's engagement, functionality, aesthetics, information, and subjective quality. After screening the titles and descriptions, 101 apps were found to be eligible out of a total of 1189 apps after the initial search. That study included 41 apps that met the eligibility criteria. A single cancer patient-friendly app has been tested in their review study [15]. The above facts from existing knowledge determine that there is a strong need for interactive and persuasive designing strategies to develop HMAs. The PSD model is also providing design strategies for designing such apps, so this study focuses on the evaluation of the PSD model.

3. The Persuasive System Design (PSD) Model

A model named PSD [16] was proposed which discusses the process of designing and evaluating persuasive systems and describes what kind of content and software functionality may be found in the final product. It also highlights seven underlying postulates behind persuasive systems and ways to analyze the persuasion context (the intent, the event, and the strategy). In light of the PSD model, Ref. [2] lists 28 design principles for persuasive system content and functionality, describing example software requirements and implementations. The basic design principles consist of four categories. These categories are primary task, dialogue, system credibility, and social support. For a better representation of the model, we have designed Figure 2 which will help the audience to easily understand the principles of the PSD Model.

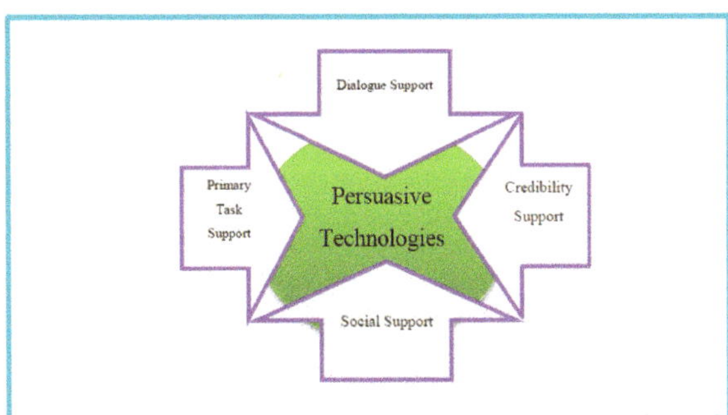

Figure 2. Basic Principles of the PSD Model.

Each category for the PSD principle has seven postulates. This study presents all twenty-eight postulates for a better understanding of the literature [3].

3.1. Primary Task Support

Primary task support facilitates users to interact with a system and helps them to track their performance through features such as self-monitoring. The design principles in primary task support are as below:

i.   Reduction: Making simpler tasks by reducing the complexity of the system design.
ii.   Tunneling: Guiding users through a process or experience.
iii.   Tailoring: The system must provide appropriate information for its user groups.
iv.   Personalization: The system must offer personalized content and services for its users.
v.   Self-monitoring: The system must provide means for users to track their routine or status.
vi.   Simulation: Immediately observe the link between cause and effect.
vii.   Rehearsal: The system must offer means to practice.

3.2. Dialogue Support

Dialogue support features improve dialogue between the user and the system, especially in terms of system feedback to better guide the user through the intended behavior/attitude change process. The seven design principles in system dialogue support are as follows:

i.   Praise: The system must allow for criticism in order to have user feedback.
ii.   Rewards: Providing a virtual environment to give credit for performing target behavior.
iii.   Reminders: Reminders should be allowed to achieve targeted behavior.
iv.   Suggestions: The system should suggest that users carry out behaviors during the system use process.
v.   Similarity: The system must follow its users by some particular method.
vi.   Liking: Visually attractive content that feels appealing to its users.
vii.   Social support: The system should adopt a social role to provide a virtual environment.

3.3. System Credibility Support

Features such as authority, expertise, a real-world feel, and verifiability promote the credibility of a persuasive system.

i.   Trustworthiness: The system must provide information that is true, fair, and unbiased.
ii.   Expertise: Must provide knowledge, experience, and competence.
iii.   Surface credibility: This should be a firsthand inspection.
iv.   Real-world feel: The system must provide information about the organization and/or actual people behind its content and services.
v.   Authority: the system should refer the inquiries to authorized powers.
vi.   Third-party endorsements: Feedback from well-known and credible sources.
vii.   Verifiability: Must offer means to verify the accuracy of system content via outside sources.

3.4. Social Support

Social support features foster user motivation through components such as cooperation, normative influence, social comparison, and social learning.

i.   Social learning: The system must offer to have information from others.
ii.   Social comparison: The system must offer an element of comparison on social forums.
iii.   Normative influence: The system must gather peoples who have the same goals.
iv.   Social facilitation: The system should provide means for recognizing other users who are performing the same behavior.
v.   Cooperation: The system should offer a cooperative platform.
vi.   Competition: The system should provide means for competing with other users.

vii. Recognition: The system should provide public recognition for users who perform their target behavior.

4. Short Summary of the PSD Model

Heading III summarizes the PSD Model, persuasive technologies, and health monitoring apps in light of the existing literature. The principles of the PSD model have been focused on. Each category of PSD principle has seven sub-principles. All twenty-eight principles are used to design a persuasive system. From the literature, it has been derived that information technology is never neutral. People like to share their views about the world for organizing and improving the promotion of their products. The hidden agenda behind persuasion is to gain maximum benefit from the targeted audience. Therefore, it will never be affordable for persuasive system developers to evaluate health monitoring technologies with the same generalized model which is designed to gain users' attention to adopting the system. During the design and analysis of any health-related application, developers should adopt moderate and selected principles to persuade users. Because these types of applications are directly involved in human life, any unintentional usage of health-related apps can even threaten the user's life.

5. Design and Extension of the PSD Model

During the writing of this study, a conference paper has been published in *"Proceedings of the 2021 International Symposium on Human Factors and Ergonomics in Health Care"* [17] that claims that users of fitness apps belong to different social and cultural involvements. That empirical study [17] focuses on two different social and cultural groups, which are discussed below.

5.1. Individualist

The Individualist group preferred more Primary Task Support, focusing on the basic targeted objectives of fitness apps which are Self-Monitoring and Goal Setting.

5.2. Collectivist

The collectivist group preferred more Dialogue Support which focuses on the basic targeted objectives of fitness apps, which are Reminders and Suggestions.

That study also recommends presenting an extended PSD model by including the following additional features for fitness application development.

- Goal Setting
- Verbal Persuasion

In 2019, the *"Gallup Global Emotions Report"* showed that 55 percent of America's population is observing stress, which is the highest stress level in the world. In addition, the same issue is observed in 35 percent of the world [18]. That study also recommends that design strategies can improve persuasiveness and ultimately increase the positive effects of stress management among users of HMAs [18].

The above [17,18] study endorses the need for improvements in the PSD model, but this study specifically evaluates the PSD model and elaborates on the need for presenting customized persuasive models for each category of applications. In addition, this study highlights the sensitivity of mobile health apps and the importance of the provision of moral support to users through mobile health apps.

6. User Studies

The results and findings have been presented from three consecutive user studies. Considering the length and conciseness of this paper, only the pertinent findings of the studies have been incorporated. The targeted study area is persuasion in mobile HMAs and smartphone users. These are planned user studies to extract appropriate data for further analysis. User studies have been conducted using the questionnaire method. The questionnaires have been designed according to the design rules of a popular HCI book,

"Human-Computer Interaction An Empirical Research Perspective" [19]. There are four major portions in each user study. Detailed descriptions regarding these user studies are described on page eight. The four major parts of the studies are:

- Consent Form
- Pre-Study Questionnaire
- Tasks to be performed
- Post-study Questionnaire

In this paper, the results of the user study are included based on the researcher's observation to present the evaluation of the PSD model.

6.1. Introduction to User Study

The purpose of this study is to determine users' motivation and behavior by analyzing the PSD model and to evaluate HMA. The case study is based on a mobile health monitoring app (iCare Health Monitor). The iCare Health Monitor (Figure 1) is a mobile health monitoring app available in the Play Store for smartphone users which can be used as a health assistant. A questionnaire was used to generate quantitative results from participants after using this app. This mobile health app is recommended to users, and they are encouraged to perform some specific tasks and evaluate the system by answering questions. Users are also encouraged to share their points of view regarding system design and their behavior/motivation level after using the application.

The following basic tasks/tests were given to participants:

1. Vision test
2. Hearing test
3. Blood pressure test
4. Psychological test, etc.

6.2. Background

It is a mobile health monitoring system that uses wireless body sensors and smartphones to monitor the health of commoners and the elderly for general wellbeing. The system enables the user to check their health conditions at any time from anywhere. The system is also enabled with tailored functions like a vise for each individual. The system is a real-time living assistant which can help users to live a convenient and comfortable life [20]. If we look at mobile applications, the most important thing with the respective user is persuasion. How many users are persuaded by the use of the application? This is the key thing for the success of the developer as well as the authority of that application. Either the user's behavior is changed or not. If it is changed what is the level of behavioral/motivational change? All these things are the ultimate demands of persuasive applications. This study aims to improve the healthcare system of the current world. Preventing health monitoring systems from persuasion will be a big failure of the research, developers, and health service providers, as it is already proven that a framework PSD model provides theories and methods to analyze the persuasive contents of technologies [2].

After investigation, Ref. [21] declared that persuasive technologies are supposed to change the behavior of patients with the help of technology available at home. The basic reason behind this is that those investigated technologies have incompetent persuasive design considerations. Our selected persuasive application is named the iCare Health Monitor (Figure 1) for health care and is a popular and well-known health monitoring technology that is web-based and available on smartphone application-based platforms like the Play Store, etc. [20]. To check the success level of designers concerning persuasion, it is necessary to check persuasive features in such a specific, highly available, and used app in the current era of ubiquitous computing and technology.

As the PSD Model is a well-known and recent persuasive design model [13], the iCare Health Monitor is designed to keep the PSD model in view [20]. In light of the above facts, investigating real-time users of iCare can be the most credible study for the evaluation of the PSD Model as well as persuasive principles. Why do users quit the application or why is the user not ready/motivated to use such apps in practical life? Of course, a very big population uses these applications, but the aim of the research, and specifically this user study, is to take the technology to the level where it actually should be. While conducting and designing a research study on HCI, it is very important to keep the persuasive context (the event, intent, and persuasive strategy) in view [2].

It is a basic characteristic of persuasive technology to influence the behavior of users through information and feedback [22]. All smartphone applications that are designed to change the attitudes or behaviors of their users through persuasion and social influence rather than force are termed persuasive technologies.

Mobile health is the creative use of emerging mobile devices to deliver and improve healthcare, health delivery, health communication, public health, health promotion, and self-management [23]. The terms mHealth, eHealth, and digital health are directly or indirectly linked with the smart technologies and tools which help users with real-time health management and monitoring.

According to [23], there are currently more than 165,000 mobile health applications (apps) publicly available in major app stores.

There is a strong need for the perfect integration of persuasive features for improving the usefulness of mobile health (mHealth) results [24]. Therefore, it remains a perpetually challenging task for mobile health application developers to balance the required persuasive features in health monitoring apps. For the ease of developers, a review study of four persuasive design models was previously conducted.

The following four models are taken by [25] in his review process:

1. Persuasive Systems Design (PSD) Process Model [2]
2. Design with Intent (DwI) Method
3. Behavior Wizard Model
4. Eight-Step Design Process

All the above models are generalized. Designers/developers of applications face tremendous problems attempting to maximize persuasion because of their general principles. For example, the expectations and needs of different categories/application users are never the same. There is a strong need for time to categorize applications with respective users' needs/expectations. There is also a need for time to develop specialized persuasive design models for each category by focusing on the expectations of users. For example, the needs and expectations of social media application (Facebook, Instagram, Twitter) are different from respective health monitoring application (iCare Health Monitor, Connected Living, CureDiva) users. We kept some questions for future researchers to answer, such as "How can designers embed arguments into designs? And if they are not embedding arguments, how can we speak of persuasion? None of the analyzed PD models for good reasons—help us answer these problems". To answer the above question [25], a user study has been planned. For this purpose, we designed a questionnaire by which questions were asked of the targeted users of one category as there is a need for proper categorization of applications. Questions are related to the needs and expectations of the selected application/technology/smartphone app and the targeted user. During the design of the questionnaire, the twenty-eight basic principles of the PSD model have been kept in view.

Questionnaires are the primary instrument for survey research, a form of research seeking to solicit a large number of people for their opinions and behaviors on a subject such as politics, spending habits, or the use of technology [19].

6.3. Problem Statement and Research Objectives

The PSD model is found to be the most recent and precise model with respect to the other frameworks for persuasive design. At the same time, [26] states that the "PSD model does not yet provide a comprehensive list of persuasive features". The facts motivated me to work on the PSD model for determining more persuasive features. That was the problem specification stage of research. The following are the main objectives of this research study:

- To find persuasion gaps in health monitoring apps.
- To overcome persuasion gaps in HMAs.

6.4. Methodology

For the sake of gathering appropriate research data for quantitative analysis, a hypothesis questionnaire is used as a key tool in this study. Questions have been derived from the literature and different research articles [10]. To measure and analyze the end-user persuasive level of any specific application, it is necessary to provide an environment where users can use an application in a free and unbiased way so that he/she can share his/her experience of behavior change regarding the application. The ultimate goal of gathering quantitative data through this questionnaire is to reach a decision to present any improved (specialized) PSD model for health monitoring apps. The PSD Model is the generalized model and, as discussed earlier, specialized models for each application category or targeted group of users need to be presented. The generated set of numerical data has been used for validating and upgrading the model.

The participants are workers in IT, professionals, and students. Keeping this in mind, we tried to get positive feedback by asking one qualitative question as well.

6.5. Study Design

The questionnaire of each study has six steps/parts as described below. Study two is discussed in detail but only the relevant results from the first and third studies have been included.

i.   Consent form
ii.   Pre-study questionnaire
iii.   Installation of a smartphone application (iCare Health Monitor) (Figure 1)
iv.   Specific tasks to be performed
v.   Post-study questionnaire
vi.   Questions regarding system design
vii.   Questions regarding the user's persuasion level/behavior
viii.   Feedback from participants/users' opinions

There are eleven questions in the pre-study questionnaire and the post-study questionnaire consists of twelve questions. Questions are designed by keeping the persuasive principles in mind. The personal data and feedback regarding the persuasion level of the user have been gathered through a questionnaire. The Likert-scale mechanism has been adopted to answer the post-study questions. The very last question aimed to get suggestions/feedback from users as suggestions/feedback will be helpful to analyze the overall persuasion level of participants.

6.6. Study Participants

As detailed in Section 6, our research encompassed a series of three user studies. Each study contains twenty-six participants. For study two and study three, we purposefully selected random participants from diverse age groups and educational backgrounds to ensure comprehensive insights into the effectiveness of our research. In the interest of clarity and focus, we have decided to provide a more in-depth elaboration of the design and procedure of user study two. This allows us to offer a comprehensive understanding of the conduct, planning, and findings of the whole user study. By judiciously presenting the details of user study two, we aim to convey a clear and concise message about our research methodology and acknowledge the relevance and significance of the other two studies within the broader context of our investigation. In the second user study, a total of twenty-six participants were involved, with an equal gender distribution of thirteen males and thirteen females. The study aimed to evaluate the participants' level of persuasion toward health monitoring apps. We have selected well-educated participants purposefully in the first user study who are familiar with the proper and effective usage of smartphones and applications. Notably, in the second study, 74% of the participants held graduate degrees, while the remaining 27% possessed higher educational qualifications such as MS/M-Phil degrees. It is worth mentioning that almost all participants in study one reported being daily users of the Internet, indicating their familiarity with digital technologies. Figure 3 provides a visual representation of this observation, reaffirming the widespread use of the Internet among the study participants. These factors collectively contribute to a well-informed and engaged participant group, ensuring valuable insights into their attitudes and responses toward health monitoring applications. All the participants of the three user studies are general users of smartphones who use mobile HMA for self-health monitoring.

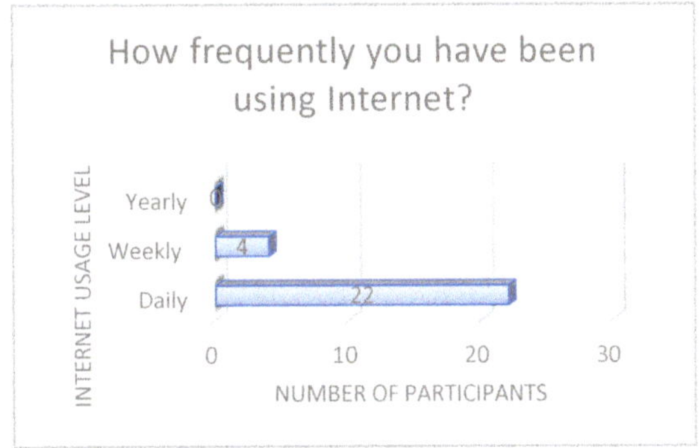

Figure 3. Participants' Internet Usage Level.

We have purposefully selected participants who frequently use the Internet as well as mHealth applications.

A total of 19 (73%) are IT professionals who now work in different organizations, and 7 (27%) of them are students. All of them have been using computers for more than ten years and smartphones for more than five years. All of them are using the Internet for education, jobs, social media, health, and news. More interestingly, Figure 4 shows that all participants are already aware of health monitoring apps.

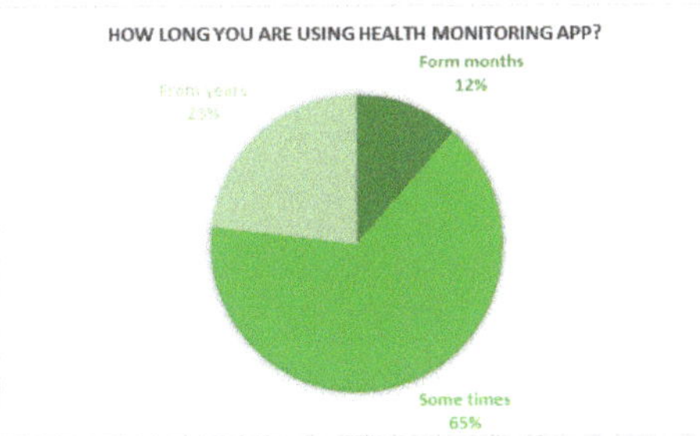

Figure 4. Participant App Usage Duration.

A total of 65% of participants use the app sometimes when needed. Collected data show that 23% of participants used these apps for a few years and 12% of participants used these apps for a few months.

6.7. Experimental Study Design

To obtain exact data, a sound and peaceful environment was provided for the duration of the study. A proper introduction to the application (iCare Health Monitor) was given to participants (Figure 1). During usage of the app, no time frame was defined. The user was open to taking his time accordingly. To aid with the research process and for the sake of contribution to the body of knowledge, a humble request was made to participants regarding their moral duties.

6.8. Study Procedure

First of all, the questionnaire was provided to the participant. After showing willingness, participants signed the consent form to become part of this research user study. After that, they filled out the pre-study questionnaire, which is about their personal information regarding their profession, education, age group, Internet usage, smartphone usage, the purpose of using the Internet, how long he is using Internet services, for what purpose he/she is using the Internet, etc. After that, participants were asked to download the app (iCare Health Monitor). If he is aware of the smartphone health monitoring app, then he/she had to perform different tasks using the app. Next, they filled out the post-study questionnaire. Suggestions/feedback regarding the app was also taken from participants. All participants are aware of health monitoring apps, while the specific app is new for various users. He/she is requested to fill out the post-study questionnaire followed by a five-point Likert scale. Finally, written feedback/suggestions were taken from the user to check the general persuasion level. The feedback question is about the worst feature of the app and any features they thought would be a necessary part of the app.

6.9. Data Analysis

There is a total of 26 participants who participated in this user study. There are 13 male and 13 female participants. Concerning their qualifications, seven out of 26 hold MS/MPhil/Ph.D. degrees, and 19 out of 26 are graduates. Almost all participants belong to the same age group, (20–35). Nineteen participants are students, while the remaining seven are professionals in different departments. Almost all participants are daily (regular) users of the Internet (Figure 3). The purpose of using the Internet has been observed for different needs accordingly, such as education, social networking, news, jobs, health, and entertainment purposes. All the participants are regular users of smartphones and have been using smartphones for more than five years and computer technology for more than ten years. Figure 5 illustrates that all the participants are already aware of health monitoring apps.

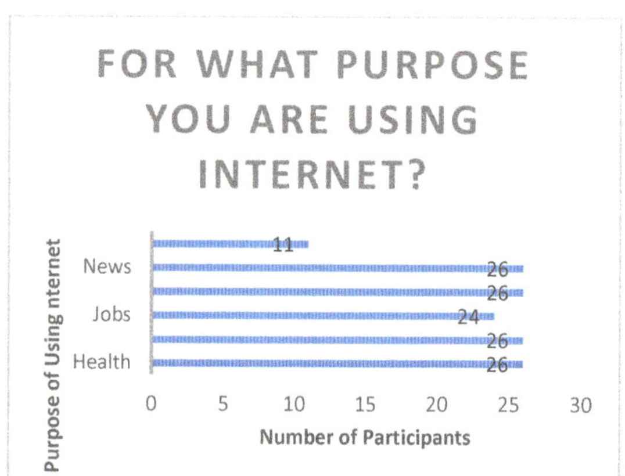

Figure 5. Purpose of using the Internet.

Figure 5 of this study shows that almost all participants are using the Internet for adopting a healthy lifestyle and looking at health-related problems.

The description of HMA usage duration (Figure 4) for participants is as follows: 65% of participants use mobile HMAs sometimes when needed, while 23% of participants used mobile HMAs for a few months and 12% of participants used mobile HMAs for years.

6.10. Results and Discussion

The mean and SD of the post-study questions are shown in Figure 6. As the study questions are divided into two categories, the first four questions are about system design and the next eight questions are about the impact of the application on the user's behavior/persuasion/motivation level. The means and SDs of the post-study questions are shown in Figure 6. The questions help us to decide whether the PSD strategy has been followed or not. There is a total of twelve questions, which have been asked of the participants by keeping the basic principles of persuasion in mind. The current technological hike gives rise to new potentials in developing more progressive and improved medical equipment as well as health applications. If we look at the rapidly growing population of elderly people, there is a strong need for time to develop a more advanced and personalized medical system that can be equally applicable for each individual to adopt a healthy life [27]. The division of the post-study questionnaire is as follows:

- Questions regarding PSD
- Questions regarding participant's behavior change
- Feedback/Suggestions from the participant
- **Result of Questions Regarding System Design**

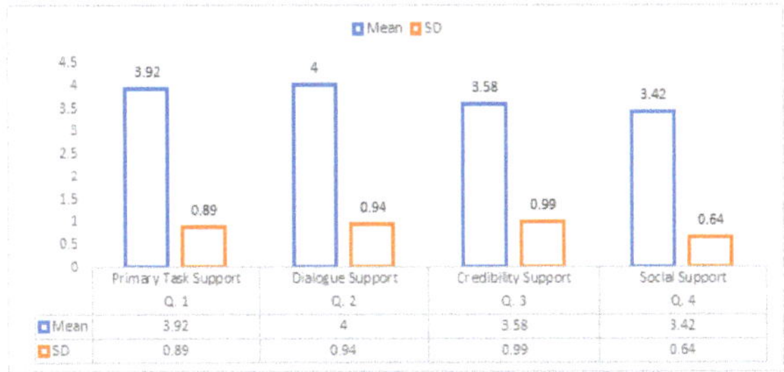

Figure 6. Post-Study Result of System Design.

The graph of basic PSD principles with respective SDs and means is shown below in Figure 6. The graph of basic PSD principles with respective Likert scales is also shown below in Figure 7.

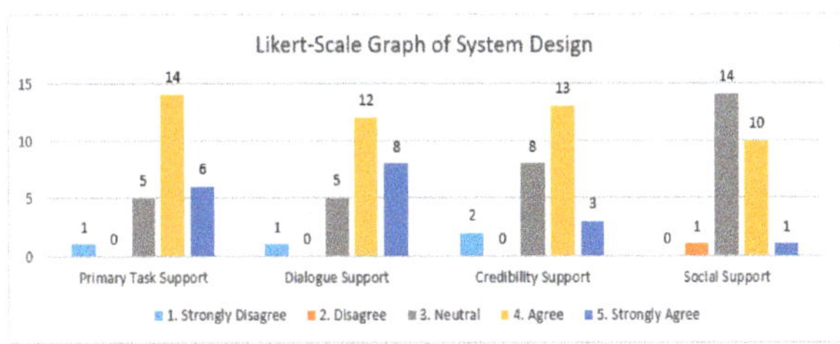

Figure 7. Likert-Scale Graph of System Design.

a. SD and Mean Graph of System Design

The results of the system design support questions are shown in Figure 6. The first question is about system primary task support, with a mean= 3.92 and SD= 0.89. The second question is about system dialogue support, with a mean = 4.00 and SD = 0.94. The third question is about system credibility support, with a mean = 3.58 and SD = 0.99. The fourth question is about system social support, with a mean = 3.42 and SD = 0.64.

The mean and SD of questions regarding PSD basic principles are almost strongly agreed by participants, which indicates that the app design is proper and not questionable in light of the PSD model.

- **Liker-Scale Graph of System Design**

Figure 7 shows the verbal response of participants using an interval scale (Likert scale). The result of questions regarding system design support has been presented graphically.

In Figures 6 and 7, the results of questions regarding system design have been shown. The values of the SD and mean in Figure 6 show that the system has been designed by following the PSD model. Figure 7 (Likert-Scale Graph of System Design) also illustrates that the system design seems to be effective with respect to the PSD model.

The results in Table 1 show that the app has been designed by following PSD model strategies. Social support must be included more but as per the author's observation, the app must persuade the user in a better way. Primary Task Support and Dialogue Support have high performance. A total of 2 out of 26 (7.69%) participants raised a question on systems credibility and 8 out of 26 (30.76%) participants showed a neutral response to the system credibility question. It means that 30.76% are not sure whether the system is credible or not.

Table 1. App Test Results in the Context of the PSD Model.

Questions about Application Test					
Q. No	Questions	1. Strongly Disagree	2.Disagree	3.Neutral	4.Agree	Strongly Agree
Q.1	Primary Task support	1	0	5	14	6
Q.2	Dialog Support	1	0	5	12	8
Q.3	Credibility Support	2	0	8	13	3
Q.4	Social Support	0	1	14	10	1
Sum		4	1	32	49	18

- **Results of Questions Regarding Participants' Behavior**

Results of questions regarding participants' behavior levels after using the application are shown in Figures 8 and 9.

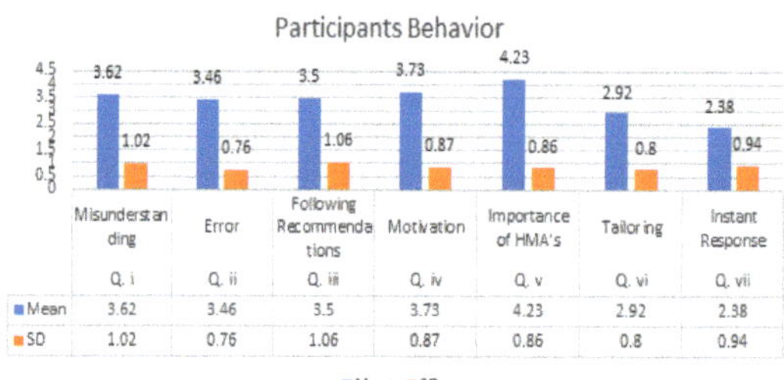

Figure 8. SD and Mean of User Behavior.

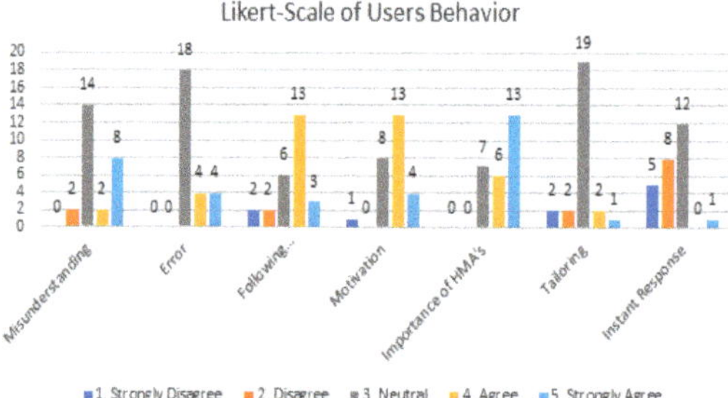

Figure 9. Likert-Scale Graph of Participant's Behavior Change.

b.   SD and Mean Graph of Participants' Behavior

The first question is about misunderstandings, where the mean = 3.62, SD = 1.02. The second question is about errors, where the mean = 3.46, and SD = 0.76. The third question is about following the recommendations of the app, where the mean = 3.5, SD = 1.06. The fourth question is about motivation, where the mean = 3.73, SD = 0.87. The fifth question is about the importance of HMAs, where the mean = 4.23, SD = 0.86. The sixth question is about tailoring, where the mean = 2.92, SD = 0.8, and the seventh question is about instant response, where the mean = 2.38 and SD = 0.94.

Figure 8 illustrates that participants observing that app can lead to misunderstandings regarding self-health management and monitoring. In addition, participants faced low instant responses. However, the participants showed great interest in following the recommendations of the app. It means that the user does not get any moral support from the app.

c.   Likert-Scale Graph of Participants' Behavioral Change

Figure 9 shows the verbal responses of participants using an interval scale (Likert scale). These are the results (Figure 9) of questions regarding participants' behavioral changes after using the app.

Figure 10 from the first user study shows the post-study results regarding the participant's level of behavioral change, which have been described below.

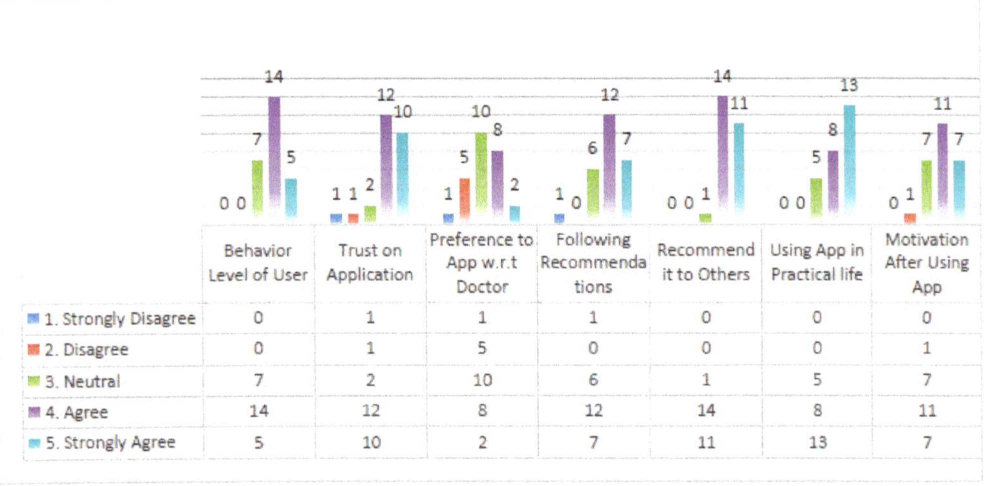

	Behavior Level of User	Trust on Application	Preference to App w.r.t Doctor	Following Recommendations	Recommend it to Others	Using App in Practical life	Motivation After Using App
1. Strongly Disagree	0	1	1	1	0	0	0
2. Disagree	0	1	5	0	0	0	1
3. Neutral	7	2	10	6	1	5	7
4. Agree	14	12	8	12	14	8	11
5. Strongly Agree	5	10	2	7	11	13	7

Figure 10. Likert-Scale Result of Questions Regarding Participant's Behavior Change.

1.   Participants showed interest in using the application and also recommended it to others.
2.   In some cases, participants preferred to use the app when visiting the doctor and in their practical life, along with following the recommendations of the app.

The above results show that the application has a great influence on the user's life but Figure 11 shows that, in the third user study, the application also creates misunderstandings regarding health-related problems. The results also show that app design can be improved and suggested that some results of health-related problems should be hidden from the user for the sake of his/her safety. Therefore, the results of such problems could be appropriately given to users so that he/she can be focused on adopting a healthy lifestyle instead of being disturbed by harsh results.

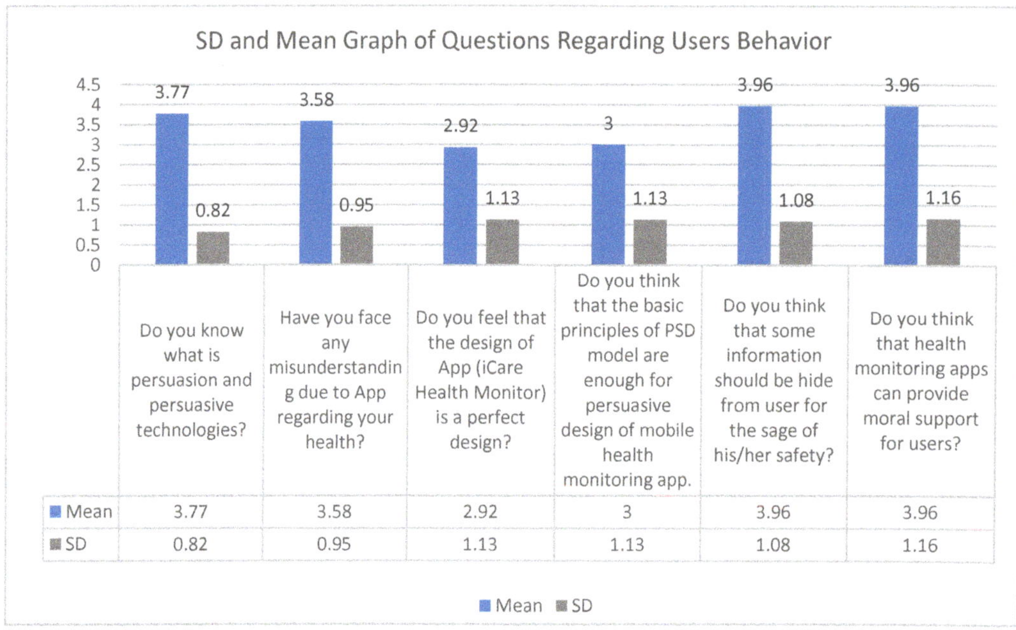

Figure 11. SD and Mean Graph of Questions Regarding User's Behavior.

Figure 11, from the third user study, shows that users have been asked some questions to learn about the actual problems behind persuasion gaps in health monitoring applications. They have also been asked about possible remedies to improve the effectiveness and usefulness of such apps.

During study two, we discovered that the application was designed with enough consideration of the PSD model's basic principles. However, the post-study results of question three in Figure 11 show some evidence that the app design can be improved. The results of post-study questions five and six reflect that this improvement can be added to the apps through providing moral support for users by hiding some particular types of test results from users, keeping his/her safety in mind. Essentially, not every result of the test should be displayed directly to the user.

6.11. Feedback/Suggestions from Participants

Fourteen participants (53.84%) out of twenty-six provided feedback/suggestions. That was a good sign because all the participants are well educated, and their suggestions/feedback will be highly encouraged. The feedback/suggestions regarding users' overall experience of the app are also shown below in Table 2.

Table 2. Participants' Feedback/Suggestions.

S No	Feedback/Suggestions of Participants after 2nd User Study
1	The app must be linked with well-known health experts.
2	For system credibility, there must be a linkage with some real-time health centers.
3	The system provides feedback on the user's interaction while the system is unaware of the user's previous health records.
4	Who is the owner of the app? For system credibility, it is very important to know who the owner is.

Table 2. *Cont.*

S No	Feedback/Suggestions of Participants after 2nd User Study
5	General tips for a healthy life are appreciable but monitoring any real health issue may not be possible.
6	Sometimes systems offer irrelevant predictions. To improve, the system should record all previous health records first.
7	The system should be linked with any hospital.
8	Some people have psychological problems. We should hide their issues in front of them but this app gives results directly.
9	The app should hide the direct result of psychological problems but, provide tips, on how to get rid of issues.
10	The app should advise on a healthy lifestyle. Giving direct feedback to the psychological user is not a good sign.
11	I look at these applications neutrally, but these apps could be more helpful if they help users subconsciously.
12	Some people have issues, but the app treats everyone equally.
13	Good application.
14	Personalized behavior is necessary for women.

7. Conclusions and Future Work

Although the PSD model provides a comprehensive framework for designing and evaluating persuasive systems and describes the content and software functionality, it does not provide specific guidance on how to balance the need for persuasion with the need to respect users' autonomy and privacy. The model assumes that persuasion is always a positive force and does not consider the potential negative consequences of persuasive systems. The model is relatively complex and may be difficult to apply in practice without significant expertise in persuasive design. This study analyzes and evaluates the user's behavior change level and system design in the context of mobile health monitoring app users. According to the post-study results, some system designing features do not align with the user's motivation level. These studies find that users face misunderstandings regarding health-related problems. According to Figure 8, the application's social support is low, and the application can create misunderstandings regarding health-related issues. Furthermore, Figure 11 illustrates that the recommendations and test results of users should be presented differently and not show serious harmful results regarding life directly to users. Instead of this, the app must suggest excellent healthcare techniques which can help the user to get out of the situation. The application must hide some results for the sake of user safety and must provide moral support to seek medical attention immediately to ensure that the user receives the proper care and treatment to manage the condition. Figures 6 and 7 and Table 1 demonstrate that the application's design adheres to the PSD model without any problems. Therefore, we can say that either the design of the application is wrong or the PSD model fails to provide the perfect persuasive design strategies for users of the application. On the other hand, Figure 8 shows that users strongly agreed to follow the recommendations of the app regarding their health-related issues. However, the basic principles of the PSD model have been followed with enough consideration by designers. This means that the app does not provide any moral support to users while using the app. Despite this, HMAs need more steps to improve user behavior and motivational levels. As mentioned previously, the majority of the participants in the first study are HCI researchers and the rest are well-educated. With this in mind, the feedback and suggestions provided by participants after the study are also not ignorable. A tailored persuasive system design model for HMAs must be presented to analyze and assess the user's behavior change level and system design principles in the context of the user's point of view. This study also investigated how mHealth apps can address the widespread issues of the emotional and mental health of users. The study also recommends designing an improved PSD model for each category of persuasive technologies so that it will be possible to get maximum reliable results and feedback. To develop a benchmark persuasive framework for HMAs, a special study is needed. As a result of the PSD model's generalized design strategies, we cannot promote these strategies for medical and health-related applications. These types of applications are directly related to human life, so any misunderstanding can affect human life. This study suggests that mHealth apps must provide moral support for their users. In short, this study has found that the PSD model fails to provide moral support for their users and strongly recommends that "technology must provide moral support for their users".

Author Contributions: Conceptualization, A.H.; methodology, A.M.; software, M.A.A.; validation, F.A.; formal analysis, S.U.; investigation, S.U.; resources, M.A.A.; data curation, S.U.; writing—original draft preparation, A.H.; writing—review and editing, A.H.; visualization, S.U. and F.A.; supervision, A.M.; project administration, F.A.; funding acquisition, F.A. All authors have read and agreed to the published version of the manuscript.

Funding: This research received no external funding.

Data Availability Statement: Not applicable.

Acknowledgments: We thank our families and colleagues who provided us with moral support.

Conflicts of Interest: The authors declare no conflict of interest.

References

1. Chittaro, L. Mobile persuasion for health and safety promotion. In Proceedings of the 17th International Conference on Human-Computer Interaction with Mobile Devices and Services Adjunct, Copenhagen, Denmark, 24–27 August 2015; ACM: New York, NY, USA, 2015; pp. 878–882. [CrossRef]
2. Oinas-Kukkonen, H.; Harjumaa, M. Persuasive systems design: Key issues, process model, and system features. *Commun. Assoc. Inf. Syst.* **2009**, *24*, 28. [CrossRef]
3. Matthews, J.; Win, K.T.; Oinas-Kukkonen, H.; Freeman, M. Persuasive Technology in Mobile Applications Promoting Physical Activity: A Systematic Review. *J. Med. Syst.* **2016**, *40*, 72. [CrossRef] [PubMed]
4. Zhao, J.; Freeman, B.; Li, M. Can mobile phone apps influence people's health behavior change? An evidence review. *J. Med. Internet Res.* **2016**, *18*, e287. [CrossRef] [PubMed]
5. Bardhan, R.; Bahuman, C.; Pathan, I.; Ramamritham, K. Designing a game based persuasive technology to promote pro-environmental behaviour (PEB). In Proceedings of the 2015 IEEE Region 10 Humanitarian Technology Conference (R10-HTC), Cebu City, Philippines, 9–12 December 2015; IEEE: Piscataway, NJ, USA, 2015; pp. 1–8.
6. Osman, A.; Mohd Zam, M.; Ibrahim, R. The mHealth: A Review of Current Persuasive Technology Design Strategies. *IOSR J. Mob. Comput. Appl. (IOSR-JMCA) E-ISSN* **2016**, *3*, 2394-0050.
7. Shati, A. Mhealth applications developed by the Ministry of Health for public users in KSA: A persuasive systems design evaluation. *Health Inform. Int. J.* **2020**, *9*, 1–13. [CrossRef]
8. Hamari, J.; Koivisto, J.; Pakkanen, T. Do persuasive technologies persuade?—A review of empirical studies. In Proceedings of the International Conference on Persuasive Technology, Padua, Italy, 21–23 May 2014; Springer: Berlin/Heidelberg, Germany, 2014; pp. 118–136.
9. McKay, F.H.; Wright, A.; Shill, J.; Stephens, H.; Uccellini, M. Using Health and Well-Being Apps for Behavior Change: A Systematic Search and Rating of Apps. *JMIR Mhealth Uhealth* **2019**, *7*, e11926. [CrossRef] [PubMed]
10. Langrial, S.; Lehto, T.; Oinas-Kukkonen, H.; Harjumaa, M.; Karppinen, P. Native Mobile Applications for Personal Well-Being: A Persuasive Systems Design Evaluation. In Proceedings of the PACIS, Ho Chi Min, Vietnam, 13 July 2012; p. 93.
11. Balcombe, L.; De Leo, D. Human-computer interaction in digital mental health. *Informatics* **2022**, *9*, 14. [CrossRef]
12. Lindemann, P.; Koelle, M.; Kranz, M. Persuasive technologies and applications. *Adv. Embed. Interact. Syst* **2015**, *3*, 46.
13. Kegel, R.H.; Wieringa, R.J. Persuasive technologies: A systematic literature review and application to pisa. *Cent. Telemat. Inf. Technol. Univ. Twente Enschede Tech. Rep. TR-CTIT-14–07* **2014**, *6*, 1–36.
14. Guillaume Faddoul, S.C. A Quantitative Measurement Model for Persuasive Technologies Using Storytelling via a Virtual Narrator. *Int. J. Hum. Comput. Interact.* **2020**, *21*, 1585–1604. [CrossRef]
15. Lu, D.J.; Girgis, M.; David, J.M.; Chung, E.M.; Atkins, K.M.; Kamrava, M. Evaluation of mobile health applications to track patient-reported outcomes for oncology patients: A systematic review. *Adv. Radiat. Oncol.* **2021**, *6*, 100576. [CrossRef] [PubMed]
16. Torning, K.; Oinas-Kukkonen, H. Persuasive system design: State of the art and future directions. In Proceedings of the 4th International Conference on Persuasive Technology, Claremont, CA, USA, 26–29 April 2009; ACM: New York, NY, USA, 2009; p. 30. [CrossRef]

17. Oyibo, K. Investigating the key persuasive features for fitness app design and extending the persuasive system design model: A qualitative approach. In Proceedings of the International Symposium on Human Factors and Ergonomics in Health Care, Baltimore, MD, USA, 12–16 April 2021; SAGE Publications Sage: Los Angeles, CA, USA; pp. 47–53.

18. Alhasani, M.; Mulchandani, D.; Oyebode, O.; Orji, R. A Systematic Review of Persuasive Strategies in Stress Management Apps. In Proceedings of the BCSS@ PERSUASIVE, Aalborg, Denmark, 21 April 2020.

19. MacKenzie, I.S. *Human-Computer Interaction: An Empirical Research Perspective*; Newnes: London, UK, 2012.

20. Lv, Z.; Xia, F.; Wu, G.; Yao, L.; Chen, Z. iCare: A Mobile Health Monitoring System for the Elderly. In Proceedings of the 2010 IEEE/ACM Int'l Conference on Green Computing and Communications & Int'l Conference on Cyber, Physical and Social Computing, IEEE Computer Society, Hangzhou, China, 18–20 December 2010; pp. 699–705. [CrossRef]

21. Jalil, S. Persuasion for In-home Technology Intervened Healthcare of Chronic Disease: Case of Diabetes Type 2. In Proceedings of the PERSUASIVE, Adjunct Proceedings, Zurich, Switzerland , 8–12 September 2013.

22. Knaus, M. Persuasive technologies and applications in health and fitness. *Persuas. Technol. Appl.* **2015**, *3*, 5–11.

23. Sadiku, M.N.O.; Shadare, A.E.; Musa, S.M. Mobile Health. *Int. J. Eng. Res.* **2017**, *6*, 4. [CrossRef]

24. Pinzon, O.E.; Iyengar, M.S. Persuasive Technology and Mobile Health: A Systematic Review. In Proceedings of the Persuasive Technology: Design for Health and Safety, the 7th International Conference on Persuasive Technology, PERSUASIVE 2012, Linköping, Sweden, 6–8 June 2012; Adjunct Proceedings. Linköping University Electronic Press: Linköping, Sweden, 2012; pp. 45–48.

25. Torning, K. A Review of Four Persuasive Design Models. *Int. J. Concept. Struct. Smart Appl.* **2013**, *1*, 17–27. [CrossRef]

26. Daud, N.A.; Sahari, N.; Muda, Z. An initial model of persuasive design in web based learning environment. *Procedia Technol.* **2013**, *11*, 895–902. [CrossRef]

27. Venckauskas, A.; Stuikys, V.; Toldinas, J.; Jusas, N. A Model-Driven Framework to Develop Personalized Health Monitoring. *Symmetry* **2016**, *8*, 65. [CrossRef]

Article

Tokenized Markets Using Blockchain Technology: Exploring Recent Developments and Opportunities

Angel A. Juan [1,*], Elena Perez-Bernabeu [1], Yuda Li [1], Xabier A. Martin [1], Majsa Ammouriova [2] and Barry B. Barrios [2]

[1] Center for Research in Production Management and Engineering, Universitat Politècnica de València, 03801 Alcoy, Spain; elenapb@upv.es (E.P.-B.); yudali@upv.es (Y.L.); xamarsol@upv.es (X.A.M.)
[2] Department of Computer Science, Multimedia and Telecommunication, Universitat Oberta de Catalunya, 08018 Barcelona, Spain; mammouriova@uoc.edu (M.A.)
* Correspondence: ajuanp@upv.es

Abstract: The popularity of blockchain technology stems largely from its association with cryptocurrencies, but its potential applications extend beyond this. Fungible tokens, which are interchangeable, can facilitate value transactions, while smart contracts using non-fungible tokens enable the exchange of digital assets. Utilizing blockchain technology, tokenized platforms can create virtual markets that operate without the need for a central authority. In principle, blockchain technology provides these markets with a high degree of security, trustworthiness, and dependability. This article surveys recent developments in these areas, including examples of architectures, designs, challenges, and best practices (case studies) for the design and implementation of tokenized platforms for exchanging digital assets.

Keywords: blockchain; tokenized markets; tokenized platforms; digital assets; cybersecurity; virtual economy; European dataspace; extended reality

Citation: Juan, A.A.; Perez-Bernabeu, E.; Li, Y.; Martin, X.A.; Ammouriova, M.; Barrios, B.B. Tokenized Markets Using Blockchain Technology: Exploring Recent Developments and Opportunities. *Information* **2023**, *14*, 347. https://doi.org/10.3390/info14060347

Academic Editors: Amar Ramdane-Cherif, Ravi Tomar and TP Singh

Received: 24 May 2023
Revised: 10 June 2023
Accepted: 16 June 2023
Published: 17 June 2023

1. Introduction

A blockchain is a sequence of interconnected blocks that contain data with digital signatures within a decentralized and distributed network. As these blockchains are managed by a group of nodes in a decentralized network, they form a public and decentralized digital ledger (or record book) that is difficult for any one node to corrupt or influence. This creates a secure method for peer-to-peer transfer of digital assets without the need for a central authority or intermediary [1]. The chain of blocks continuously grows as new blocks are appended to the existing chain. Although cryptocurrencies are the most widely known application of blockchain technology, it has numerous other applications. Surveys exploring these applications, challenges, opportunities, and cryptocurrencies based on blockchain technology have been done [2–4]. This article focuses on one of these alternative applications, which involves using blockchain technology to create virtual markets where individuals can exchange tokenized digital assets.

Utilizing the public, decentralized, and distributed ledger that is provided by a blockchain, it is possible to create virtual representations of nearly any physical asset, such as a house, painting, or jewel. This process, known as asset tokenization or digitization, involves transforming asset rights into digital tokens that, in theory, can be securely and reliably bought, sold, and traded using blockchain technology. However, in practice, significant challenges remain in the development of these "tokenized virtual markets", such as cybersecurity threats and the absence of government and industry regulation. The tokenization process involves four primary steps [5]: (i) identifying the physical asset to be tokenized; (ii) assessing the asset's true value; (iii) determining the parameters that define the tokenized asset, such as the number and value of tokens; and (iv) creating and auditing smart contracts that govern the exchange of tokenized assets.

The Ethereum token standards ERC-20 and ERC-721 created two types of tokens on the blockchain (Figure 1): (i) fungible tokens that are identical and can be replaced (such as cryptocurrencies and carbon credits), and (ii) non-fungible tokens (NFTs) that represent a unique asset and cannot be easily exchanged for other tokens (such as music files or copyright certificates). NFTs are like digital IDs registered on the blockchain and cannot be divided further. Despite this, they can be traded using smart contracts on the blockchain, which supports decentralized and secure virtual markets.

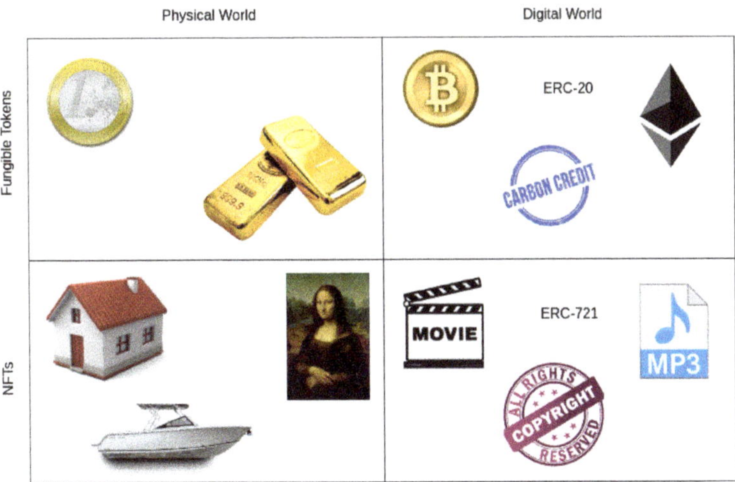

Figure 1. Fungible tokens vs. NFTs.

In this paper, we examine recent developments related to the design and implementation of tokenized platforms for exchanging digital assets. The paper covers a variety of topics, such as architectures, designs, challenges, and best practices (case studies). It is organized into several sections for easy reading. Section 2 provides an introduction to the basic concepts of tokenized platforms, using diagrams and other visual aids to make the material more accessible to non-experts. Section 3 outlines the methodology we used to conduct our review. Section 4 reviews recent research on the design of tokenized platforms for exchanging digital assets. Section 5 discusses the latest research on distributed data management infrastructure, with a focus on the European data strategy. Section 6 presents best practices for implementing tokenized platforms. Section 7 explores how artificial intelligence (AI) can be used to detect and combat cyber threats. Section 8 presents two case studies of platforms that use the blockchain and NFTs. Finally, Section 9 provides an overview of the key trends and challenges facing tokenized markets.

2. Basic Concepts and Conceptual Schemas

This section is intended to provide non-expert readers with a basic understanding of some technical concepts that will be discussed throughout the manuscript. The overview is designed to be accessible and avoids unnecessary technical details while providing all the necessary information to fully comprehend the rest of the paper.

2.1. Blockchain

As explained in Section 1, a blockchain is a chain of linked blocks in a distributed and decentralized network. Each block contains data associated with a transaction operation, a cryptographic hash of the previous block (a unique identifier generated using advanced cryptographic methods), and a timestamp of the operation (Figure 2). Since each new block is linked to the previous one, the data in any registered block cannot be changed without affecting its hash code, which would break the link connection with the next block, rendering the change invalid to other nodes in the distributed network. Consequently, a blockchain constitutes a secure and public digital ledger.

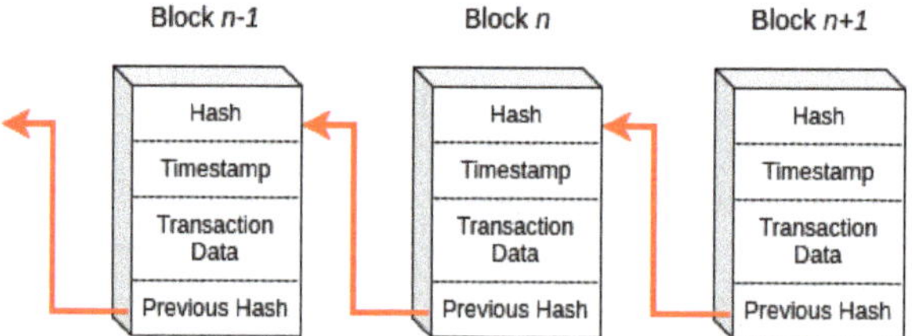

Figure 2. A blockchain as a series of linked blocks.

The process of how blockchain technology works is illustrated in Figure 3. When a new transaction request is made (step 1), a new block containing the transaction data is created (step 2). To be registered in the public ledger, the block must first undergo a process called "proof of work" (PoW) in some blockchains (e.g., Bitcoin-like ones) and "proof of stake" (PoS) in others (e.g., Ethereum-like ones), which involves analysis by the nodes in the distributed network (step 3). Nodes that successfully complete the PoW/PoS process receive a reward in the form of cryptocurrencies. The block must then be validated by achieving consensus among the nodes in the network (step 4). If a block is found to be invalid or modified, it will not be added to the blockchain, as it would differ from the copy of the valid blockchain that each node stores. The combination of secure hashing, a PoW/PoS process, and a consensus mechanism in a peer-to-peer network makes it extremely difficult to alter a block within a blockchain. Once validated, the new block is added to the blockchain and distributed throughout the network (step 5), and the transaction is considered completed (step 6).

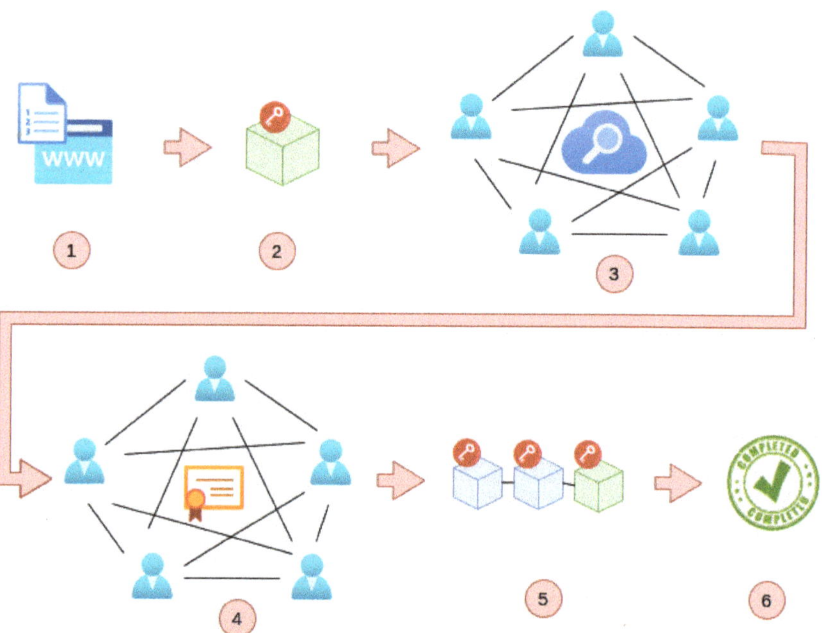

Figure 3. A 6-step process summarizing how blockchain works.

2.2. Smart Contracts and Tokenization

A smart contract is a type of digital contract that is securely and immutably stored within a blockchain network. It contains pre-agreed clauses and conditions that are automatically executed when specific predetermined criteria are met. This distributed nature of the smart contract ensures that it is secure and transparent for all parties involved.

The concept of 'tokenization' is the process of representing an asset (either digital or physical) with digital tokens that can be bought, sold, and traded on blockchains using smart contracts [6,7]. This process involves first creating a digital representation of the asset and then splitting it into individual and non-divisible parts called tokens. Virtually every physical asset could be tokenized, from a house to a boat or a painting. Ownership or other property rights over the asset can be transferred between parties without the participation of intermediaries or central authorities when a digital token is exchanged via a smart contract. However, legal issues may need to be addressed before these transactions can take place in a digital market [8].

3. Review Methodology

The use of blockchain technology is a promising approach to maintaining the traceability and immutability of data. This study aims to answer a series of questions about using blockchain technology in tokenized platforms. To achieve this, we followed the preferred reporting items for systematic reviews and meta-analyses (PRISMA) methodology, a well-established approach for conducting systematic reviews. The PRISMA methodology involves a comprehensive checklist of elements, such as search strategy, study selection criteria, data extraction, and risk of bias assessment, which ensures that systematic reviews are conducted consistently and transparently. By adhering to this methodology, researchers can enhance the quality and credibility of their findings [9].

We simplified and applied the PRISMA methodology to investigate questions pertaining to tokenized platforms. Our approach involved defining research questions, selecting inclusion and exclusion criteria, conducting a thorough search of pertinent databases, screening search results, extracting data from relevant studies, evaluating study quality, synthesizing results, and presenting findings according to PRISMA guidelines.

Our search strategy was designed to identify the latest and most relevant research on blockchain-based markets and their efficiency and included three major databases: Web of Science, Scopus, and IEEE Xplore. We limited our inclusion criteria to English-language documents published between 2013 and 2023, ensuring that our review included the most up-to-date and comprehensive research. We focused on identifying relevant books, journal articles, and conference proceedings, both final and in press, to provide a broad and accurate view of the topic. Following our search strategy, we found 1527 articles that contain the keywords "blockchain" and "tokens" in their title or abstract. The vast number of articles makes it difficult to review them manually, so we utilized state-of-the-art topic modeling techniques to automatically extract a selection of topics from this collection of documents (corpus). Specifically, we used the Non-negative Matrix Factorization (NMF) algorithm [10], one of the most popular approaches in topic modeling, to gain a better understanding of the underlying themes and patterns present within the corpus, which can be used for further analysis or generate new insights. Figures 4–6 graphically represent the information related to this first step of the content analysis of the literature review.

We set the number of selected topics to four. The ten most weighted descriptors of the four topics identified by the NMF model from the abstracts of the selected papers are: (i) blockchain, token, smart, system, technology, based, Ethereum, transaction, paper, platform; (ii) data, access, IoT, scheme, based, authentication, security, privacy, control, system; (iii) NFT, non, fungible, digital, art, ownership, asset, metaverse, data, research; and (iv) ICO, crypto, coin, initial, financial, offering, market, regulation, legal, investor.

Based on the identified topics, Topic 1 seems to focus on the technology behind blockchain and tokens, while Topic 2 is related to security and privacy. Topic 3 is related to non-fungible tokens (NFTs) and their applications, and Topic 4 is related to the financial aspect of blockchain and tokens.

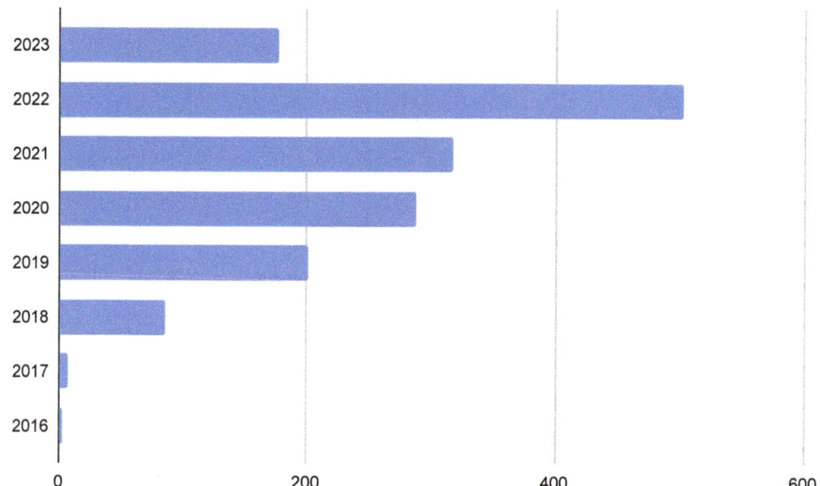

Figure 4. Annual distribution of publications.

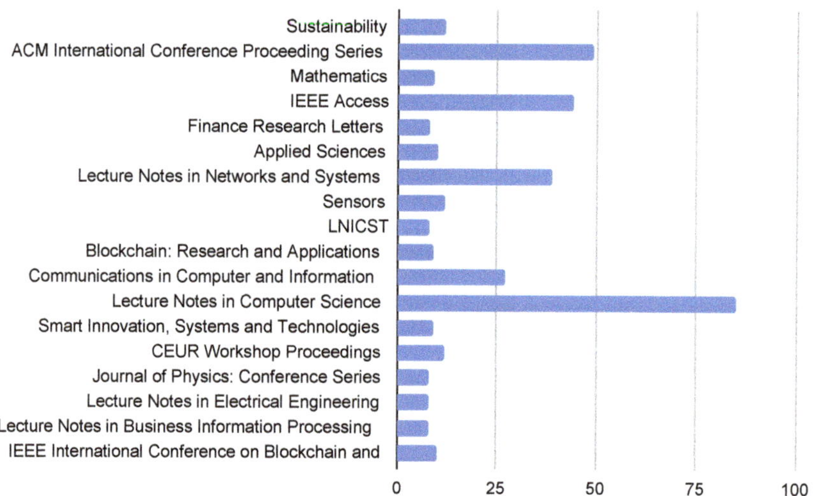

Figure 5. Articles per journal.

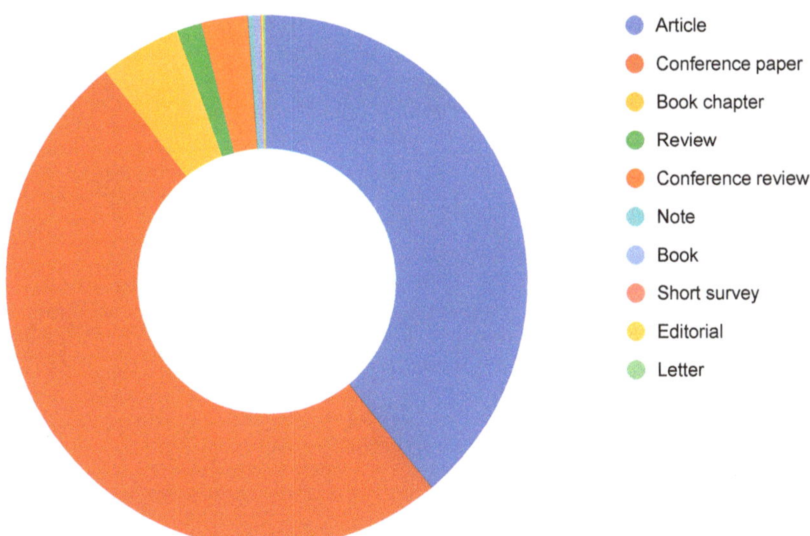

Figure 6. Article types.

To narrow down the scope of our research, we formulated five specific research questions and established corresponding criteria for each one. These questions center around the advantages and disadvantages of using tokenized platforms for digital asset exchange, the essential elements of the decentralized data management infrastructure, the optimal approaches for implementing tokenized platforms, the techniques for mitigating cybersecurity threats, and the principal developments and difficulties in tokenized markets. Additionally, for each question, its corresponding semantic fields are defined. Semantic fields refer to clusters of words or phrases that share a common meaning and are commonly used within a particular context or subject matter. These semantic fields are tailored to each research question in order to facilitate the discovery of relevant answers. Thus, the main questions and semantic fields employed in our review are provided next:

- What are the benefits and drawbacks of tokenized platforms for exchanging digital assets? Semantic fields: Design, token, digital assets.
- What are the key factors of a distributed data management infrastructure? Semantic fields: Data management, distributed systems, infrastructure.
- What are the best practices associated with the implementation of tokenized platforms? Semantic fields: Implementation, tokenization, best practices.
- What are the tools for fighting cyber threats? Semantic fields: Machine learning, cyber-threat detection, cybersecurity tools.
- What are the main trends and challenges associated with tokenized markets? Semantic fields: Market trends, tokenization, challenges.

Our focus was on tokenized platforms, which simplify the exchange and trade of tokens and have a significant impact on the liquidity and accessibility of tokenized assets. Through a comprehensive search and screening process, we identified a number of studies that met our inclusion criteria. These studies covered a broad range of topics related to blockchain technology's use in market tokenization, including its impact on market efficiency, challenges, opportunities, and emerging trends. We conducted a systematic review that followed PRISMA guidelines, ensuring that the findings were reliable and valid. Our study provides valuable insights and data-driven recommendations for researchers interested in understanding tokenized markets' mechanics and identifying opportunities for further research in this fast-evolving field.

4. Design of Tokenized Platforms for Exchange of Digital Assets

As already mentioned, tokenization is the conversion of an asset or its ownership rights into unique units called tokens. This term is often associated with blockchain technology, where tokens are utilized to represent the ownership of valuable assets. Various tokenized platforms exist, and a few examples are provided below:

- Bitbond (https://www.bitbond.com/, accessed on 2 May 2023): A peer-to-peer lending platform that uses blockchain technology to facilitate cross-border lending. It allows borrowers to access loans from investors globally using digital currencies such as Bitcoin.
- TrustToken (https://www.trusttoken.com/, accessed on 2 May 2023): A platform that enables the creation of asset-backed tokens. These tokens are backed by real-world assets, such as currencies or commodities, and can be traded on various exchanges.
- Harbor (https://goharbor.io/, accessed on 2 May 2023): A platform that streamlines the process of investing in private securities by using blockchain technology. It allows issuers to digitize their securities and investors to access and trade these securities in a secure and compliant manner.
- Polymath (https://polymath.network/, accessed on 2 May 2023): A platform that enables the creation of security tokens, which are digital tokens that represent ownership of a traditional security, such as stocks or bonds. It provides tools for issuers to create, issue, and manage security tokens in compliance with relevant regulations.
- Securitize (https://securitize.io/, accessed on 2 May 2023): A platform that specializes in the issuance and management of security tokens. It enables issuers to create and manage compliant security tokens and provides investors with a transparent and secure platform to buy and sell these tokens.

These platforms allow people and companies to establish a digital record of ownership for tangible assets that are easily tradable, such as real estate or funds. The tokens that represent these assets are distinct from security tokens because they represent actual capital and liquid value. The concept of tokenization has become a crucial topic in blockchain technology. By using blockchain, assets or rights can be tokenized and represented on a digital ledger. This connection between the off-chain and on-chain world is explored in Heines et al. [11]. Tokenization leverages blockchain technology to securitize both traded and non-traded assets, providing benefits such as increased liquidity, faster settlement, lower costs, and bolstered risk management, as explained in another article [11]. J.P. Morgan's Onyx Digital Assets is an example of a tokenization platform that aims to bring traditional assets into the blockchain ecosystem [11]. In the design field, design tokens have been introduced as a new paradigm for design deliverables, creating more efficient and consistent design systems, as described in Freni et al. [12] and Guggenberger et al. [13].

Tokenized platforms have emerged as a solution to the need for secure, transparent, and efficient methods of exchanging value. These platforms allow for the exchange of various types of digital assets, including cryptocurrencies, securities, and real-world assets, in a peer-to-peer manner, without intermediaries. These platforms have several advantages over traditional exchange platforms, including increased transparency and security. This is due to the use of blockchain technology, which records every transaction on a decentralized ledger accessible to all participants, ensuring that all transactions are secure, tamper-proof, and transparent. Another key feature of tokenized platforms is the use of smart contracts, which automatically enforce the terms of a transaction, enabling users to exchange digital assets in a trustless manner without the need for intermediaries such as banks. Smart contracts can also be customized to include various conditions, such as price limits, expiration dates, and other terms, providing a high degree of flexibility and control to users. However, designing a tokenized platform is challenging, as it requires careful consideration of various factors, such as security, scalability, and user experience. One of the main challenges is ensuring the platform is secure and resistant to attacks, which involves implementing various security measures, such as multi-factor authentication, encryption, and other mechanisms to prevent unauthorized access.

Designing a tokenized platform involves several critical factors, including scalability, security, and user experience. Scalability is vital as the platform should be able to handle a vast number of transactions without experiencing performance issues. High-performance computing infrastructure, such as cloud computing services and efficient consensus algorithms such as proof-of-stake or sharding are crucial in achieving this. Moreover, creating a user-centric design is essential to provide an intuitive, easy-to-use, and accessible platform for a wide range of users. This requires conducting user research, creating user personas, and implementing user feedback. In conclusion, building tokenized platforms to exchange digital assets is a complex task that demands careful consideration of various factors. Still, with blockchain technology, smart contracts, and advanced technologies, it is possible to develop secure, efficient, and transparent platforms that facilitate peer-to-peer asset exchange without intermediaries.

5. Distributed Data Management Infrastructure

In order to handle databases associated with real systems, scalable data models that can utilize distributed systems are required [14]. A distributed system is a collection of autonomous computing elements that appears to its users as a single coherent system [15] (p. 2). For these systems, an effective distributed data management system is necessary to manage, store, and process data across different locations. According to Moysiadis et al. [16], such systems must guarantee no loss of stored data in case of failure, process data from different sources (heterogeneity), be scalable, adaptive in bandwidth consumption, have low latency, be efficient in energy consumption, and comply with security and privacy standards.

To establish a distributed data management infrastructure, various frameworks such as Apache Hadoop, Apache Spark, and Apache Cassandra can be adopted. These frameworks provide fundamental tools, such as file systems and data processing engines, that are necessary for managing data across different locations. For example, the Hadoop distributed file system (HDFS) is used to store data as different blocks. Apache Hadoop and Spark are suitable for handling big data, and they rely on HDFS and MapReduce. In this system, the files are stored as small blocks at different nodes, classified as data nodes and master nodes. The master nodes request permission to access a file and receive a list of data nodes that store the different blocks of the file. MapReduce is used to split files into blocks. According to Ahmed et al. [17], Spark outperforms Hadoop in word count work and is more stable and faster. This advantage is due to the processing of data in memory before storage, although this performance degrades with large amounts of input data. On the other hand, Apache Cassandra has no defined master or data nodes, and all nodes in the network are equal. This framework has high update throughput and low latency, but it does not support privacy [16]. In general, when establishing a distributed data management system, it is crucial to use a framework that is scalable, adaptable in bandwidth consumption, efficient in energy consumption, and compliant with security and privacy standards while ensuring no loss of stored data in case of failure.

The European Union has initiated the European Dataspace for Digital Assets (EDSA) to enable the secure sharing and exchange of data, including digital assets, across Europe. As discussed by Scerri et al. [18], the establishment of a European digital market presents opportunities for businesses, citizens, science, and government to improve services and collect real-time statistics. However, technical, legal compliance, organizational, and national challenges must also be addressed, including issues related to data protection and digital transformation. The data strategy aims to facilitate the flow of data across different sectors in the EU while ensuring high-quality data and compliance with European rules and values, as well as fair and practical rules for accessing, sharing, and using the data. As a way to promote the EDSA, various measures have been proposed for establishing a common data space in the EU [19]. The aim is to unlock the potential benefits of the data economy and innovation opportunities while respecting European values and rules. The proposed actions include securing access to health data, promoting research, empowering citizens, and achieving person-centered care. The common data space is based on two types of data: public and publicly funded data and private sector data. For private sector data, key principles for contractual agreements have been defined, including transparency, shared value creation, respect for commercial interests, undistorted competition, and minimizing data lock-in, especially for business-to-business sharing.

6. Implementing Tokenized Platforms: Best Practices

Tokenization using blockchain technology and smart contracts enables any type of asset, including physical, intellectual, and creative property, to be converted into digital assets. This globalizes liquidity for all assets by creating a platform where these assets can be exchanged without the need for a central authority. However, it is crucial to design and implement these platforms with great care. To ensure the best practices for implementing such platforms, it is important to review case studies from the literature that illustrate the design and deployment of tokenized platforms.

In their article, Khan et al. [20] provided an exploratory analysis of the tokenization of sukuk, a financial instrument similar to bonds. The authors discussed the challenges involved in issuing sukuk and how blockchain technology can be used to resolve them. They reviewed different blockchain architectures and implemented a basic smart contract for Sukuk al-Murabaha on Ethereum. Finally, they conducted a cost-benefit analysis comparing conventional sukuk issuance to sukuk tokenization. Meanwhile, Tian et al. [21] explored how asset tokenization can create a new economic model that integrates non-financial values, such as social and environmental impacts, into tradable tokens. They analyzed SolarCoin, WePower Token, and ZiyenCoin as examples of how blockchain-enabled asset tokenization can be applied to support the economy and build social resilience. The authors identified that tokenization promoted inclusiveness and sustainability through shareholder empowerment, incentive monetization, and finance optimization, and they discussed obstacles to broader adoption and policy implications. Zarifis and Cheng [22] studied the business models focused on NFTs and identified four NFT business models, including NFT creator, NFT marketplace, a company offering their own NFT, and a computer game with NFT sales. Finally, Calandra et al. [23] took a multiple case study approach to explore the relationship between blockchain technology and sustainable business models (SBMs). Through an analysis of various databases, the authors demonstrated how blockchain could be used for environmental management and highlighted the main application of blockchain in relation to SBMs, which is supply chain cost reduction.

Recent case studies suggest some best practices for implementing asset tokenization platforms. Firstly, tokenized platforms must comply with relevant securities laws and regulations, including conducting KYC and AML checks on investors and ensuring proper classification of tokens as securities or utility tokens [24]. Secondly, these platforms rely on smart contracts to manage ownership and transfer of tokens, which requires careful design to protect against potential hacks or exploits. Using established standards such as ERC-20 for security tokens or ERC-721 for non-fungible tokens is recommended [25]. Thirdly, tokenized platforms must be scalable to handle large volumes of transactions, which can be achieved through selecting appropriate consensus mechanisms such as PoW or PoS, implementing off-chain solutions such as sidechains or state channels, and optimizing transaction fees to pay network validators for their services to the blockchain [26]. Lastly, tokenized platforms must be user-friendly and accessible to a broad range of investors. This requires a well-designed user interface, clear instructions for buying and selling tokens, and adequate support for investors and issuers [27]. Table 1 identifies which of the inferred best practices follow the tokenization platforms listed in Section 4. Based on publicly available information, it appears that all of these tokenized platforms follow the best practices listed in the table: require users to undergo KYC/AML checks in order to participate in token offerings on their platforms, use Ethereum-based smart contract standards such as ERC-20 and ERC-721 to create and manage tokens, use proof-of-stake (PoS) as their consensus mechanism, and offer user-friendly interfaces and resources, such as documentation and support, to help users participate in token offerings on their platforms. Additionally, some of these platforms offer features that make it easy for users to invest in tokens, such as integration with popular wallets or investor portals.

Table 1. Summary of best practices provided tokenized platforms.

Tokenized Platform	Compliance	Contract Standards	Consensus	User Experience
Bitbond	✓	✓	PoS	✓
TrustToken	✓	✓	PoS	✓
Harbor	✓	✓	PoS	✓
Polymath	✓	✓	PoS	✓
Securitize	✓	✓	PoS	✓

By examining recent case studies, best practices for developing asset tokenization platforms have been identified. These practices include ensuring compliance with regulations, carefully designing smart contracts, scaling the platform appropriately, and prioritizing user experience [24–27]. Asset tokenization platforms hold significant potential as a use case for blockchain technology.

7. AI Tools for Cyber-Threat Detection

Today's interconnected world faces increasing worry with the rise of cyber threats. Businesses, organizations, and individuals are at a greater risk of cyber attacks, which can lead to severe consequences such as sensitive data theft, critical infrastructure disruption, and financial loss. Cybersecurity or cyber-threat detection has emerged to address this concern, aiming to develop effective tools and methods to identify potential cyber threats and prevent them from causing damage. However, traditional approaches such as signature-based detection and rule-based systems are losing their effectiveness as cyber threats become more complex, serious, and sophisticated [28]. The exponential growth of data in cyberspace and the increasing computing power have made machine learning (ML) the most efficient and essential approach to counter cyber threats and overcome the restrictions of traditional security systems. Various ML techniques have been implemented to detect and categorize different types of cyber threats, including decision trees, ensemble methods, Bayesian networks, support vector machines, K-nearest neighbor (k-NN), and artificial neural networks. These are just a few examples of commonly used ML algorithms in cybersecurity, as documented by research surveys [29,30].

ML has been widely used in cybersecurity, particularly in applications such as intrusion detection, malware analysis, and spam detection [31]. Intrusion detection involves monitoring a network or computer system for unauthorized access or harmful activities, aiming to identify potential security breaches so that necessary measures can be taken to prevent damage. Malware analysis, on the other hand, entails examining malicious software to understand its behavior, purpose, and potential impact on a system or network. It is the responsibility of security researchers to conduct malware analysis to develop effective defenses against further infection or damage. Spam detection is a critical component of email and messaging security, which involves distinguishing and filtering unwanted or unsolicited messages from legitimate ones, such as those promoting a product or service, distributing malware, or phishing for sensitive information. Numerous ML algorithms, including random forest [32,33], support vector machine [34–36], and k-NN [37–40], have been investigated in the literature for developing intrusion detection models, malware analysis, and spam detection.

Over the past twenty years, ML has played an increasingly crucial role in the battle against cyber attacks. However, ML is not a panacea in cybersecurity. Cyber attackers are constantly evolving their methods and tactics, requiring ML algorithms to be retrained continuously on constantly changing data to keep up with these changes. Moreover, ML algorithms are also being employed for malicious purposes, such as developing more sophisticated and convincing social engineering attacks, creating malware that is harder to detect and analyze, and automating the process of identifying and exploiting system and network vulnerabilities. A study by Kaloudi and Li [41] examined prior research on AI-based cyber attacks and proposed a framework for categorizing several aspects of malicious AI use throughout the cyber attack life cycle, providing a basis for detecting and predicting future threats. For a more comprehensive review of ML for cybersecurity, interested readers can refer to [29,42,43].

8. Case Study: Axie Infinity and Solbeatz—Leveraging Blockchain Technology for Tokenized Markets

This case study explores the applications and recent developments of blockchain-based tokenized markets by examining two promising platforms: Axie Infinity (https://axieinfinity.com/, accessed on 2 May 2023) in the gaming industry and Solbeatz (https://www.solbeatz.xyz/, accessed on 2 May 2023) in the music industry. Axie Infinity, founded in 2018, is a pioneering play-to-earn gaming platform that has gained significant popularity. Solbeatz, on the other hand, is a recently founded platform with early access, showing promising potential in revolutionizing the music industry through tokenization.

The traditional music industry has long faced issues such as limited access for independent artists, and a lack of transparency in the revenue system. Solbeatz, founded in 2022, aims to address these issues by leveraging blockchain and tokenization. By utilizing the decentralized nature of blockchain and creating a marketplace for music creators and consumers, Solbeatz provides a transparent and fair ecosystem that empowers artists and rewards creativity. It connects music creators directly with their audience, enabling them to share their music, collaborate, and monetize their work. Artists can tokenize their music and associated rights, creating unique digital assets that can be bought, sold, and traded on the Solbeatz marketplace. Users can purchase and stream music using cryptocurrency, providing a seamless and transparent transaction experience.

Axie Infinity, on the other hand, is currently the most popular play-to-earn gaming platform that occupies a large portion of the online gaming market [44]. Axie Infinity is a game centered around digital creatures called Axies, which are unique NFTs that can be bought, sold, and bred on the Ethereum blockchain. Each Axie possesses different traits and abilities, players can build a team of Axies and engage in strategic battles with other players, earning in-game tokens called Smooth Love Potions (SLP) and Axie Infinity Shards (AXS). By combining elements of gaming, NFTs, and decentralized finance, Axie Infinity has created an innovative ecosystem that enables players to earn real money from their gameplay. Through its marketplace and breeding mechanics, Axie Infinity facilitates the buying, selling, and trading of NFTs, enabling players to build valuable collections and participate in a vibrant secondary market. The platform's success demonstrates the potential for tokenization to revolutionize the gaming industry, providing players with true ownership, transparent money exchange, and new economic opportunities.

The success of Axie Infinity and the emergence of Solbeatz showcase the potential of blockchain-based tokenized markets in different sectors. Axie Infinity provides players with economic opportunities and genuine ownership of their in-game assets by integrating tokenization, play-to-earn mechanics, and a thriving NFT marketplace. On the other hand, Solbeatz empowers artists in the music industry by enabling them to tokenize their music and associated rights. Artists have control over pricing, fan interactions, and royalties, fostering a direct and transparent connection between artists and their audience. Table 2 illustrates the main characteristics of Axie Infinity and Solbeatz. By leveraging blockchain technology and embracing tokenization, these platforms redefine traditional industries, offering participants enhanced ownership, transparent value exchange, and innovative economic models. As the blockchain landscape matures, the ongoing exploration and implementation of tokenized markets will likely unlock new frontiers, transforming industries and fostering novel possibilities within the global economy.

Table 2. Comparative table of Axie Infinity and Solbeatz.

Aspect	Axie Infinity	Solbeatz
Founding Year	2018	2022
Industry	Gaming	Music
Technology	Blockchain	Blockchain
Tokenization	Game creatures (Axies), SLP, AXS	Music and associated rights
Use Cases	Play-to-earn gaming, NFT marketplace	Tokenized music marketplace
Economic Model	In-game cryptocurrency rewards	Music rights transactions
User Community	Global community of players	Global community of artists and audience
Main Features	Collectible digital creatures, battles, marketplace	Tokenized music
Primary Focus	Gaming and play-to-earn model	Music industry disruption

9. Trends and Challenges of Tokenized Markets

Tokenized markets create and trade digital tokens representing a variety of assets, including real estate, commodities, currencies, and more. These markets offer numerous potential use cases:

- Real estate: Tokenized real estate allows for fractional ownership of property, increasing liquidity for an otherwise non-liquid asset. By buying and selling tokens representing a share of the property, investors can bypass intermediaries such as brokers and reduce transaction costs.
- Commodities: Tokenized commodities enable investors to trade fractions of assets such as gold, silver, and oil, providing easier exposure to these assets without the need for physical ownership.
- Currencies: Tokenized currencies represent fiat currencies in digital form and can be used for cross-border payments, reducing the time and costs associated with traditional currency exchanges.
- Loyalty programs: Loyalty programs can use blockchain to offer more flexible and valuable rewards, such as those offered by Loyyal (https://loyyal.com/, accessed on 2 May 2023). Tokens allow customers to use their points across multiple programs and earn rewards beyond just purchases.
- Identity verification: Civic's blockchain platform (https://www.civic.com/, accessed on 2 May 2023) allows individuals to securely prove their identity without sharing sensitive information with third parties. This has the potential to revolutionize industries such as finance, healthcare, and government.
- Digital gold: Platforms such as Rush Gold (https://rush.gold/, accessed on 2 May 2023) use blockchain to enable users to invest in digital gold, offering a secure and transparent way to invest in this traditional store of value and hedge against inflation.

Overall, tokenized markets have the potential to democratize access to assets and markets, increase transparency, and reduce transaction costs. However, as with any new technology, there are potential risks and challenges, such as regulatory uncertainty and technical issues with blockchain implementations.

10. Conclusions and Future Work

Blockchain technology has enabled secure and decentralized peer-to-peer transfer of digital assets without intermediaries, and asset tokenization has opened up opportunities for virtual markets. Li et al. [45] found that fintechs that use big data, cloud computing, blockchain, and other technical innovations can simplify consumer financial transactions, increase user satisfaction, and enhance enterprise marketing, especially on user word of mouth communication and promotion of the platform. However, challenges such as cybersecurity threats and the lack of government and industry regulation must be addressed. Tokenized platforms have advantages over traditional exchange platforms, such as increased transparency and security, using blockchain technology to record transactions on a decentralized ledger. Designing a tokenized platform requires consideration of security, scalability, and user experience.

Frameworks such as Apache Hadoop, Apache Spark, and Apache Cassandra can be adopted to manage distributed data for handling databases associated with real systems. The European dataspace for digital assets has been initiated to facilitate the secure sharing and exchange of data across Europe while ensuring high-quality data and compliance with European rules and values. Machine learning has become an effective approach to counter cyber threats, but attackers are constantly evolving their methods, requiring ML algorithms to be retrained continuously. Tokenized markets allow for the creation and trading of digital tokens representing various assets, providing benefits such as fractional ownership, cross-border payments, and flexible loyalty program rewards. However, as tokenized platforms continue to evolve, regulatory uncertainty and technical issues with blockchain implementations must be considered, and mechanisms to strike a balance between innovation and consumer protection, ensuring fair and transparent operations within tokenized markets, should also be studied. Additionally, to mitigate the massive energy consumed by blockchain operations, scalable blockchain solutions, such as sharding and layer-two protocols, and other energy-efficient consensus mechanisms to minimize the environmental impact of blockchain operations need to be explored.

Future research should focus on developing and implementing blockchain-based tokenized platforms that securely and efficiently exchange digital assets, improving distributed data management systems, continued research and development of machine learning techniques to detect and categorize cyber threats, and exploring the potential benefits and risks of tokenized markets for various asset types. Furthermore, while the initial focus of blockchain technology has been on cryptocurrencies and financial assets, there is significant potential for its application across various industries, such as supply chain management, intellectual property rights, healthcare, and voting systems.

Overall, future research should continue to explore the potential benefits and risks associated with tokenized markets for various asset types and industries. By addressing these research areas, we can unlock the full potential of blockchain-based tokenized platforms, enabling the secure, transparent, and efficient exchange of digital assets while promoting economic growth, innovation, and compliance with legal and ethical standards across industries.

Author Contributions: Conceptualization, A.A.J. and E.P.-B.; methodology, Y.L.; validation, A.A.J. and M.A.; writing—original draft preparation, Y.L., E.P.-B., X.A.M., M.A. and B.B.B.; writing—review and editing, A.A.J.; supervision, A.A.J. All authors have read and agreed to the published version of the manuscript.

Funding: This work has received financial support from the Horizon Europe Research & Innovation Programme under Grant agreement N. 101092612 (Social and hUman ceNtered XR—SUN project), as well as from the Regional Department of Innovation, Universities, Science and Digital Society of the Generalitat Valenciana "Programa Investigo" (INVEST/2022/342), within the framework of the Plan de Recuperación, Transformación y Resiliencia funded by the European Union—NextGenerationEU.

Institutional Review Board Statement: Not applicable.

Informed Consent Statement: Not applicable.

Data Availability Statement: Not applicable.

Conflicts of Interest: The authors declare no conflict of interest.

References

1. Di Pierro, M. What is the blockchain? *Comput. Sci. Eng.* **2017**, *19*, 92–95. [CrossRef]
2. Zheng, Z.; Xie, S.; Dai, H.N.; Chen, X.; Wang, H. Blockchain challenges and opportunities: A survey. *Int. J. Web Grid Serv.* **2018**, *14*, 352–375. [CrossRef]

3. Monrat, A.A.; Schelén, O.; Andersson, K. A survey of blockchain from the perspectives of applications, challenges, and opportunities. *IEEE Access* **2019**, *7*, 117134–117151. [CrossRef]
4. Steinmetz, F.; Von Meduna, M.; Ante, L.; Fiedler, I. Ownership, uses and perceptions of cryptocurrency: Results from a population survey. *Technol. Forecast. Soc. Chang.* **2021**, *173*, 121073. [CrossRef]
5. Sazandrishvili, G. Asset tokenization in plain English. *J. Corp. Account. Financ.* **2020**, *31*, 68–73. [CrossRef]
6. Bala, R. Tokenization of Assets. In *Handbook on Blockchain*; Springer: Berlin/Heidelberg, Germany, 2022; pp. 577–602.
7. Zheng, M.; Sandner, P. Asset Tokenization of Real Estate in Europe. In *Blockchains and the Token Economy: Theory and Practice*; Springer: Berlin/Heidelberg, Germany, 2022; pp. 179–211.
8. Garcia-Teruel, R.M.; Simón-Moreno, H. The digital tokenization of property rights. A comparative perspective. *Comput. Law Secur. Rev.* **2021**, *41*, 105543. [CrossRef]
9. Moher, D.; Liberati, A.; Tetzlaff, J.; Altman, D.G.; Group, P. Reprint—Preferred reporting items for systematic reviews and meta-analyses: The PRISMA statement. *Phys. Ther.* **2009**, *89*, 873–880. [CrossRef]
10. Lee, D.D.; Seung, H.S. Learning the parts of objects by non-negative matrix factorization. *Nature* **1999**, *401*, 788–791. [CrossRef]
11. Heines, R.; Dick, C.; Pohle, C.; Jung, R. The Tokenization of Everything: Towards a Framework for Understanding the Potentials of Tokenized Assets. In Proceedings of the PACIS, Virtual, 12–14 July 2021; p. 40.
12. Freni, P.; Ferro, E.; Moncada, R. Tokenomics and blockchain tokens: A design-oriented morphological framework. *Blockchain Res. Appl.* **2022**, *3*, 100069. [CrossRef]
13. Guggenberger, T.; Schellinger, B.; von Wachter, V.; Urbach, N. Kickstarting blockchain: Designing blockchain-based tokens for equity crowdfunding. *Electron. Commer. Res.* **2023**, 1–35. [CrossRef]
14. Warren, J.; Marz, N. *Big Data: Principles and Best Practices of Scalable Realtime Data Systems*; Simon and Schuster: New York, NY, USA, 2015.
15. Van Steen, M.; Tanenbaum, A.S. *Distributed Systems*; Maarten van Steen: Leiden, The Netherlands, 2017.
16. Moysiadis, V.; Sarigiannidis, P.; Moscholios, I. Towards distributed data management in fog computing. *Wirel. Commun. Mob. Comput.* **2018**, *2018*, 7597686 . [CrossRef]
17. Ahmed, N.; Barczak, A.L.; Susnjak, T.; Rashid, M.A. A comprehensive performance analysis of Apache Hadoop and Apache Spark for large scale data sets using HiBench. *J. Big Data* **2020**, *7*, 110. [CrossRef]
18. Scerri, S.; Tuikka, T.; de Vallejo, I.L.; Curry, E. Common European Data Spaces: Challenges and Opportunities. *Data Spaces Des. Deploy. Future Dir.* **2022**, 337–357. [CrossRef]
19. European Commission. *Towards a Common European Data Space*; European Commission: Brussels, Belgium, 2018.
20. Khan, N.; Kchouri, B.; Yatoo, N.A.; Kräussl, Z.; Patel, A.; State, R. Tokenization of sukuk: Ethereum case study. *Glob. Financ. J.* **2022**, *51*, 100539. [CrossRef]
21. Tian, Y.; Minchin, R.; Chung, K.; Woo, J.; Adriaens, P. Towards Inclusive and Sustainable Infrastructure Development through Blockchain-enabled Asset Tokenization: An Exploratory Case Study. *IOP Conf. Ser. Mater. Sci. Eng.* **2022**, *1218*, 012040.
22. Zarifis, A.; Cheng, X. The business models of NFTs and fan tokens and how they build trust. *J. Electron. Bus. Digit. Econ.* **2022**, *ahead-of-print*. [CrossRef]
23. Calandra, D.; Secinaro, S.; Massaro, M.; Dal Mas, F.; Bagnoli, C. The link between sustainable business models and Blockchain: A multiple case study approach. *Bus. Strategy Environ.* **2022**, *32*, 1403–1417. [CrossRef]
24. Benedetti, H.; Rodríguez-Garnica, G. Tokenized Assets and Securities. In *The Emerald Handbook on Cryptoassets: Investment Opportunities and Challenges*; Emerald Publishing Limited: Bingley, UK, 2023; pp. 107–121.
25. Kopp, A.; Orlovskyi, D. Towards the Tokenization of Business Process Models using the Blockchain Technology and Smart Contracts. *CMIS* **2022**, *3137*, 274–287.
26. Buldas, A.; Draheim, D.; Gault, M.; Laanoja, R.; Nagumo, T.; Saarepera, M.; Shah, S.A.; Simm, J.; Steiner, J.; Tammet, T.; et al. An ultra-scalable blockchain platform for universal asset tokenization: Design and implementation. *IEEE Access* **2022**, *10*, 77284–77322. [CrossRef]
27. Mazzei, D.; Baldi, G.; Fantoni, G.; Montelisciani, G.; Pitasi, A.; Ricci, L.; Rizzello, L. A Blockchain Tokenizer for Industrial IOT trustless applications. *Future Gener. Comput. Syst.* **2020**, *105*, 432–445. [CrossRef]
28. Hubballi, N.; Suryanarayanan, V. False alarm minimization techniques in signature-based intrusion detection systems: A survey. *Comput. Commun.* **2014**, *49*, 1–17. [CrossRef]
29. Buczak, A.L.; Guven, E. A survey of data mining and machine learning methods for cyber security intrusion detection. *IEEE Commun. Surv. Tutor.* **2015**, *18*, 1153–1176. [CrossRef]
30. Shaukat, K.; Luo, S.; Chen, S.; Liu, D. Cyber threat detection using machine learning techniques: A performance evaluation perspective. In Proceedings of the 2020 International Conference on Cyber Warfare and Security (ICCWS), Norfolk, VA, USA, 12–13 March 2020; pp. 1–6.
31. Apruzzese, G.; Colajanni, M.; Ferretti, L.; Guido, A.; Marchetti, M. On the effectiveness of machine and deep learning for cyber security. In Proceedings of the 2018 10th International Conference on Cyber Conflict (CyCon), Tallinn, Estonia, 29 May–1 June 2018; pp. 371–390.
32. Miller, S.T.; Busby-Earle, C. Multi-perspective machine learning a classifier ensemble method for intrusion detection. In Proceedings of the 2017 International Conference on Machine Learning and Soft Computing, Ho Chi Minh City, Vietnam, 13–16 January 2017; pp. 7–12.

33. Narudin, F.A.; Feizollah, A.; Anuar, N.B.; Gani, A. Evaluation of machine learning classifiers for mobile malware detection. *Soft Comput.* **2016**, *20*, 343–357. [CrossRef]
34. Gauthama Raman, M.; Somu, N.; Jagarapu, S.; Manghnani, T.; Selvam, T.; Krithivasan, K.; Shankar Sriram, V. An efficient intrusion detection technique based on support vector machine and improved binary gravitational search algorithm. *Artif. Intell. Rev.* **2020**, *53*, 3255–3286. [CrossRef]
35. Ghanem, K.; Aparicio-Navarro, F.J.; Kyriakopoulos, K.G.; Lambotharan, S.; Chambers, J.A. Support vector machine for network intrusion and cyber-attack detection. In Proceedings of the 2017 Sensor Signal Processing for Defence Conference (SSPD), London, UK, 6–7 December 2017; pp. 1–5.
36. Li, Y.; Xia, J.; Zhang, S.; Yan, J.; Ai, X.; Dai, K. An efficient intrusion detection system based on support vector machines and gradually feature removal method. *Expert Syst. Appl.* **2012**, *39*, 424–430. [CrossRef]
37. GuangJun, L.; Nazir, S.; Khan, H.U.; Haq, A.U. Spam detection approach for secure mobile message communication using machine learning algorithms. *Secur. Commun. Netw.* **2020**, *2020*, 8873639. [CrossRef]
38. Lin, W.C.; Ke, S.W.; Tsai, C.F. CANN: An intrusion detection system based on combining cluster centers and nearest neighbors. *Knowl.-Based Syst.* **2015**, *78*, 13–21. [CrossRef]
39. Meng, W.; Li, W.; Kwok, L.F. Design of intelligent KNN-based alarm filter using knowledge-based alert verification in intrusion detection. *Secur. Commun. Netw.* **2015**, *8*, 3883–3895. [CrossRef]
40. Shapoorifard, H.; Shamsinejad, P. Intrusion detection using a novel hybrid method incorporating an improved KNN. *Int. J. Comput. Appl.* **2017**, *173*, 5–9. [CrossRef]
41. Kaloudi, N.; Li, J. The ai-based cyber threat landscape: A survey. *ACM Comput. Surv. (CSUR)* **2020**, *53*, 1–34. [CrossRef]
42. Martínez Torres, J.; Iglesias Comesaña, C.; García-Nieto, P.J. Machine learning techniques applied to cybersecurity. *Int. J. Mach. Learn. Cybern.* **2019**, *10*, 2823–2836. [CrossRef]
43. Sarker, I.H.; Kayes, A.; Badsha, S.; Alqahtani, H.; Watters, P.; Ng, A. Cybersecurity data science: An overview from machine learning perspective. *J. Big Data* **2020**, *7*, 41. [CrossRef]
44. Delic, A.J.; Delfabbro, P.H. Profiling the Potential Risks and Benefits of Emerging "Play to Earn" Games: A Qualitative Analysis of Players' Experiences with Axie Infinity. *Int. J. Ment. Health Addict.* **2022**, 1–14. [CrossRef]
45. Li, Y.; Ma, X.; Li, Y.; Li, R.; Liu, H. How does platform's fintech level affect its word of mouth from the perspective of user psychology? *Front. Psychol.* **2023**, *14*, 1085587. [CrossRef]

 *information*

Article

A Robust Hybrid Deep Convolutional Neural Network for COVID-19 Disease Identification from Chest X-ray Images

Theodora Sanida [1,*], Irene-Maria Tabakis [1], Maria Vasiliki Sanida [2], Argyrios Sideris [1] and Minas Dasygenis [1]

[1] Department of Electrical and Computer Engineering, University of Western Macedonia, 50131 Kozani, Greece
[2] Department of Digital Systems, University of Piraeus, 18534 Piraeus, Greece
* Correspondence: thsanida@uowm.gr; Tel.: +30-24610-56534

Abstract: The prompt and accurate identification of the causes of pneumonia is necessary to implement rapid treatment and preventative approaches, reduce the burden of infections, and develop more successful intervention strategies. There has been an increase in the number of new pneumonia cases and diseases known as acute respiratory distress syndrome (ARDS) as a direct consequence of the spread of COVID-19. Chest radiography has evolved to the point that it is now an indispensable diagnostic tool for COVID-19 infection pneumonia in hospitals. To fully exploit the technique, it is crucial to design a computer-aided diagnostic (CAD) system to assist doctors and other medical professionals in establishing an accurate and rapid diagnosis of pneumonia. This article presents a robust hybrid deep convolutional neural network (DCNN) for rapidly identifying three categories (normal, COVID-19 and pneumonia (viral or bacterial)) using X-ray image data sourced from the COVID-QU-Ex dataset. The proposed approach on the test set achieved a rate of 99.25% accuracy, 99.10% Kappa-score, 99.43% AUC, 99.24% F1-score, 99.25% recall, and 99.23% precision, respectively. The outcomes of the experiments demonstrate that the presented hybrid DCNN mechanism for identifying three categories utilising X-ray images is robust and effective.

Keywords: hybrid DCNN mechanism; diagnosis; chest X-ray images; radiography images; lung opacity; pneumonia; COVID-19

Citation: Sanida, T.; Tabakis, I.-M.; Sanida, M.V.; Sideris, A.; Dasygenis, M. A Robust Hybrid Deep Convolutional Neural Network for COVID-19 Disease Identification from Chest X-ray Images. *Information* 2023, 14, 310. https://doi.org/10.3390/info14060310

Academic Editor: Gholamreza Anbarjafari (Shahab)

Received: 28 February 2023
Revised: 20 May 2023
Accepted: 25 May 2023
Published: 29 May 2023

1. Introduction

The COVID-19 pandemic has led to a considerable increase in pneumonia patients worldwide. The most prominent indications and symptoms of COVID-19 are chest discomfort, cough, sore throat, fever, and shortness of breath, similar to other pneumonia types. COVID-19 pneumonia presents unique challenges as it can cause severe respiratory distress, which can advance rapidly to acute respiratory distress syndrome (ARDS) [1]. So, to successfully combat the disease and implement preventative measures, it is necessary to differentiate between COVID-19 infection and other bacterial or viral pneumonias. A delay in treatment may result in mortality, or other problems, including impaired lung function and chronic non-communicable respiratory infections, such as asthma or chronic obstructive pulmonary disease (COPD) [2,3].

In diagnosing a broad range of lung-related disorders, chest X-rays (CXR) are often used as one of the diagnostic techniques [4]. Chest X-rays are inexpensive, widely available, and can be performed quickly, making them an effective diagnostic tool for pneumonia. In contrast, computed tomography (CT) and magnetic resonance imaging (MRI) are higher-cost techniques and require more time. For this reason, chest radiography has developed into an important diagnostic technique for COVID-19 infection pneumonia in hospitals. However, when performing chest X-rays in the early stages of pneumonia, the radiographic features may not be distinct, making their interpretation more difficult. As a result, especially during the COVID-19 pandemic, using a CAD system for pneumonia diagnosis can help to manage the high volume of patients presenting with respiratory symptoms [5].

Consequently, designing a computer-aided diagnostic (CAD) system has become essential in supporting medical practitioners in establishing an accurate diagnosis of pneumonia on time [6,7].

Lately, increased attention has been paid to deep learning (DL) methods, particularly deep convolutional neural networks (DCNNs), in CAD systems based on computer vision methodologies so as to identify diseases using chest X-ray images. The use of DCNNs in chest X-rays for COVID-19 detection has gained significant traction due to its potential to provide a fast and accurate diagnosis, which is crucial for handling the COVID-19 pandemic [8]. DCNNs are a kind of artificial intelligence (AI) often utilised in image categorisation tasks because they can extract characteristics from images and categorise them based on these features. In the medical imaging domain [9], it has been shown that identifying DCNNs in chest X-rays plays a crucial role in diagnosing bacterial pneumonia, viral pneumonia, and other chest disorders. Therefore, the development of such models to identify radiographs contaminated with COVID-19 is urgently required to be able to make suitable clinical decisions to assist radiologists, medical experts, practitioners and doctors [10,11].

This work proposes a CAD system with a hybrid identification strategy that uses chest X-ray image data to categorize three distinct diseases that might cause pneumonia. The hybrid DCNN consists of a combination of VGG [12] blocks and an inception [13,14] module. Compared to previous methods that have already been established, our novel network achieves a higher rate of accurate identification using a large collection. Thus, by employing X-ray images, this image-based pneumonia disease diagnosis method will assist medical professionals in the early and rapid identification of pneumonia.

The most significant contributions of this work are as follows:

- The identification of pneumonia is performed using a hybrid DCNN mechanism. The modified VGG19 model includes two inception blocks to take advantage of simultaneous feature extraction capabilities. The hybrid DCNN is equipped with powerful feature extraction capabilities.
- We conducted exhaustive high-level simulations to assess the effectiveness of the presented hybrid DCNN. The proposed hybrid DCNN mechanism findings were compared to those obtained from the most current and advanced networks.

The remaining structure of this work is as follows: In Section 2, we address similar articles on pneumonia and the diagnosis of COVID-19 that have been published in the literature. Section 3 analyzes the materials and methods used in our experimental work. The experiment outcomes are discussed in Section 4, along with assessment metrics, and then the accuracy rate is compared with existing identification techniques. Finally, the study's conclusion is in presented in Section 5, which includes some predictions for the future.

2. Related Work

The automated investigation and analysis of an extensive collection of image data create new and exciting challenges that call for state-of-the-art computational strategies and classic machine learning (ML), deep learning (DL), or computational intelligence (CI) approaches that can provide high-performance and specialized medical services [15]. In the last two years, a significant number of investigators from all over the world have developed and published many studies to detect and slow the spread of the COVID-19 virus. A substantial number of these researchers have used a variety of AI methodologies to analyze and diagnose X-ray images to identify various diseases. The capacity of DL techniques [16] to generate better results than typical ML approaches has made them the most popular methods for identifying images. In this part, we will concentrate on research that uses novel methodologies to identify COVID-19 based on DL methods.

COVID-AleXception is proposed in [17], which is a concatenation of the features from two pre-trained CNN methods, Xception and AlexNet. The dataset comprises 15,153 X-ray images (1345 pneumonia, 3616 COVID-19 and 10,192 normal). Each CNN method was trained for 100 epochs with the Adam optimisation algorithm. The COVID-

AleXception method achieved an identification accuracy rate of 98.68% over Xception and AlexNet, which yielded an identification accuracy of 95.63% and 94.86%, respectively. Hafeez et al. [18] designed a customised CNN prediction system for chest X-rays and compared it with two pre-trained CNN methods (VGG16 and AlexNet). The accuracy of the proposed system for the three categories (normal, COVID-19, and virus bacteria) is 89.855%, 89.015% for VGG16 and 89.155% for AlexNet.

In [19], the authors suggested a lightweight CNN technique for COVID-19 identification utilising X-ray images and evaluated it with seven pre-trained CNN systems (InceptionV3, Xception, ResNet50V2, MobileNetV2, DenseNet121, EfficientNet-B0, and EfficientNetV2). The dataset comprised 600 COVID-19, 600 normal, and 600 pneumonia images. Each CNN method was trained for 50 epochs. The rate of the accuracy of the proposed method for the three categories is 98.33% and 97.73% from EfficientNetV2. CoroNet is proposed in [20], based on the Xception method. The utilised collection comprised 330 bacterial pneumonia, 327 viral pneumonia, 284 COVID-19, and 310 normal X-ray images. The CoroNet method was trained for 80 epochs, and four categories reached a rate of accuracy of 89.60%.

Ghose et al. [21] designed a customised CNN automatic diagnosis system. The dataset comprises 10,293 X-ray images, including 4200 pneumonia, 2875 COVID-19, and 3218 normal images. The customised CNN was trained for 25 epochs with the Adam optimisation algorithm. The proposed method attained 98.50% accuracy, a 98.30% F1-score, and 99.20% precision. In [22], the authors suggested a DL diagnosis system to quickly detect pneumonia using X-ray images. They compare the VGG19 and ResNet50 methods for three distinct diseases of lung detection. The dataset comprises 11,263 pneumonia, 11,956 COVID-19 and 10,701 normal images. Each CNN method was trained for 180 epochs. The accuracy of the proposed diagnosis system for the three categories is 96.60% for the VGG19 method and 95.80% for ResNet50.

Furthermore, in [23], the authors compare four DL methods (VGG16, ResNet50, DenseNet121, and VGG19) to diagnose X-ray images as COVID-19 or normal. The dataset comprises 1592 X-ray images (802 normal, 790 COVID-19). Each CNN method was trained for 30 epochs. The VGG16 method for the two categories achieved an accuracy rate of 99.33%, ResNet50 achieved 97.00%, DenseNet121 achieved 96.66%, and VGG19 achieved 96.66%. In [24], the authors suggested a DL model based on MobileNetV2 to identify COVID-19 infection. The dataset comprises 1576 normal, 3616 COVID-19 and 4265 pneumonia X-ray images. Each CNN strategy was trained for 80 epochs. The accuracy rate of the suggested diagnosis approach for the three categories is 97.61%.

Nayak et al. [25] designed a CNN technique called LW-CORONet. The suggested method is evaluated by employing two datasets where dataset-1 has 2250 images (750 pneumonia, 750 normal, and 750 COVID-19) and dataset-2 has 15,999 images (5575 pneumonia, 8066 normal, and 2358 COVID-19). The customised CNN was trained for 100 epochs with the Adam optimisation algorithm. The identification accuracy obtained is 98.67% on dataset-1 and 95.67% on dataset-2 for three category cases, respectively. In [26], the authors suggested a CNN model for medical diagnostic image analysis to identify COVID-19. The proposed approach is based on the MobileNetV2 method. The dataset comprises 10.192 normal, 3616 COVID-19, 6012 lung opacity and 1345 viral pneumonia images. The proposed diagnosis method achieves an identification accuracy rate of 95.80%.

Most researchers fed their identification networks with data from relatively small collections. Consequently, most of the networks reached high levels of accuracy; however, the prediction results based on those networks cannot be generalised owing to the small number of image data on which the networks were trained [27]. Table 1 presents a comprehensive description of the categorisation of the above systems for identifying COVID-19 and analyzes the model employed and the accuracy rate achieved.

In our work, the collection included 33,920 chest X-ray image data, balanced with around 10,500 images belonging to each category. Thus, a hybrid DCNN identification mechanism was created for diagnosing pneumonia and COVID-19 disease based on image evidence from a much larger collection. Our suggested design has a primary key goal to improve disease detection accuracy and reduce the frequency of inaccurate identifications. The hybrid DCNN network was trained and tested using X-ray image data that included three distinct types of pneumonia. According to the experiments' findings, the model's categorization accuracy is 99.23%. Since it has a high accuracy rate, the recommended strategy may be of assistance to those operating in the medical industry.

Table 1. A summary of studies using CNN methods for COVID-19 identification.

Study	Best Method	Accuracy (%)
[17]	COVID-AleXception	98.68
[18]	Custom CNN	89.855
[19]	Lightweight CNN	98.33
[20]	CoroNet	89.60
[21]	Custom CNN	98.50
[22]	VGG19	96.60
[23]	VGG16	99.33
[24]	MobileNetV2	97.61
[25]	LW-CORONet	98.67
[26]	MobileNetV2	95.80

3. Materials and Methods

3.1. Dataset Collection

There are 33,920 chest X-ray image data in the collection COVID-QU-Ex [28], all of which are available to the public.

The COVID-QU-Ex collection consists of three categories: normal, non-COVID infection, and COVID-19. Patients with normal (healthy) situations represent 32% of the total collection with 10,701 instances, non-COVID infection situations represent 33% with 11,263 instances, and COVID-19 situations represent 35% with 11,956 instances. These images represent two different diseases and one healthy state. Each image's resolution in the collection, which is in a PNG file format, is 256 pixels per flank. Figure 1 illustrates a sample of a normal instance and two distinct disorders that may damage the lungs. Since the collection is already large and relatively well-balanced, as shown in Figure 2, there is no need to use data augmentation techniques to make it more balanced.

From the radiographic findings in Figure 1, a normal lung X-ray typically shows clear lung fields without any significant opacities or abnormalities. The lung markings appear normal, with the blood vessels and airway passages clearly visible. In cases of viral or bacterial pneumonia, the X-ray image often reveals areas of opacity or consolidation. These areas appear as dense, cloudy regions within the lung fields, indicating the presence of inflammation, fluid, or pus. The opacities can be patchy, focal, or lobar, depending on the severity and extent of the infection. X-ray findings in COVID-19 pneumonia show ground glass opacities (blurry areas) in multiple areas of the lungs. These opacities often have a peripheral distribution and can affect both lungs symmetrically [29,30].

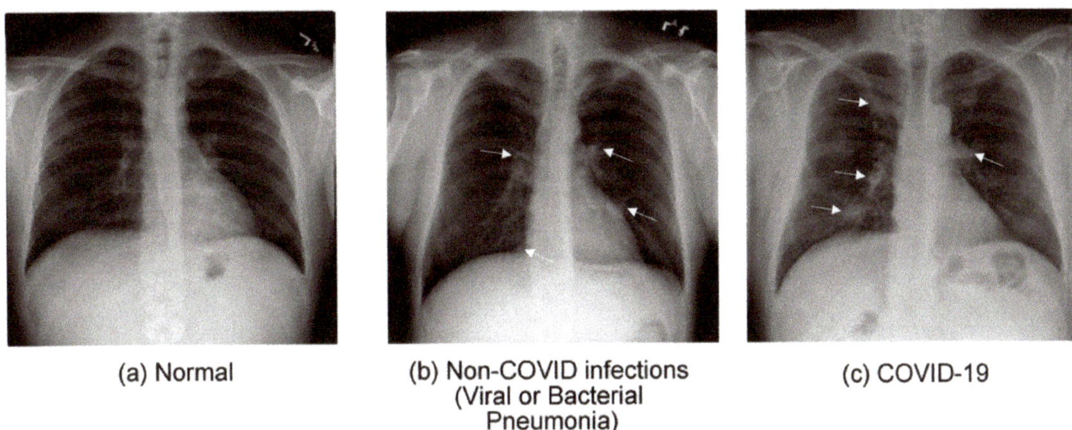

(a) Normal (b) Non-COVID infections (Viral or Bacterial Pneumonia) (c) COVID-19

Figure 1. X-ray samples by category from the COVID-QU-Ex collection (the white markers indicate infected areas).

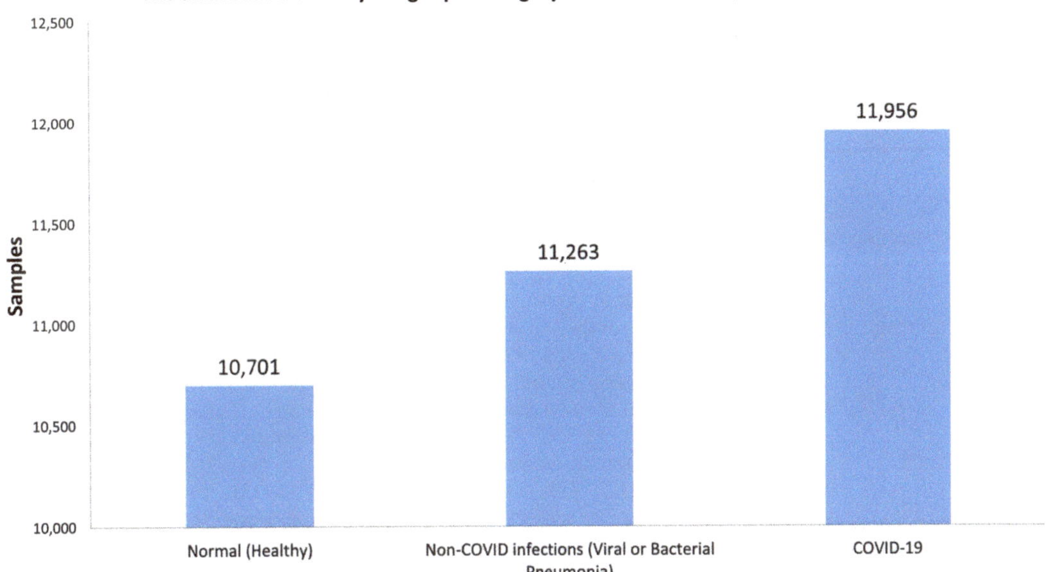

Figure 2. The allocation of X-ray image data per category from the COVID-QU-Ex collection is balanced; thus, no augmentation techniques are required.

3.2. Split Collection

In DCNN, a collection is divided into three parts: training/validation/testing. So, the training set optimizes the network's weights by reducing the predicted and actual output differences. The validation set is utilised to estimate the network's effectiveness on unknown data during training to enhance the network's performance. The test set provides a final, objective assessment of the network's performance after training on unseen data [31].

The COVID-QU-Ex collection split ratio is 80:20 for train and test objectives. Additionally, 20% of training images were used as validation data for the network during the training phase. This rate is typically used when there is a large collection, adequate data is available to train stage the network, and sufficient image data remains for validation and testing of the network [32]. The number of image data per type utilized for training/validation/testing is outlined in Table 2.

Table 2. Number of image data per type for training/validation/testing in the COVID-QU-Ex collection.

Category	Number of Images	Training Images	Validation Images	Test Images
Normal (Healthy)	10,701	6849	1712	2140
Non-COVID infections (Viral or Bacterial Pneumonia)	11,263	7208	1802	2253
COVID-19	11,956	7658	1903	2395
Total	33,920	21,715	5417	6788

3.3. Hybrid DCNN for Diagnosing Pneumonia and COVID-19 Disease

We developed a hybrid DCNN mechanism that is effective in distinguishing between the three distinct categories that have the potential to have an impact on the lungs. The hybrid DCNN network was based on combining VGG blocks and the inception module. So, combining VGG19 with the inception module increases accuracy, improves feature extraction, and improves computing efficiency.

The VGG19 [12] network comprises a total of 19 layers, 16 of which are convolutional layers and 3 of which are completely connected. It was developed specifically to perform well on image categorization tasks, making it a popular option for various computer vision applications due to its architecture. The 16 convolutional layers are separated into five blocks of two or three convolutional layers followed by a max pooling layer. Additionally, the blocks use small filters (3×3) with a stride of 1, and as the network becomes deeper, the number of filters gradually increases. Each of the three fully connected layers has 4096 neurons and uses a softmax activation function to perform the final categorization.

The inception [13] module is composed of several parallel branches that each have a different size filter, including: Convolutional layers use filters of varying sizes to extract characteristics from the input image $224 \times 224 \times 3$. The max pooling layer minimizes the spatial dimensionality of the feature maps generated by convolutional layers. The concatenation layer merges the outputs from multiple branches of the inception module into a single multi-scale representation of the input image. The inception module is widely employed in modern DCNN designs for computer vision and has demonstrated exemplary performance in various image category tasks.

The hybrid DCNN mechanism for COVID-19 disease identification has the following elements: ten convolutional layers for feature extraction, four max-pooling layers for spatial dimension of the feature maps, two inception modules, a global average pooling (GAP) layer and a fully connected (FC) layer to conduct the categorization. The hybrid DCNN mechanism takes an input size image (224, 224, 3) and passes it through the network to identify the disease categories in the image. The initial VGG block utilizes 64 filters, which results in a feature map that is (224, 224, 64) in size. The output shape produced as a consequence of this process is (112, 112, 64). The second VGG block utilizes 128 filters and produces a feature map that is (112, 112, 128) in size, and the resulting shape of the output is (56, 56, 128). The following third VGG block utilizes 256 filters and generates a feature map (56, 56, 256) in size; the output shape this creates is a rectangle (28, 28, 256). The final VGG block utilizes 512 filters and generates a feature map with dimensions of (28, 28, 512), and the shape of the output is (14, 14, 512). The first inception module utilizes 512 filters, and the shape of the output produced as a consequence is (7, 7, 512). The second inception module utilizes 512 filters, and the shape of the output that this generates is (7, 7, 512). An output shape is possessed by the GAP layer (1, 1, 512). Finally, the FC layer has an output shape that consists of (1, 1, 3). Figure 3 depicts the diagram of the inception module, whereas Figure 4 illustrates the hybrid DCNN mechanism diagram.

Our hybrid DCNN approach combines the strengths of the VGG19 architecture and the inception module to achieve improved accuracy and speed while reducing computational complexity over existing methods. Specifically, the first four blocks of the VGG19 architecture form the backbone of our network, which are highly efficient in extracting low-level features in images. However, the later blocks of VGG19 are computationally expensive, particularly when working with large datasets. To address this, we add two inception modules of the VGG19 architecture to improve the network's ability to learn more valuable features, leading to an even higher accuracy. The inclusion of two inception modules provides additional flexibility and complexity in the network, enabling it to capture a broader range of image features. So, by combining these two architectures, we leverage both strengths to achieve improved accuracy and speed while reducing computational complexity.

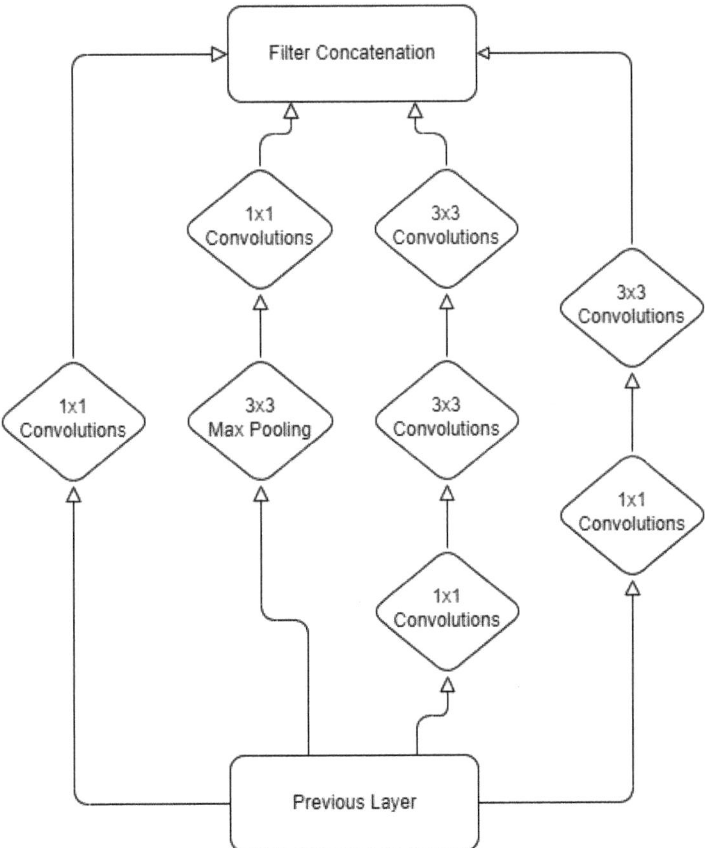

Figure 3. The block diagram of the inception module.

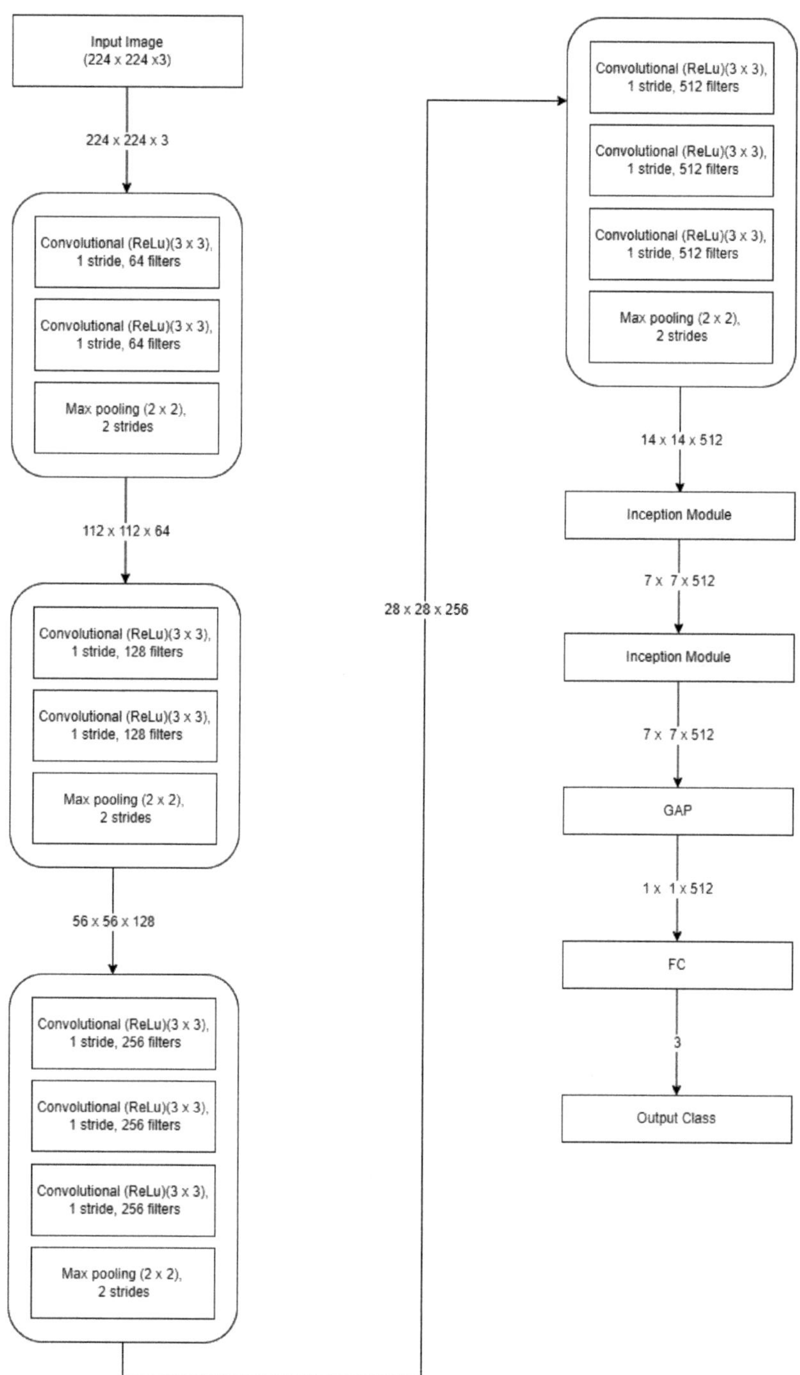

Figure 4. The block diagram of proposed hybrid DCNN for COVID-19 disease identification.

3.4. Implementation Description

All experimentations were performed utilizing a GPU (NVIDIA RTX 3050 with 8 GB RAM). Python 3, CUDA, the Keras package, CuDNN, Matplotlib and NumPy were the main libraries used to implement all networks. All networks were optimized using the Adam [33] optimizer with a learning rate of 0.0001, a number of epochs of 30 and categorical cross-entropy as a loss function. Table 3 displays the specific training parameters for all networks.

Table 3. Configurations of the training parameters for all networks.

Name of Parameter	Value for Training
Optimizer	Adam
Number of epochs	30
Learning rate	0.0001
Mini batch size	32
Loss function	Cross-entropy

3.5. Performance Measures

Accuracy, precision (specificity), recall (sensitivity), and F1-score are the most popular measures to evaluate deep learning networks [34]. In addition, the Kappa score [35] coefficient is used to assess the level of agreement between the predicted labels and the actual labels in the test data. Consequently, these measures were selected for this work. All measures are based on the number of true negative (TN), true positive (TP), false positive (FP), and false negative (FN) cases. Furthermore, the confusion matrix is used to evaluate the performance of networks during categorization tasks. Finally, the ROC curve demonstrates how effectively the network can discriminate between various kinds of image data; when the indicator is increased, the network can satisfactorily distinguish between the type with the infection and without infection. The formulas for the measures above are provided using Equations (1)–(6):

$$\text{Accuracy} = ((TP + TN)/(TP + FN + TN + FP)) \times 100\% \tag{1}$$

$$\text{Precision} = (TP/(TP + FP)) \times 100\% \tag{2}$$

$$\text{Recall} = (TP/(TP + FN)) \times 100\% \tag{3}$$

$$\text{F1-score} = 2 \times ((\text{Precision} \times \text{Recall})/(\text{Precision} + \text{Recall})) \times 100\% \tag{4}$$

$$\text{Random Accuracy} = (((TN + FP) \times (TN + FN) + (FN + TP) \times (FP + TP))/ \\ ((TP + FN + TN + FP) \times (TP + FN + TN + FP))) \times 100\% \tag{5}$$

$$\text{Kappa-score} = ((\text{ Accuracy } - \text{ Random Accuracy })/(1 - \text{ Random Accuracy })) \times 100\% \tag{6}$$

4. Experimental Results

The immediate purpose of our suggested network is to sweeten the identification accuracy of the COVID-19 disease and reduce miscategorization. Figures 5 and 6 illustrate the accuracy and loss curves for 30 epochs during the training and validation stages. The highest training and validation accuracy is shown in the hybrid DCNN with 99.32% and 97.60%, and loss is 0.1062 and 0.9260. On the contrary, the lowest training and validation accuracy is obtained at 98.97% and 91.57%, and loss is 0.1548 and 0.9157 for the ResNet50 network. Analyzing the accuracy curve, it is seen that the accuracy values of the hybrid DCNN are statble without showing overfitting over the other networks.

As can be seen in Figure 5, the Hybrid DCNN outperforms the other popular CNN architectures, starting from more than 0.85 when epochs are 0. The main reason for the superior performance is that the proposed approach uses the generic characteristics of images extracted from the ImageNet [36] dataset by the VGG19. So, our model has already learned to recognize a great deal of visually valuable elements, such as edges, textures, and shapes, which can be used for identification, and learns specific features of the COVID-19 disease identification task due to the two newly added inception modules. On the other hand, the other CNN architectures are also initialized with pre-trained weights from ImageNet, but they do not achieve optimal results for the COVID-19 identification task. Thus, the Hybrid DCNN is a powerful model that combines the advantages of both generic and specific feature extraction capabilities, resulting in top performance for the COVID-19 identification task.

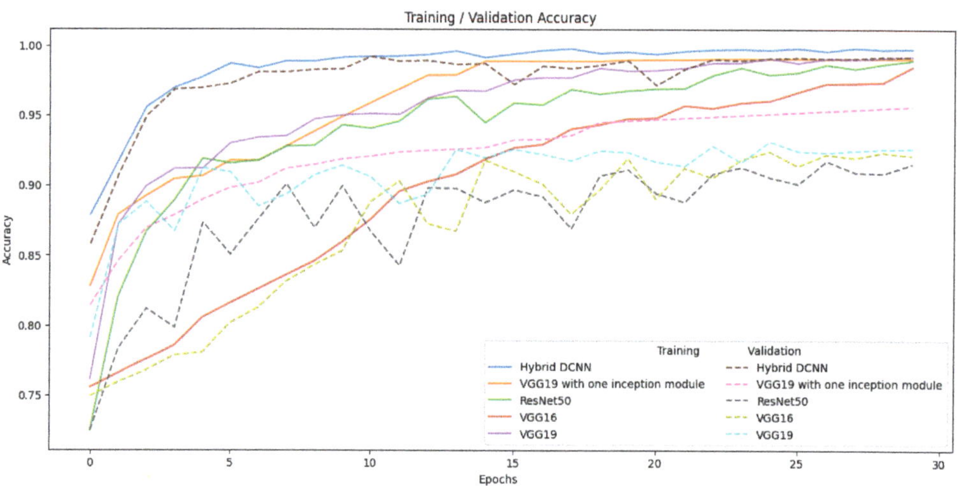

Figure 5. Comparison plot of the accuracy curves of the training/validation for each network.

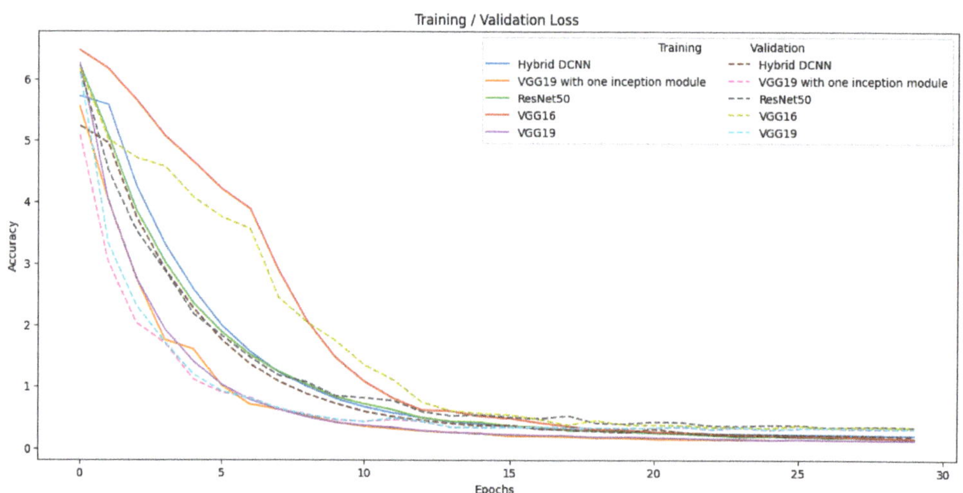

Figure 6. Comparison plot of the loss curves of the training/validation for each network.

Figures 7–11, demonstrate the confusion matrix and ROC curve plots for all networks. Among the 6788 instances, 51 were miscategorised by the proposed hybrid DCNN, the VGG19 with one inception module miscategorised 96, the VGG19 network miscategorised 124, the VGG16 network miscategorised 181, and the ResNet50 network misclassified 241 instances.

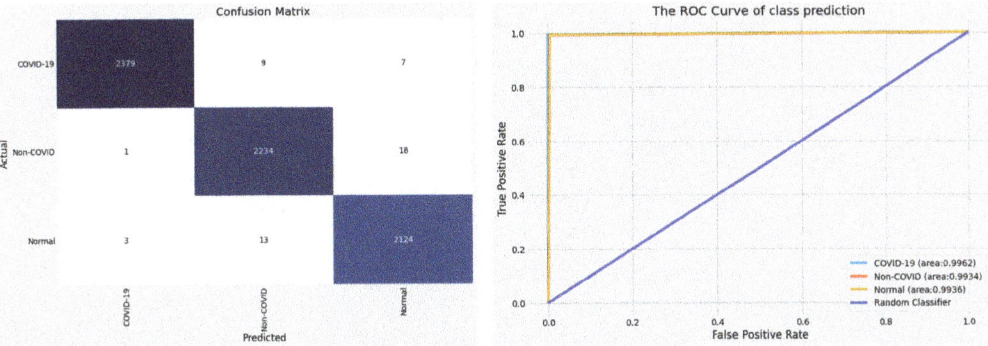

Figure 7. Results of the confusion matrix and ROC curve for the hybrid DCNN on the test dataset.

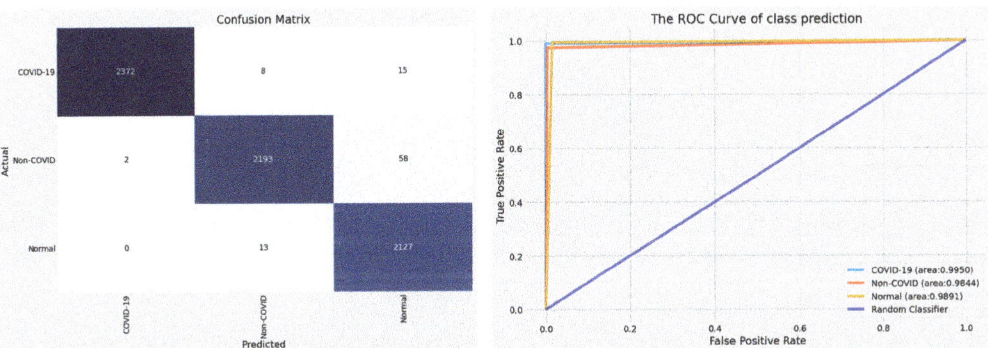

Figure 8. Results of the confusion matrix and ROC curve for the VGG19 with one inception module on the test dataset.

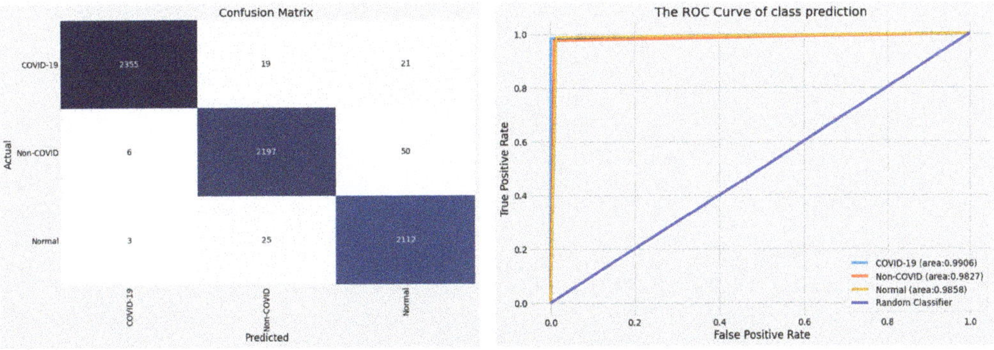

Figure 9. Results of the confusion matrix and ROC curve for the VGG19 network on the test dataset.

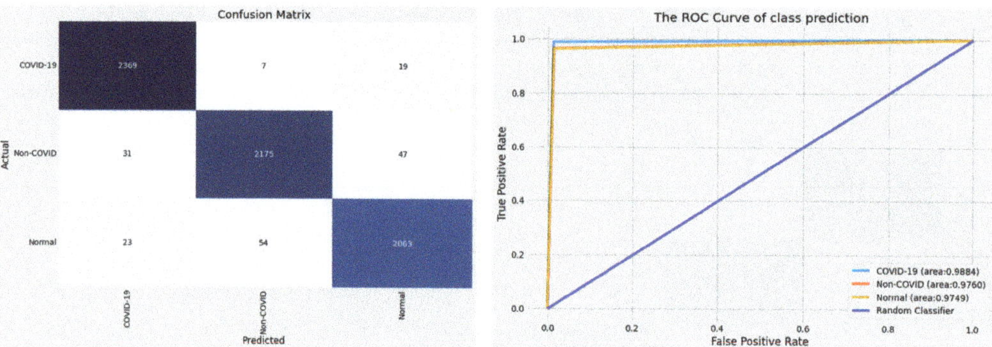

Figure 10. Results of the confusion matrix and ROC curve for the VGG16 network on the test dataset.

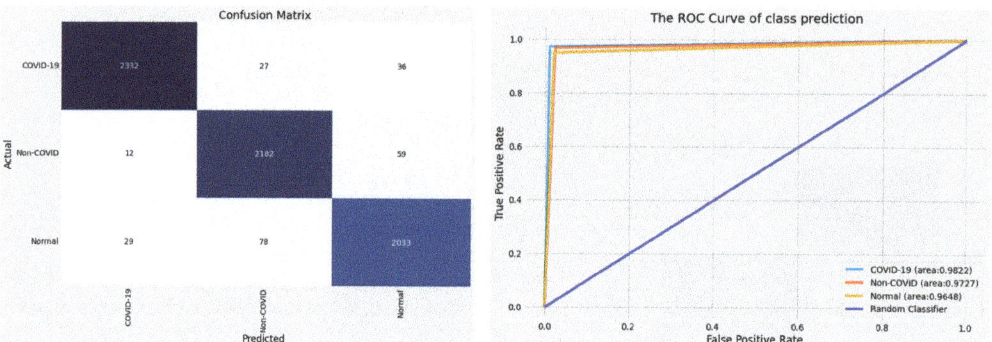

Figure 11. Results of the confusion matrix and ROC curve for the ResNet50 network on the test dataset.

The performance of the five networks is outlined in Table 4; the standard deviation is included in parentheses. The proposed hybrid DCNN mechanism has the best performance with 99.25% accuracy, 99.23% precision, 99.25% recall, a 99.24% F1-score, 99.43% AUC, and 99.10% Kappa score. On the contrary, a comparatively low performance was obtained by the ResNet50 network with 96.45% accuracy, 96.41% precision, 96.41% recall, 96.40% F1-score, 97.32% AUC, and 95.17% Kappa score. Hence, as shown in Table 5, it was revealed that the proposed hybrid DCNN is superior to other networks.

Figure 12 shows results from the hybrid DCNN mechanism on some sample instances from the test set. For example, the proper category kind, shown in Figure 12a top/bottom, is accurately diagnosed with a probability greater than 98.82% as "Normal". Moreover, the suggested strategy accurately identifies each instance in Figure 12b (top/bottom images). The proposed mechanism is accurately identified, as shown in the top image of Figure 12c. In contrast, the irregular opacity of the lungs affects the feature extraction process. So, erroneous lung disease identifications may arise, as illustrated in Figure 12c (bottom image). Considering the outcomes, it can be deduced that the recommended hybrid DCNN mechanism enhances the accuracy of COVID-19 disease identification. Specifically, combining the highly effective first four blocks of the VGG19 architecture with the efficient inception modules allows our network to capture useful features missed by other methods, thereby reducing the frequency of inaccurate identifications.

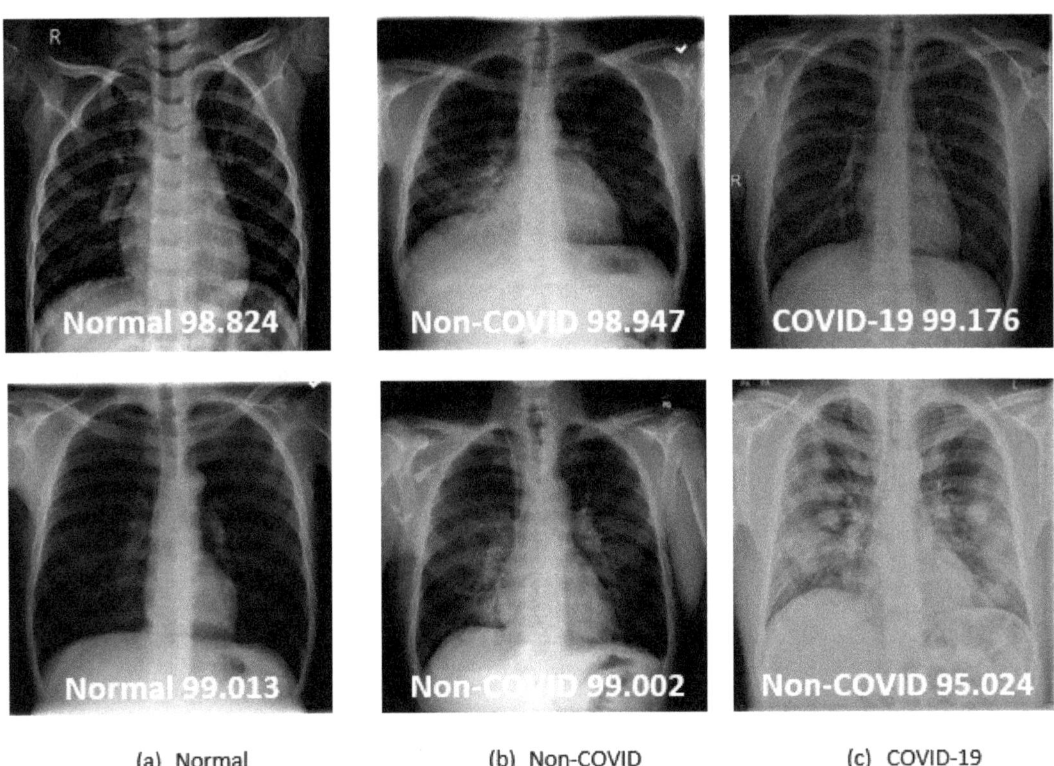

(a) Normal   (b) Non-COVID   (c) COVID-19

Figure 12. Indicative instances evaluated by hybrid DCNN mechanism.

Table 4. Performance measures of five networks, standard deviation included in parentheses.

Network	Accuracy (%)	Precision (%)	Recall (%)	F1-Score (%)	AUC (%)	Kappa-Score (%)
Hybrid DCNN	**99.25** **(0.0254)**	**99.23** **(0.0270)**	**99.25** **(0.0295)**	**99.24** **(0.0307)**	**99.43** **(0.0354)**	**99.10** **(0.0386)**
VGG19 with one inception module	98.59 (0.0427)	98.55 (0.0454)	98.59 (0.0458)	98.56 (0.0507)	98.94 (0.0492)	98.45 (0.0471)
VGG19	98.17 (0.0474)	98.13 (0.0432)	98.18 (0.0481)	98.15 (0.0507)	98.63 (0.0531)	97.84 (0.0516)
VGG16	97.33 (0.0706)	97.31 (0.0732)	97.28 (0.0713)	97.30 (0.0634)	97.97 (0.0642)	96.61 (0.0770)
ResNet50	96.45 (0.0552)	96.41 (0.0507)	96.41 (0.0587)	96.40 (0.0602)	97.32 (0.0580)	95.17 (0.0524)

Table 5. Performance evaluation of five networks.

Network	Categories	Precision	Recall	F1-Score
	COVID-19	0.9983	0.9933	0.9958
Hybrid DCNN	Non-COVID	0.9902	0.9916	0.9909
	Normal	0.9884	0.9925	0.9904
	COVID-19	0.9972	0.9901	0.9948
VGG19 with one inception module	Non-COVID	0.9901	0.9734	0.9819
	Normal	0.9668	0.9919	0.9802
	COVID-19	0.9962	0.9833	0.9897
VGG19	Non-COVID	0.9804	0.9751	0.9777
	Normal	0.9675	0.9869	0.9771
	COVID-19	0.9777	0.9891	0.9834
VGG16	Non-COVID	0.9727	0.9654	0.9690
	Normal	0.9690	0.9640	0.9665
	COVID-19	0.9827	0.9737	0.9782
ResNet50	Non-COVID	0.9541	0.9685	0.9612
	Normal	0.9554	0.9500	0.9527

5. Conclusions and Future Work

The COVID-19 pandemic has created a global health concern, with millions of individuals infected worldwide. The rapid spread of the disease has made early detection and accurate diagnosis crucial to prevent its further spread. This work presents a CAD system with a hybrid identification strategy that uses chest X-ray image data to categorize three distinct diseases. The hybrid DCNN identification mechanism consists of a combination of VGG blocks and three inception modules. Our network mechanism achieves 99.25% accuracy, a 99.10% Kappa score, 99.43% AUC, and 99.24% F1-score. These results demonstrate that the proposed strategy can effectively distinguish between pneumonia, COVID-19, and typical chest X-ray images. In further research, the diagnostic accuracy of large-scale medical datasets has to be investigated, and appropriate experiments must be conducted to verify our hybrid DCNN identification strategy in specialized services such as service-oriented networks (SONs).

Author Contributions: Methodology, T.S.; conceptualization, T.S.; formal analysis, T.S., I.-M.T., M.V.S. and A.S.; investigation, T.S., M.V.S. and I.-M.T.; software, T.S.; project administration, T.S.; resources, T.S., I.-M.T. and A.S.; validation, T.S., M.V.S., I.-M.T. and A.S.; visualization, T.S. and A.S.; supervision, M.D.; writing—original draft preparation, T.S. and M.V.S.; writing—review and editing, M.D. All authors have read and agreed to the published version of the manuscript.

Funding: This research received no external funding.

Data Availability Statement: This article uses the COVID-QU-Ex collection, which is fully available in [28].

Conflicts of Interest: The authors declare no conflict of interest.

Abbreviations

The following abbreviations are used in this manuscript:

AI	Artificial Intelligence
ARDS	Acute Respiratory Distress Syndrome
CAD	Computer-Aided Diagnostic
COPD	Chronic Obstructive Pulmonary Disease
CT	Computed Tomography
CXR	Chest X-Ray

DCNN	Deep Convolutional Neural Network
DL	Deep Learning
GPU	Graphics processing unit
ML	Machine Learning
MRI	Magnetic Resonance Imaging
ReLU	Rectified linear unit
SONs	Service-Oriented Networks
VGG	Visual geometry group

References

1. Lerner, D.K.; Garvey, K.L.; Arrighi-Allisan, A.E.; Filimonov, A.; Filip, P.; Shah, J.; Tweel, B.; Del Signore, A.; Schaberg, M.; Colley, P.; et al. Clinical features of parosmia associated with COVID-19 infection. *Laryngoscope* **2022**, *132*, 633–639. [CrossRef] [PubMed]
2. Mollarasouli, F.; Zare-Shehneh, N.; Ghaedi, M. A review on corona virus disease 2019 (COVID-19): Current progress, clinical features and bioanalytical diagnostic methods. *Microchim. Acta* **2022**, *189*, 103. [CrossRef]
3. Watanabe, A.; So, M.; Mitaka, H.; Ishisaka, Y.; Takagi, H.; Inokuchi, R.; Iwagami, M.; Kuno, T. Clinical features and mortality of COVID-19-associated mucormycosis: A systematic review and meta-analysis. *Mycopathologia* **2022**, *187*, 271–289. [CrossRef]
4. Irmici, G.; Cè, M.; Caloro, E.; Khenkina, N.; Della Pepa, G.; Ascenti, V.; Martinenghi, C.; Papa, S.; Oliva, G.; Cellina, M. Chest X-ray in Emergency Radiology: What Artificial Intelligence Applications Are Available? *Diagnostics* **2023**, *13*, 216. [CrossRef] [PubMed]
5. Taleghani, N.; Taghipour, F. Diagnosis of COVID-19 for controlling the pandemic: A review of the state-of-the-art. *Biosens. Bioelectron.* **2021**, *174*, 112830. [CrossRef]
6. Ravi, V.; Narasimhan, H.; Pham, T.D. A cost-sensitive deep learning-based meta-classifier for pediatric pneumonia classification using chest X-rays. *Expert Syst.* **2022**, *39*, e12966. [CrossRef]
7. Rajaraman, S.; Guo, P.; Xue, Z.; Antani, S.K. A Deep Modality-Specific Ensemble for Improving Pneumonia Detection in Chest X-rays. *Diagnostics* **2022**, *12*, 1442. [CrossRef]
8. Hasan, M.M.; Islam, M.U.; Sadeq, M.J.; Fung, W.K.; Uddin, J. Review on the Evaluation and Development of Artificial Intelligence for COVID-19 Containment. *Sensors* **2023**, *23*, 527. [CrossRef]
9. Soomro, T.A.; Zheng, L.; Afifi, A.J.; Ali, A.; Yin, M.; Gao, J. Artificial intelligence (AI) for medical imaging to combat coronavirus disease (COVID-19): A detailed review with direction for future research. *Artif. Intell. Rev.* **2022**, *55*, 1409–1439. [CrossRef]
10. Pfaff, E.R.; Girvin, A.T.; Bennett, T.D.; Bhatia, A.; Brooks, I.M.; Deer, R.R.; Dekermanjian, J.P.; Jolley, S.E.; Kahn, M.G.; Kostka, K.; et al. Identifying who has long COVID in the USA: A machine learning approach using N3C data. *Lancet Digit. Health* **2022**, *4*, e532–e541. [CrossRef]
11. Ahsan, M.M.; Luna, S.A.; Siddique, Z. Machine-learning-based disease diagnosis: A comprehensive review. *Healthcare* **2022**, *10*, 541. [CrossRef] [PubMed]
12. Simonyan, K.; Zisserman, A. Very deep convolutional networks for large-scale image recognition. *arXiv* **2014**, arXiv:1409.1556. [CrossRef].
13. Szegedy, C.; Vanhoucke, V.; Ioffe, S.; Shlens, J.; Wojna, Z. Rethinking the inception architecture for computer vision. In Proceedings of the IEEE Conference on Computer Vision and Pattern Recognition, Las Vegas, NV, USA, 27–30 June 2016; pp. 2818–2826. [CrossRef]
14. Szegedy, C.; Liu, W.; Jia, Y.; Sermanet, P.; Reed, S.; Anguelov, D.; Erhan, D.; Vanhoucke, V.; Rabinovich, A. Going deeper with convolutions. In Proceedings of the IEEE Conference on Computer Vision and Pattern Recognition, Boston, MA, USA, 7–12 June 2015; pp. 1–9.
15. Conti, V.; Militello, C.; Rundo, L.; Vitabile, S. A novel bio-inspired approach for high-performance management in service-oriented networks. *IEEE Trans. Emerg. Top. Comput.* **2020**, *9*, 1709–1722. [CrossRef]
16. Han, X.; Hu, Z.; Wang, S.; Zhang, Y. A Survey on Deep Learning in COVID-19 Diagnosis. *J. Imaging* **2022**, *9*, 1. [CrossRef]
17. Ayadi, M.; Ksibi, A.; Al-Rasheed, A.; Soufiene, B.O. COVID-AleXception: A Deep Learning Model Based on a Deep Feature Concatenation Approach for the Detection of COVID-19 from Chest X-ray Images. *Healthcare* **2022**, *10*, 2072. [CrossRef] [PubMed]
18. Hafeez, U.; Umer, M.; Hameed, A.; Mustafa, H.; Sohaib, A.; Nappi, M.; Madni, H.A. A CNN based coronavirus disease prediction system for chest X-rays. *J. Ambient. Intell. Humaniz. Comput.* **2022**, 1–15. [CrossRef] [PubMed]
19. Huang, M.L.; Liao, Y.C. A lightweight CNN-based network on COVID-19 detection using X-ray and CT images. *Comput. Biol. Med.* **2022**, *146*, 105604. [CrossRef] [PubMed]
20. Khan, A.I.; Shah, J.L.; Bhat, M.M. CoroNet: A deep neural network for detection and diagnosis of COVID-19 from chest X-ray images. *Comput. Methods Programs Biomed.* **2020**, *196*, 105581. [CrossRef] [PubMed]
21. Ghose, P.; Uddin, M.A.; Acharjee, U.K.; Sharmin, S. Deep viewing for the identification of covid-19 infection status from chest X-ray image using cnn based architecture. *Intell. Syst. Appl.* **2022**, *16*, 200130. [CrossRef]
22. Ibrokhimov, B.; Kang, J.Y. Deep Learning Model for COVID-19-Infected Pneumonia Diagnosis Using Chest Radiography Images. *BioMedInformatics* **2022**, *2*, 654–670. [CrossRef]
23. Khan, I.U.; Aslam, N. A deep-learning-based framework for automated diagnosis of COVID-19 using X-ray images. *Information* **2020**, *11*, 419. [CrossRef]

24. Kaya, Y.; Gürsoy, E. A MobileNet-based CNN model with a novel fine-tuning mechanism for COVID-19 infection detection. *Soft Comput.* **2023**, *27*, 5521–5535. [CrossRef] [PubMed]

25. Nayak, S.R.; Nayak, D.R.; Sinha, U.; Arora, V.; Pachori, R.B. An Efficient Deep Learning Method for Detection of COVID-19 Infection Using Chest X-ray Images. *Diagnostics* **2023**, *13*, 131. [CrossRef] [PubMed]

26. Sanida, T.; Sideris, A.; Tsiktsiris, D.; Dasygenis, M. Lightweight neural network for COVID-19 detection from chest X-ray images implemented on an embedded system. *Technologies* **2022**, *10*, 37. [CrossRef]

27. Sanida, T.; Sideris, A.; Chatzisavvas, A.; Dossis, M.; Dasygenis, M. Radiography Images with Transfer Learning on Embedded System. In Proceedings of the 2022 7th South-East Europe Design Automation, Computer Engineering, Computer Networks and Social Media Conference (SEEDA-CECNSM), Ioannina, Greece, 23–25 September 2022; IEEE: Piscataway, NJ, USA, 2022; pp. 1–4. [CrossRef]

28. Tahir, A.M.; Chowdhury, M.E.; Khandakar, A.; Rahman, T.; Qiblawey, Y.; Khurshid, U.; Kiranyaz, S.; Ibtehaz, N.; Rahman, M.S.; Al-Maadeed, S.; et al. COVID-19 infection localization and severity grading from chest X-ray images. *Comput. Biol. Med.* **2021**, *139*, 105002. [CrossRef]

29. Yasin, R.; Gouda, W. Chest X-ray findings monitoring COVID-19 disease course and severity. *Egypt. J. Radiol. Nucl. Med.* **2020**, *51*, 193. [CrossRef]

30. Rousan, L.A.; Elobeid, E.; Karrar, M.; Khader, Y. Chest X-ray findings and temporal lung changes in patients with COVID-19 pneumonia. *BMC Pulm. Med.* **2020**, *20*, 245. [CrossRef]

31. Sanida, M.V.; Sanida, T.; Sideris, A.; Dasygenis, M. An Efficient Hybrid CNN Classification Model for Tomato Crop Disease. *Technologies* **2023**, *11*, 10. [CrossRef]

32. Mohanty, S.P.; Hughes, D.P.; Salathé, M. Using deep learning for image-based plant disease detection. *Front. Plant Sci.* **2016**, *7*, 1419. [CrossRef]

33. Sanida, T.; Tsiktsiris, D.; Sideris, A.; Dasygenis, M. A heterogeneous implementation for plant disease identification using deep learning. *Multimed. Tools Appl.* **2022**, *81*, 15041–15059. [CrossRef]

34. Tharwat, A. Classification assessment methods. *Appl. Comput. Inform.* **2020**, *17*, 168–192. [CrossRef]

35. Delgado, R.; Tibau, X.A. Why Cohen's Kappa should be avoided as performance measure in classification. *PLoS ONE* **2019**, *14*, e0222916. [CrossRef] [PubMed]

36. Deng, J.; Dong, W.; Socher, R.; Li, L.J.; Li, K.; Fei-Fei, L. Imagenet: A large-scale hierarchical image database. In Proceedings of the 2009 IEEE Conference on Computer Vision and Pattern Recognition, Miami, FL, USA, 20–25 June 2009; IEEE: Piscataway, NJ, USA, 2009; pp. 248–255. [CrossRef]

 *information*

Article

A Double-Stage 3D U-Net for On-Cloud Brain Extraction and Multi-Structure Segmentation from 7T MR Volumes

Selene Tomassini, Haidar Anbar, Agnese Sbrollini, MHD Jafar Mortada, Laura Burattini * and Micaela Morettini *

Department of Information Engineering, Università Politecnica delle Marche, Via Brecce Bianche 12, 60131 Ancona, Italy; s.tomassini@pm.univpm.it (S.T.); s1101956@studenti.univpm.it (H.A.); a.sbrollini@staff.univpm.it (A.S.); s1101958@studenti.univpm.it (M.J.M.)
* Correspondence: l.burattini@univpm.it (L.B.); m.morettini@univpm.it (M.M.)

Abstract: The brain is the organ most studied using Magnetic Resonance (MR). The emergence of 7T scanners has increased MR imaging resolution to a sub-millimeter level. However, there is a lack of automatic segmentation techniques for 7T MR volumes. This research aims to develop a novel deep learning-based algorithm for on-cloud brain extraction and multi-structure segmentation from unenhanced 7T MR volumes. To this aim, a double-stage 3D U-Net was implemented in a cloud service, directing its first stage to the automatic extraction of the brain and its second stage to the automatic segmentation of the grey matter, basal ganglia, white matter, ventricles, cerebellum, and brain stem. The training was performed on the 90% (the 10% of which served for validation) and the test on the 10% of the Glasgow database. A mean test Dice Similarity Coefficient (DSC) of 96.33% was achieved for the brain class. Mean test DSCs of 90.24%, 87.55%, 93.82%, 85.77%, 91.53%, and 89.95% were achieved for the brain structure classes, respectively. Therefore, the proposed double-stage 3D U-Net is effective in brain extraction and multi-structure segmentation from 7T MR volumes without any preprocessing and training data augmentation strategy while ensuring its machine-independent reproducibility.

Keywords: brain extraction; brain multi-structure segmentation; cloud computing; deep learning; double-stage 3D U-Net; neuroradiology; 7T magnetic resonance; volume measure analysis

Citation: Tomassini, S.; Anbar, H.; Sbrollini, A.; Mortada, M.J.; Burattini, L.; Morettini, M. A Double-Stage 3D U-Net for On-Cloud Brain Extraction and Multi-Structure Segmentation from 7T MR Volumes. *Information* **2023**, *14*, 282. https://doi.org/10.3390/info14050282

Academic Editor: Amar Ramdane-Cherif

Received: 31 March 2023
Revised: 5 May 2023
Accepted: 8 May 2023
Published: 10 May 2023

1. Introduction

Magnetic Resonance (MR) is a radiological imaging modality of pivotal importance in diagnostics. The brain is the organ most frequently studied by MR, as it allows to obtain the greatest sensitivity for the characterization of brain structures by combining radio waves and strong magnetic fields [1,2]. MR imaging is an effective way to diagnose neurological diseases and conditions by detecting structural or connectivity alterations, such as the Grey Matter (GM) atrophy in Alzheimer's disease, the shrinkage of the Brain Stem (BS) structures (e.g., substantia nigra) in Parkinson's disease, the presence of lesions in the White Matter (WM) in multiple sclerosis and the abnormal connectivity of the cortico-cerebellar-striatal-thalamic loop in schizophrenia [2,3]. Nowadays, the advancements in instrumentation technology join improved acquisition methodologies. The emergence of 7T scanners, in particular, has increased the MR imaging resolution to a sub-millimeter level, making the visualization of such brain structures more evident [4,5]. Despite the potentialities, these innovative technologies come with new technical challenges, such as more pronounced radiofrequency field non-uniformities, larger spatial distortion near air-tissue interfaces, and more susceptibility artifacts.

Brain structure segmentation is an important step in MR imaging diagnostics for monitoring the presence of anatomical alterations by isolating specific brain areas and, thus, allowing a region-by-region quantitative analysis [2]. Manual segmentation is the gold standard for brain structure segmentation in MR [6]. It is necessary for providing the Ground Truth (GT), requiring experienced operators to first define the region of interest and then draw boundaries surrounding it [7]. Although being the most accurate, manual segmentation is only feasible for small collections of data, as it is time-consuming, because performed slice-by-slice, and labor-intensive, due to the noisy yet complex tissue edges [8]. Moreover, its results are difficult to reproduce, as even experienced operators exhibit significant variability with respect to their previous delineations [6,9]. It may also happen that high-resolution MR images, such as 7T ones, no longer have a crisp boundary of the region of interest. As a consequence, slight variations in the selection of pixels may lead to errors [7]. Automatic segmentation techniques have recently aroused great interest for their use in both research and clinical applications. However, most such techniques require labeled MR images obtained through manual segmentation and, thus, experience similar constraints as mentioned earlier. Additional challenges with automatic segmentation techniques include the poor contrast between brain areas, the complex anatomical environment of the brain, and the wide variations in size, shape, and texture found in the brain tissue of subjects. Lack of consistency in source data acquisition may also result in such variations. Consequently, most existing approaches based on clustering, watershed, and machine learning share the problem of a lack of global applicability, which limits their usage to a limited number of applications. Deep Learning (DL)-based algorithms are capable of processing unenhanced data by extracting the salient features automatically, thus eliminating the need for manually-extracted features [10]. DL-based brain structure segmentation seems to be currently the most promising, thanks to the rapid increase of hardware capabilities together with computational and memory resources that have largely reduced the execution time [7,9].

Over the past few years, researchers have reported in the literature on various brain structure automatic segmentation techniques of different accuracy and degrees of complexity [6,8,11–13]. Some researchers, in particular, developed DL-based algorithms for brain structure segmentation from 1.5T and 3T MR volumes, and 3D Convolutional Neural Network (CNN) was the neural architecture used most predominantly. In 2019, Sun et al. [14] proposed a spatially-weighted 3D U-Net for the automatic segmentation of the brain structures into WM, GM, and cerebrospinal fluid from T1-weighted MR volumes of the MRBrainS13 and MALC12 databases, then extended to multi-modal MR volumes. In the same year, Wang et al. [15] proposed a 3D CNN including recursive residual blocks and a pyramid pooling module for the automatic segmentation of the brain structures into WM, GM, and cerebrospinal fluid from T1-weighted MR volumes of the CANDI and IBSR databases. One year later, Bontempi et al. [16] proposed a 3D CNN trained in a weakly-supervised fashion by exploiting a large database collected from the Centre for Cognitive Neuroimaging of the University of Glasgow and composed of T1-weighted MR volumes. Again in 2020, Ramzan et al. [17] proposed a 3D CNN with residual learning and dilated convolution operations for the automatic segmentation of the brain structures into nine different classes, including WM, GM, and cerebrospinal fluid from T1-weighted MR volumes of the ADNI, MRBrainS18, and MICCAI 2012 databases. In 2022, Laiton-Bonadiez et al. [18] injected T1-weighted MR sub-volumes of the Mindboggle-101 database into a set of successive 3D CNN layers free of pooling operations for extracting the local information. Later, they sent the resulting feature maps to successive self-attention layers for obtaining the global context, whose output was later dispatched to the decoder composed mostly of up-sampling layers. However, there is still a severe lack of brain structure automatic segmentation techniques for 7T MR volumes compared to lower field MR volumes. To the authors' best knowledge, the only DL-based algorithm for brain structure segmentation from unenhanced 7T MR volumes is the one proposed by Svanera et al. [19]. They pretrained a 3D CNN on the Glasgow database in a weakly-supervised fashion by taking advantage of training data augmentation strategies. Additionally, they took into account two different collections of data for exploring the condition of limited data availability. However, they directed their research to focus more on the demonstration of the practical portability of a pretrained neural architecture with a fine-tuning procedure involving very few MR volumes rather than on effective performance evaluation and analysis.

Thus, the focus of this research is on developing a novel DL-based algorithm for on-cloud brain extraction and multi-structure segmentation from 7T MR volumes without any preprocessing and training data augmentation strategy. To this aim, a double-stage 3D U-Net was designed and implemented in a scalable GPU cloud service, directing its first stage to the automatic extraction of the brain by removing the background and stripping the skull, and its second stage to the automatic segmentation of the GM, Basal Ganglia (BG), WM, VENtricles (VEN), CereBellum (CB), and BS.

2. Data and Methodology

2.1. Data Labeling and Division

Data used in this research come from the Glasgow database (https://search.kg.ebrains. eu/instances/Dataset/2b24466d-f1cd-4b66-afa8-d70a6755ebea, accessed on 2 January 2023), which was collected at the Imaging Centre of Excellence of the Queen Elizabeth University Hospital in Glasgow and publicly released by Svanera et al. [19]. The database includes 142 out-of-the-scanner T1-weighted MR volumes of $256 \times 352 \times 224$ mm^3, obtained with an MP2RAGE sequence at 0:63 mm^3 isotropic resolution, acquired by a Siemens 7T Terra Magnetom scanner with a 32-channel head coil, and belonging to 76 healthy subjects. Neck cropping was the only preprocessing performed by the data providers using the INV2 volume obtained during the acquisition. Together with MR volumes, a multi-class segmentation mask is also included in the database. The segmented classes are (0, 1, 2, 3, 4, 5, 6) for, respectively, the background, GM, BG, WM, VEN, CB, and BS [19]. Once selected, all MR volumes were stored as compressed NIFTI files without applying any further preprocessing.

Due to the considerable time cost and expertise required to produce manual annotations on such a database, the inaccurate GTs (iGTs), made available together with MR volumes by Svanera et al. [19], were exploited, and corrections were then applied. The automatic data labeling procedure accounts for an upper branch dealing with GM and WM automatic segmentation, and a lower branch dealing with BG, VEN, CB, and BS automatic segmentation. In the upper branch, AFNI-3dSeg proposed by Cox et al. [20] was used, followed by geometric and clustering techniques as seen in Fracasso et al.'s [21] research. In the lower branch, FreeSurfer v6 proposed by Fischl et al. [22] was used, with preliminary denoising of MR volumes as in O'Brien et al. [23]. Then, the two branches were combined, and a manual correction was carried out to reduce the major errors (e.g., CB wrongly labeled as GM) using ITK-SNAP as in Yushkevich et al. [24]. In those cases in which the iGTs came with black holes (Figure 1) due to inaccuracies in the automatic data labeling procedure, an additional correction was performed by applying a morphological operation, called dilation, to both increase the object area and fill the black holes. The appropriateness of such a correction was confirmed by an expert neurosurgeon.

Data division was performed not from a data-level perspective but from a subject-level one, being careful not to include data belonging to the same subject in training, validation, and test sets, in order to avoid a biased prediction. Thus, MR volumes were partitioned into 90% (128 MR volumes, 62 subjects) for training, 10% of which served for validation, and the remaining 10% (14 MR volumes, 14 subjects) for testing.

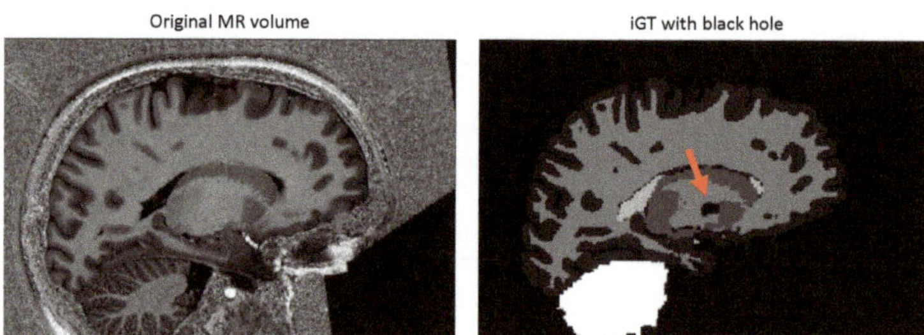

Figure 1. Example of one inaccurate Ground Truth (iGT) with a black hole (pointed out by the red arrow) to be corrected.

2.2. Double-Stage 3D U-Net

Entire unenhanced 7T MR volumes were processed to take advantage of both the global and local spatial information of MR, conducting the analysis in two learning stages, both accomplished by the same neural architecture, as displayed in Figure 2.

The first learning stage is directed to the automatic extraction of the brain by removing the background and stripping the skull. To fulfill this learning stage, the multi-class segmentation mask was adjusted by giving the background the value of 0 and giving all six brain structures the value of 1. Then, the original MR volumes were injected into the double-stage 3D U-Net.

The second learning stage is directed to the automatic segmentation of the brain structures into GM, BG, WM, VEN, CB, and BS at once. To accomplish this learning stage, the multi-class segmentation mask was kept unaltered. The original MR volumes were multiplied with the predicted masks obtained from the first learning stage, and the resulting brain MR volumes were later injected into the double-stage 3D U-Net.

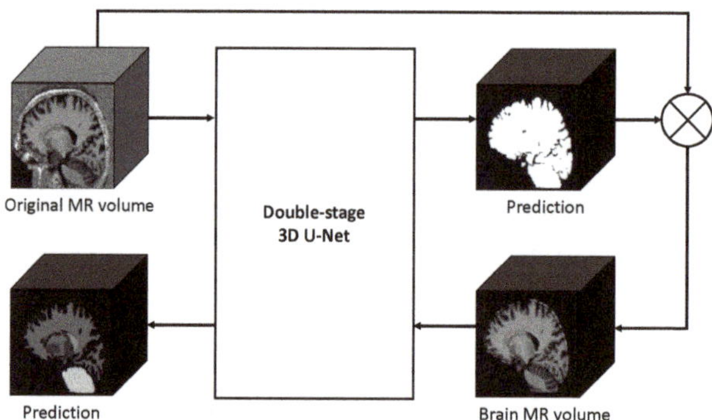

Figure 2. Workflow of the double-stage 3D U-Net.

2.2.1. Neural Architecture

A double-stage 3D U-Net based on the standard U-Net neural architecture proposed by Ronneberger et al. [25] was designed. Architecturally, it consists of a down-sampling path and an up-sampling path, as depicted in Figure 3. The down-sampling path is made up of five stadiums. The first stadium consists of two $3 \times 3 \times 3$ Convolution (Conv3D) layers with Rectified Linear Unit (ReLU) activation function followed by a Batch Normalization (BN) layer, used to accelerate the training by reducing the internal covariate shift [26]. The second stadium consists of a 3D Average Pooling (AvgPool3D) layer with a stride of 2, used to look at the complete extent of the input by smoothing it out, thus smoothly extracting the features. The third stadium consists of two Conv3D layers with a ReLU activation function followed by a BN layer. The fourth and fifth stadiums are analogous to the second and third ones with the only addition of a 3D Spatial Dropout (SpatialDrop3D) layer with a dropout rate of 0.5, in order to reduce the overfitting effect. A SpatialDrop3D layer was introduced in the neural architecture because of the reduced training size, in order to improve the generalization performance by preventing activations from becoming strongly correlated and, thus, avoiding overtraining. The SpatialDrop3D layer, indeed, drops entire 3D feature maps in place of individual elements. If adjacent voxels within 3D feature maps are strongly correlated, a 3D regular dropout will not regularize the activations. The SpatialDrop3D, instead, will help promote the independence between 3D feature maps. The filter sizes of the Conv3D layers in each stadium of the down-sampling path are 8, 16, 32, 64, and 128, respectively. The up-sampling path is made up of five stadiums as well. Differently from the standard U-Net neural architecture of [25], in the first four stadiums, a 3D Transposed Convolution (TransposeConv3D) layer was used in place of the 3D up-sampling layer, followed by one Conv3D layer with ReLU activation function. The TransposeConv3D layer served to up-sample the volumes by increasing the size, height, and width of their inputs. Then, a Concatenation (Concat) layer was added for skip connections, followed by two Conv3D layers with ReLU activation function and a BN layer. The last stadium consists of a Conv3D layer with Softmax activation function, as it assigns probabilities to each class by squashing the outputs to real values between 0 and 1, with a sum of 1 [27]. It has 2 (i.e., [0, 1]) and 7 (i.e., [0, 1, 2, 3, 4, 5, 6]) output neurons for, respectively, the first and second learning stage of the double-stage 3D U-Net. Details of the double-stage 3D U-Net neural architecture are also summarized in Table 1.

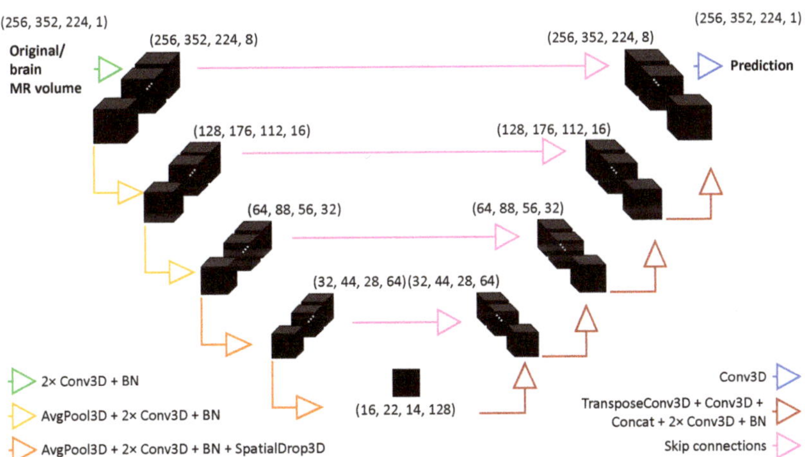

Figure 3. Neural architecture of the double-stage 3D U-Net. Conv3D refers to 3D Convolution layer, BN refers to Batch Normalization layer, AvgPool3D refers to 3D Average Pooling layer, SpatialDrop3D refers to 3D Spatial Dropout layer, TransposeConv3D refers to 3D Transposed Convolution layer, and Concat refers to Concatenation layer.

Table 1. Details of the double-stage 3D U-Net neural architecture. Conv3D refers to the 3D Convolution layer, BN refers to the Batch Normalization layer, AvgPool3D refers to the 3D Average Pooling layer, SpatialDrop3D refers to the 3D Spatial Dropout layer, TransposeConv3D refers to the 3D Transposed Convolution layer, and Concat refers to the Concatenation layer.

Layer	Output Shape	Number of Parameters
Input	(None, 256, 352, 2, 24, 1)	0
Conv3D	(None, 256, 352, 22, 4, 8)	224
Conv3D	(None, 256, 352, 22, 4, 8)	1736
BN	(None, 256, 352, 22, 4, 8)	32
AvgPool3D	(None, 128, 176, 11, 2, 8)	0
Conv3D	(None, 128, 176, 11, 2, 16)	3472
Conv3D	(None, 128, 176, 11, 2, 16)	6928
BN	(None, 128, 176, 11, 2, 16)	64
AvgPool3D	(None, 64, 88, 56, 16)	0
Conv3D	(None, 64, 88, 56, 32)	13,856
Conv3D	(None, 64, 88, 56, 32)	27,680
BN	(None, 64, 88, 56, 32)	128
AvgPool3D	(None, 32, 44, 28, 32)	0
Conv3D	(None, 32, 44, 28, 64)	55,360
Conv3D	(None, 32, 44, 28, 64)	110,656
BN	(None, 32, 44, 28, 64)	256
SpatialDrop3D	(None, 32, 44, 28, 64)	0
AvgPool3D	(None, 16, 22, 14, 64)	0
Conv3D	(None, 16, 22, 14, 128)	221,312
Conv3D	(None, 16, 22, 14, 128)	442,496
BN	(None, 16, 22, 14, 128)	512
SpatialDrop3D	(None, 16, 22, 14, 128)	0
TransposeConv3D	(None, 32, 44, 28, 64)	65,600
Conv3D	(None, 32, 44, 28, 64)	32,832
Concat	(None, 32, 44, 28, 128)	0
Conv3D	(None, 32, 44, 28, 64)	221,248
Conv3D	(None, 32, 44, 28, 64)	110,656
BN	(None, 32, 44, 28, 64)	256
TransposeConv3D	(None, 64, 88, 56, 32)	16,416
Conv3D	(None, 64, 88, 56, 32)	8224
Concat	(None, 64, 88, 56, 64)	0
Conv3D	(None, 64, 88, 56, 32)	55,328
Conv3D	(None, 64, 88, 56, 32)	27,680
BN	(None, 64, 88, 56, 32)	128
TransposeConv3D	(None, 128, 176, 11, 2, 16)	4112
Conv3D	(None, 128, 176, 11, 2, 16)	2064
Concat	(None, 128, 176, 11, 2, 32)	0
Conv3D	(None, 128, 176, 11, 2, 16)	13,840
Conv3D	(None, 128, 176, 11, 2, 16)	6928
BN	(None, 128, 176, 11, 2, 16)	64
TransposeConv3D	(None, 256, 352, 22, 4, 8)	1032
Conv3D	(None, 256, 352, 22, 4, 8)	520
Concat	(None, 256, 352, 22, 4, 16)	0
Conv3D	(None, 256, 352, 22, 4, 8)	3464
Conv3D	(None, 256, 352, 22, 4, 8)	1736
BN	(None, 256, 352, 22, 4, 8)	32
Conv3D	(None, 256, 352, 22, 4, 2/7)	18/63

Total: 1,456,890/1,456,935
Trainable: 1,456,154/1,456,199
Non-trainable: 736

2.2.2. Experimental Setup and Learning Process

The double-stage 3D U-Net was developed in Python, exploiting the Pro version of Google Colab to take advantage of the cloud storage and computing power of the

Google servers. The GPU hardware acceleration (NVIDIA Tesla P100 with 16 GB of video RAM) and high system RAM (34 GB) settings were chosen. The Keras library built on a TensorFlow backend was also used.

The combination of both neural network and training parameters that led to the best performance on validation data was considered. Thus, the double-stage 3D U-Net was trained by scratch for 50 epochs, fixing the batch size to 1 and the learning rate to 0.001. The RMSprop was utilized as an optimizer because, during training, it uses an adaptive mini-batch learning rate that changes over time. A combination of Weighted Dice Loss (WDL) and Categorical Cross Entropy (CCE) was used as loss function. The combination of WDL, which is a region-based loss, and CCE, which is a distribution-based loss, allows to simultaneously minimize dissimilarities between two distributions while minimizing the mismatch or maximizing the overlapping regions between desired and predicted outputs [12,28]. Multiple loss functions and a weighting strategy were used here to minimize the problems coming from the highly imbalanced sizes of brain structures. The early stopping callback with a patience of 5 was also used to further minimize the overfitting effect. The weights that led to the lowest validation loss were saved and then used to evaluate the performance of the double-stage 3D U-Net on test data.

2.3. Performance Evaluation and Volume Measure Analysis

A comparison between iGTs and predictions was performed to evaluate the performance of the double-stage 3D U-Net on test data. In this guise, the three metrics adopted by the MICCAI MRBrainS18 Challenge were taken into account, being the most commonly used in the context of semantic segmentation, namely Dice Similarity Coefficient (DSC), Volumetric Similarity (VS), and Hausdorff Distance 95% percentile (HD_{95}). In addition to the DSC, the ACCuracy (ACC), loss, weighted DSC, and mean DSC were monitored in both training and validation phases to provide a better perspective on the behavior of the double-stage 3D U-Net throughout the learning process. Focusing on DSC, VS, and HD_{95}, DSC is an overlap-based metric useful to predict the similarity index between iGTs and predictions by comparing the pixel-wise agreement between the couple. It is also used as an index of spatial overlap, where a value of 1 indicates the perfect overlap [13]. It is computed as in Equation (1), where L refers to the iGT pixels, and S refers to the prediction pixels:

$$DSC(S, L) = \frac{|S \cap L|}{|S| + |L|}.$$ (1)

VS is not an overlap-based metric but rather a measure that considers the volumes of the segments to indicate the similarity [29]. Despite there are several definitions for this metric, VS is typically defined as $1 - VD$, where VD is the Volumetric Distance. Mathematically, VS is defined as the absolute volume difference divided by the sum of the compared volumes, as reported in Equation (2), where FN stands for False Negative, FP stands for False Positive, and TP stands for True Positive:

$$VS = 1 - \frac{|FN - FP|}{2TP + FP + FN}.$$ (2)

HD_{95} is one of the most commonly used boundary-base metrics, essential for calculating the distance between iGTs and predictions. Basically, it calculates the maximum of all shortest distances for all points from one object boundary to the other. Small values represent a high segmentation accuracy. Specifically, 0 refers to a perfect segmentation (distance of 0 to the reference boundary), and no fixed upper bound exists. It is computed as in Equation (3), where L refers to the iGT and S refers to the prediction [29]:

$$HD_{95} = max\{K^{th}_{s \in S} \min_{l \in L} \| S - L \|, K^{th}_{s \in S} \min_{l \in L} \| L - S \|\}.$$ (3)

In addition, the volume measures of each prediction and corresponding iGT were analyzed for evaluating the goodness of the predictions of the double-stage 3D U-Net on test data. To do so, the volumes (volume$_{es}$) of each prediction and corresponding iGT were calculated by automatically counting the number of voxels (number$_{vox}$) inside, respectively, the brain, GM, BG, WM, VEN, CB, and BS, and multiplying it by the voxel volume (volume$_{vox}$), expressed in cm^3, according to Equation (4):

$$volume_{es} = number_{vox} \times volume_{vox}. \tag{4}$$

A comparison between the volume distributions computed by the iGT and the prediction related to the same subject was also performed. Specifically, the volume distributions computed by iGTs and predictions were calculated and reported in terms of 50th (median) [25th; 75th] percentiles. Then, non-normal volume distributions computed by iGTs and predictions were statistically compared by means of Mean Absolute Error (MAE, %) and paired Wilcoxon rank-sum test, setting 0.05 as the statistical level of significance (P).

3. Results

The behavior of the double-stage 3D U-Net throughout the learning process is reported in Table 2, in terms of ACC, loss, weighted DSC, mean DSC, and DSC (computed for each brain structure class). The trend of ACC and loss across the epochs in both training and validation phases is also depicted in Figure 4 for the first learning stage and in Figure 5 for the second learning stage. For the first learning stage, the lowest validation loss value (0.043) was reached in the 44th epoch. For the second learning stage, the lowest validation loss value (0.075) was reached in the 33rd epoch. Due to the highly imbalanced sizes of brain structures in the second learning stage, the trend of weighted DSC and mean DSC across the epochs in both training and validation phases was also monitored, and it is displayed in Figure 6.

Table 2. Behavior of the double-stage 3D U-Net in both training and validation phases, in terms of ACCuracy (ACC), loss, weighted Dice Score Coefficient (DSC), mean DSC, and DSC (computed for each brain structure class). GM refers to Grey Matter, BG refers to Basal Ganglia, WM refers to White Matter, VEN refers to VENtricles, CB refers to CereBellum, and BS refers to Brain Stem.

Metrics	Class	First Learning Stage		Second Learning Stage	
		Training	Validation	Training	Validation
ACC (%)	All	98.31	98.24	96.95	96.93
Loss (−)	All	0.04	0.04	0.08	0.08
Weighted DSC (%)	All	-	-	79.41	79.09
Mean DSC (%)	All	-	-	87.63	87.91
	GM	-	-	86.62	87.46
	BG	-	-	80.42	80.54
DSC (%)	WM	-	-	91.46	92.53
	VEN	-	-	82.22	82.05
	CB	-	-	88.81	88.48
	BS	-	-	87.09	86.55

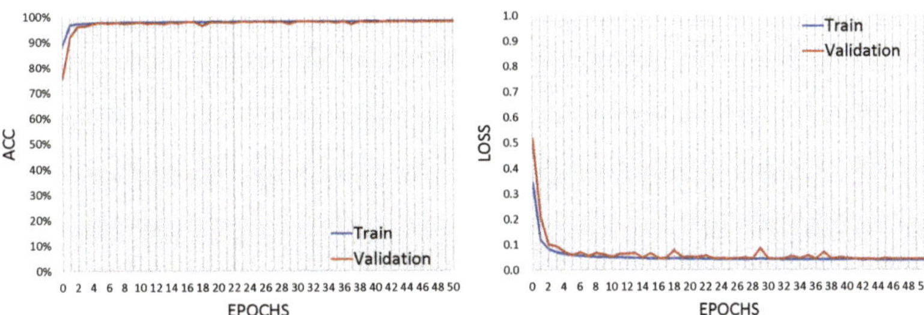

Figure 4. Trend of ACCuracy (ACC) and loss across the epochs in both training and validation phases of the first learning stage.

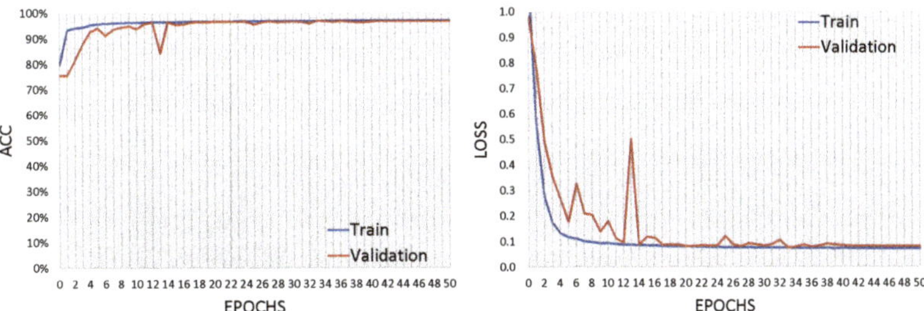

Figure 5. Trend of ACCuracy (ACC) and loss across the epochs in both training and validation phases of the second learning stage.

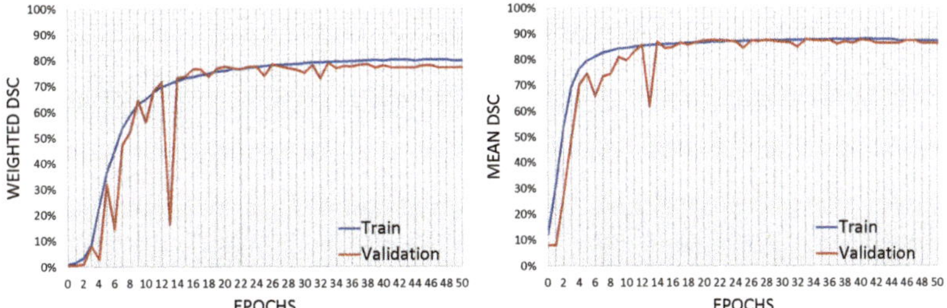

Figure 6. Trend of weighted Dice Score Coefficient (DSC) and mean DSC across the epochs in both training and validation phases of the second learning stage.

The test performance of the double-stage 3D U-Net in the automatic extraction of the brain (i.e., first learning stage) and segmentation of the brain structures into six different classes at once (i.e., second learning stage) is reported in Table 3. Since unenhanced 7T MR volumes reserved for testing are the same 14 for both learning stages, DSC, VS, and HD$_{95}$ values of the eight total classes (background, brain, GM, BG, WM, VEN, CB, and BS) were computed and expressed as mean $\pm$ standard deviation. The qualitative outcome of the double-stage 3D U-Net in the automatic extraction of the brain and segmentation of the brain structures into GM, BG, WM, VEN, CB, and BS at once is provided in Figure 7 and Figure 8, respectively. The qualitative outcome of the double-stage 3D U-Net in the automatic segmentation of the above-mentioned brain structure classes is also displayed in Figure 9, with a different color for each predicted class. Eventually, the volume measures of each prediction and corresponding iGT are reported in Table 4, together with the volume distributions computed by the iGT and the prediction related to the same subject.

Table 3. Test performance of the double-stage 3D U-Net in automatically extracting the brain and segmenting its structures into Grey Matter (GM), Basal Ganglia (BG), White Matter (WM), VENtricles (VEN), CereBellum (CB), and Brain Stem (BS) at once, in terms of Dice Score Coefficient (DSC), Volumetric Similarity (VS), and Hausdorff Distance 95% percentile (HD_{95}). Values are reported as mean ± standard deviation.

Class	Learning Stage	DSC (%)	VS (%)	HD_{95} (mm)
Background	First	98.78 ± 0.22	99.75 ± 0.25	2.74 ± 0.68
Brain	First	96.33 ± 0.51	99.27 ± 0.67	3.36 ± 0.54
GM	Second	90.24 ± 1.04	98.61 ± 1.33	1.15 ± 0.21
BG	Second	87.55 ± 0.83	94.88 ± 1.82	2.94 ± 0.31
WM	Second	93.82 ± 0.87	98.38 ± 1.51	1.03 ± 0.11
VEN	Second	85.77 ± 4.16	96.91 ± 2.11	2.15 ± 0.94
CB	Second	91.53 ± 1.96	96.87 ± 2.05	5.93 ± 1.73
BS	Second	89.95 ± 2.63	97.46 ± 1.36	2.92 ± 0.91

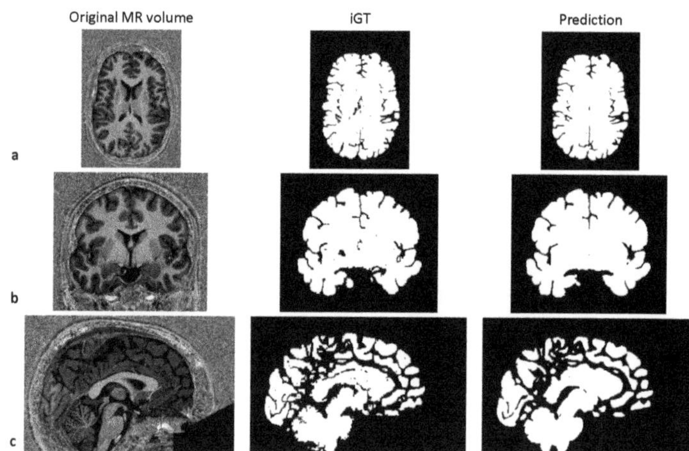

Figure 7. Qualitative outcome of the double-stage 3D U-Net in the automatic extraction of the brain. Mid-axial (**a**), mid-coronal (**b**), and mid-sagittal (**c**) slices of original MR volume together with corresponding inaccurate Ground Truth (iGT) and prediction are reported.

Table 4. Measures of the automatically extracted volume of the brain and segmented volumes of the Grey Matter (GM), Basal Ganglia (BG), White Matter (WM), VENtricles (VEN), CereBellum (CB), and Brain Stem (BS) in terms of 50th (median) [25th; 75th] percentiles. The Mean Absolute Error (MAE, %) between distributions of volumes computed by prediction and corresponding iGT is also reported.

Class	Learning Stage	iGT (cm^3)	Prediction (cm^3)	MAE (%)
Brain	First	1269 [1152; 1312]	1253 [1162; 1313]	1.02 [0.83; 1.73]
GM	Second	624 [581; 663]	630 [589; 672]	2.11 [0.55; 3.66]
BG	Second	46 [42; 47]	50 [48; 54] *	11.72 [8.69; 14.29]
WM	Second	444 [385; 461]	421 [386; 446]	2.45 [1.28; 4.39]
VEN	Second	16 [15; 21]	16 [15; 21]	6.56 [0; 8]
CB	Second	109 [104; 115]	110 [106; 113]	5.83 [4.27; 10]
BS	Second	17 [16; 18]	17 [16; 18]	5.72 [5.26; 7.14]

*: $p < 0.05$ (paired Wilcoxon rank-sum test).

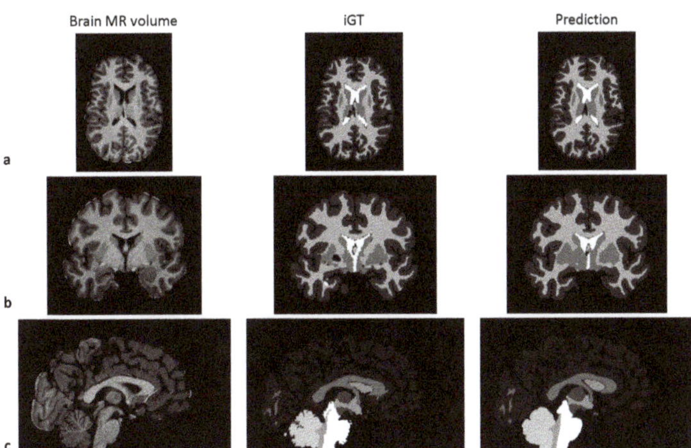

Figure 8. Qualitative outcome of the double-stage 3D U-Net in the automatic segmentation of the brain structures into six different classes at once. Mid-axial (**a**), mid-coronal (**b**), and mid-sagittal (**c**) slices of original MR volume together with corresponding inaccurate Ground Truth (iGT) and prediction are reported.

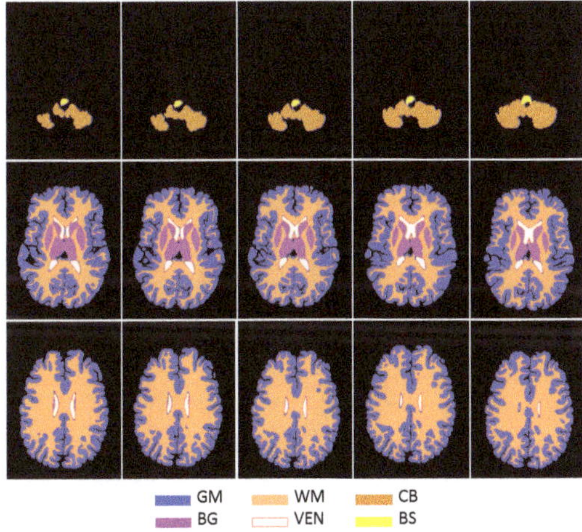

Figure 9. Qualitative outcome of the double-stage 3D U-Net in the automatic segmentation of the Grey Matter (GM), Basal Ganglia (BG), White Matter (WM), VENtricles (VEN), CereBellum (CB), and Brain Stem (BS), with a different color for each predicted class, on a bunch of axial MR slices (15 out of total 224) of the same MR volume.

4. Discussion

In this research, a novel DL-based algorithm for on-cloud brain extraction and multi-structure segmentation from unenhanced 7T MR volumes was developed by taking advantage of a double-stage 3D U-Net, the first stage of which was directed to automatically extract the brain by removing the background and stripping the skull, and the second served for the automatic segmentation of the GM, BG, WM, VEN, CB, and BS.

During the learning process, the ACC increased smoothly till it reached a value above 98% (first learning stage) and 97% (second learning stage) while the loss decreased steadily and under a value of 0.04 and 0.08 in both training and validation phases of the double-stage 3D U-Net without going to either overfitting or underfitting (Table 2, Figures 4 and 5).

The reasons for the small decrease in both training and validation ACC (Table 2) from, respectively, 98.31% and 98.24% (first learning stage) to, respectively, 96.95% and 96.93% (second learning stage) rely on the different segmentation tasks. In the first learning stage of the double-stage 3D U-Net, the task was to automatically extract the brain, thus automatically segmenting the entire MR volume into two major classes (background and brain). In the second learning stage of the double-stage 3D U-Net, instead, the task was to automatically segment the entire brain volume (as extracted in the first learning stage) into six further brain structures (GM, BG, WM, VEN, CB, and BS). Here, the complexity of the six brain structures, due to both overlapping and interference, made their segmentation much harder, especially at the boundaries and edges, which led to an expected yet quite low decrease in both training and validation ACC (-1.38% and -1.33%, respectively). Because of imbalanced data, the trend in weighted DSC and mean DSC was also monitored for the second learning stage (Table 2 and Figure 6). When monitoring weighted DSC and mean DSC in both training and validation phases of the double-stage 3D U-Net, the DSC of the GM, BG, WM, VEN, CB, and BS were also taken into account to make sure that all six brain structures were getting automatically segmented properly. All six brain structures were automatically segmented with a DSC higher than 80%, and the brain structures with the highest DSC are the WM, CB, and GM in both training and validation phases of the double-stage 3D U-Net (Table 2). These three brain structures are the ones that were segmented with the highest DSC also in the test phase of the double-stage 3D U-Net (Table 3). All eight total classes achieved DSCs higher than 86%, VS values higher than 95%, and HD_{95} values lower than 6 mm, and the brain structures with the highest DSC, highest VS, and lowest HD_{95} in the test phase of the double-stage 3D U-Net are the WM and GM (Table 3). The reason may rely on the fact that the strong T1 contrast that is present between fluid and more solid anatomical structures is likely to make the delineation of brain structures such as WM and GM easier [27,30]. Eventually, the analysis of volumes revealed the goodness of the predictions of the double-stage 3D U-Net, considering that all classes (except for the BG class) present volume distributions computed by predictions that are not statistically different from volume distributions computed by iGTs, with median MAE lower than 10%. As for the BG class, the reason for the statistical difference between volume distributions computed by predictions and corresponding iGTs may rely on its complex structure and specific location in the brain, which makes its delineation challenging. Moreover, in high-resolution MR images like 7T ones, BG no longer has a crisp boundary and, thus, slight voxel variations may lead to errors [7]. However, the associated median MAE is lower than 12% and, thus, still very close to 10%.

The proposed double-stage 3D U-Net for on-cloud brain extraction and multi-structure segmentation digests entire 7T MR volumes at once, avoiding the drawbacks of the tiling process. Moreover, it preserves the two-scale (i.e., global and local) analysis (that is a peculiar characteristic of manual segmentation [19]). Instead, the publicly-available non-DL tools for brain structure automatic segmentation, such as Statistical Parametric Mapping 12 (SPM12, www.fil.ion.ucl.ac.uk/spm accessed on 31 March 2023) [31] and FMRIB Software Library (FSL, www.fmrib.ox.ac.uk/fsl accessed on 31 March 2023) [32], emulate these two steps using atlases to gain global clues and, for most of them, gradient methods for the local processing. In SPM12, the brightness information and position of voxels, along with tissue probability maps, are considered, and the construction of appropriate priors is recommended. In FSL, the GM, WM, and cerebrospinal fluid segmentation is performed using an FMRIB Automated Segmentation Tool, which works on the extracted brain and uses the Markov random field model along with the expectation-maximization algorithm. Instead, the subcortical segmentation is performed using an FMRIB Integrated Registration and Segmentation Tool, which provides a deformation model-based segmentation. SPM12 and FSL are sufficiently resilient with respect to noise and artifacts introduced at the acquisition stage, and have performed consistently across different collections of data [6,8]. However, they are still far from being accepted at par with manual segmentation. Firstly, the quality of these approaches is limited by the accuracy of the pairwise registration method [8]. Secondly, image contrast, gross morphological deviation, high noise levels, and high spatial signal bias may lead to erroneous segmentation of brain structures [6]. Relying on priors, SPM12 and FSL are prone to erroneous results or simply fail in the presence of abnormal contrast or gross morphological alterations [8]. Thirdly, image artifacts due to poor subject compliance may systematically skew the results [8]. Eventually, despite the fact that such automated tools return a higher number of brain structures, it is often the case that, depending on the final application, having too many labels is not always useful, thus re-clustering is needed. Moreover, the higher the number of labels, the less accurate the available segmentation. Therefore, DL-based brain structure segmentation turns out to be more useful for MR protocols that are missing proper anatomical data or in those cases in which it is difficult to achieve high-quality anatomical structure registration, thanks to its applicability directly to data together with its property to automatically and adaptively learn spatial hierarchies of image features from low- to high-level patterns [33].

In the literature, Svanera et al. [19] were the only researchers to develop a DL-based algorithm for brain structure segmentation from unenhanced 7T MR volumes, as addressed in Section 1. Since they mentioned the results only graphically while reporting the overall test performance across the classes and used a different data division protocol, a fair quantitative comparison between their findings and the ones achieved in this research could not be provided, but a methodological comparison could be performed. Like Svanera et al. [19], a neural architecture able to deal with the full spatial information contained in the analyzed data was designed in this research, as 3D neural architectures can find voxel relationships in the three anatomical planes, thus maximizing the use of the intrinsic spatial nature of MR imaging. The Glasgow database was considered, and entire unenhanced 7T MR volumes, instead of sub-volumes, were processed. Additionally, the iGT automatic procedure was exploited for data labeling, although an additional neurosurgeon-approved correction was performed in this research to eliminate the inaccuracies (i.e., black holes). Differently from Svanera et al. [19], who pretrained a 3D CNN on the Glasgow database in a weakly-supervised fashion to demonstrate its practical portability with a fine-tuning procedure involving very few MR volumes of two different collections of data, this research was directed toward the effective performance evaluation and analysis of the proposed double-stage 3D U-Net by monitoring its behavior throughout the learning process (Table 2, Figures 4–6), assessing its test performance on eight total classes (Table 3, Figures 7–9), and evaluating the goodness of the predictions by means of a volume measure analysis (Table 4). The 3D U-Net was called 'double-stage' because the analysis was conducted in two learning stages for the automatic extraction of the brain and segmentation of the brain structures into six different classes at once, respectively. The 3D U-Net was also customized with respect to the standard 3D U-Net neural architecture of Ronneberger et al. [25] by adding two SpatialDrop3D layers in the fourth and fifth stadiums of the down-sampling path, in order to lift the generalization performance. It has been found that the SpatialDrop3D layer contributed to improving the performance without the need for any training data augmentation strategy by extending the dropout value across the entire feature map. Moreover, a TransposeConv3D layer was used in the first four stadiums of the up-sampling path because it is a convolution layer and, thus, it has trainable kernels. It has also been found that the combination of WDL and CCE as loss functions helped in overcoming the problem of data imbalance. In addition, the proposed double-stage 3D U-Net was developed in a scalable GPU service running entirely in the cloud. To allow so, a publicly-available database, already compliant with ethical and regulatory issues, was used. The uploading of entire 7T MR volumes in the compressed NIFTI format took just a few seconds. Although the training of the proposed double-stage 3D U-Net took approximately 6 hours to be completed, only 10 to 20 s served to generate the predictions on test data, which is a perfectly acceptable time in terms of execution efficiency. Moreover, no special network bandwidth requirements were necessary for on-cloud training, validating, testing, and visualizing. Cloud computing was chosen in this research because one of the main challenges facing radiological imaging analysis is the development of benchmarks that allow methodologies to be compared under common standards and measures. Thus, the cloud can contribute to creating such benchmarks [34]. Additionally, the scalable and distributed computational resources of cloud computing have the potential to increase the execution speed while keeping the costs low [35]. The pivotal component of the cloud, in fact, is the analysis platform, which supports a wide spectrum of data queries and cost-effectiveness solutions without the surcharge of purchasing and maintaining additional setups. In case the proposed double-stage 3D U-Net would need to be tested on a private collection of data, the neural network weights could be exploited to produce predictions in a local environment, with no need for defacing data/anonymizing facial features. Furthermore, if more brain structures other than the ones segmented in this research would need to be analyzed, the training of only the second learning stage of the proposed double-stage 3D U-Net could be re-performed after simply modifying the number of output classes (and, eventually, the loss function weights).

In this research, unenhanced 7T MR volumes were processed by performing no preliminary quality check, and the proposed double-stage 3D U-Net was demonstrated to be effective in brain extraction and multi-structure segmentation from raw, noisy input data. Accordingly, it can be assumed that it is highly generalizable to the quality of data, even though this cannot yet be demonstrated due to the lack of availability of suitable MR volumes.

In fact, MR volumes with annotations of seven classes are really difficult to achieve, and the results found can state only to the treated database. However, being a DL-based brain extraction and multi-structure segmentation algorithm, it is likely to generalize well to heterogeneous data coming from scanners of different manufacturers and/or acquired with 3T and lower field strengths, as DL is able to adaptively learn directly from data (i.e., it is fully data-driven). Moreover, in case of such heterogeneity, preliminary scan registration and intensity normalization may be enclosed in the pipeline, in order to align multiple brain structures for verifying their spatial correlation in anatomical terms and reduce the intensity variation caused by the use of different scanners [27].

One limitation is that, due to the unavailability of other openly-accessible collections of 7T MR volumes, performance was evaluated using test data selected from the same database used for training and validation. However, as mentioned before, we were careful not to mix data belonging to the same subject to avoid a biased prediction, which frequently happens when data belonging to the same subject end up in training, validation, and test sets.

Another limitation is that the choice to analyze entire unenhanced 7T MR volumes made it impossible to increase the batch size to a value greater than 1 because of the technical constraints linked to the hardware capabilities. This resulted in undercutting the advantages that larger batch sizes (without exceeding, otherwise the generalization may decrease) could carry, such as faster convergence. However, hardware capabilities are now experiencing rapid empowerment, so it will soon be possible to manage this limitation. A further limitation relies on the choice to investigate only one sequence. This choice, however, not only limited the scanning time but also avoided the need for sequence alignment while reducing distortion. Moreover, T1-weighted MR imaging is extremely useful in analyzing brain structures from an anatomical point of view [30]. For instance, the presence of brain shrinkage and anomalies on subcortical structures caused by neurodegeneration can be appreciated easily from a T1-weighted MR volume [27].

In the future, therefore, further analysis on the effective reproducibility of the proposed double-stage 3D U-Net for multi-site data will be conducted. In addition, future investigations would also demonstrate an optimization in the delineation of BG boundaries and edges. Eventually, it might be appropriate to extend the analysis to pathological MR volumes to monitor the anatomical abnormalities of the brain structures involved the most in neurological pathologies, especially neurodegenerative ones.

5. Conclusions

The double-stage optimized 3D U-Net proposed in this research is powerful for use in brain extraction and multi-structure segmentation from 7T MR volumes without any preprocessing and training data augmentation strategy. Furthermore, it ensures a machine-independent reproducibility of its implementation and has the potential to be integrated into any decision-support system, thanks to the cloud nature and, thus, machine-independent reproducibility.

Author Contributions: Conceptualization, S.T., H.A. and A.S.; methodology, S.T., H.A. and A.S.; software, S.T., H.A. and M.J.M.; validation, S.T., H.A., A.S., M.J.M., L.B. and M.M.; formal analysis, S.T.; investigation, S.T. and H.A.; resources, L.B.; data curation, S.T. and H.A.; writing—original draft preparation, S.T.; writing—review and editing, H.A., A.S., M.J.M., L.B. and M.M.; visualization, S.T.; supervision, L.B. and M.M.; project administration, L.B. and M.M. All authors have read and agreed to the published version of the manuscript.

Funding: This research received no external funding.

Informed Consent Statement: Informed consent was obtained from all subjects employed in collecting data for the Glasgow database [19].

Data Availability Statement: Data analyzed in this research are openly available in https://search.kg.ebrains.eu/instances/Dataset/2b24466d-f1cd-4b66-afa8-d70a6755ebea (accessed on 2 January 2023) under the need for account generation. For research purposes, the developed algorithm will be released free of charge to the scientific community by contacting the corresponding authors (L.B. and M.M.).

Acknowledgments: The authors thank Svanera et al. [19] for sharing data. The authors also thank Consortium GARR, the Italian National Research & Education Network, for promoting this research.

Conflicts of Interest: The authors declare no conflict of interest.

References

1. Haq, E.U.; Huang, J.; Kang, L.; Haq, H.U.; Zhan, T. Image-based state-of-the-art techniques for the identification and classification of brain diseases: A review. *Med Biol. Eng. Comput.* **2020**, *58*, 2603–2620. [CrossRef] [PubMed]
2. Zhao, X.; Zhao, X.M. Deep learning of brain magnetic resonance images: A brief review. *Methods* **2021**, *192*, 131–140. [CrossRef] [PubMed]
3. Tomassini, S.; Sernani, P.; Falcionelli, N.; Dragoni, A.F. CASPAR: Cloud-based Alzheimer's, schizophrenia and Parkinson's automatic recognizer. In Proceedings of the IEEE International Conference on Metrology for Extended Reality, Artificial Intelligence and Neural Engineering, Rome, Italy, 26–28 October 2022; pp. 6–10.
4. Keuken, M.C.; Isaacs, B.R.; Trampel, R.; Van Der Zwaag, W.; Forstmann, B. Visualizing the human subcortex using ultra-high field magnetic resonance imaging. *Brain Topogr.* **2018**, *31*, 513–545. [CrossRef] [PubMed]
5. Helms, G. Segmentation of human brain using structural MRI. *Magn. Reson. Mater. Phys. Biol. Med.* **2016**, *29*, 111–124. [CrossRef]
6. González-Villà, S.; Oliver, A.; Valverde, S.; Wang, L.; Zwiggelaar, R.; Lladó, X. A review on brain structures segmentation in magnetic resonance imaging. *Artif. Intell. Med.* **2016**, *73*, 45–69. [CrossRef]
7. Haque, I.R.I.; Neubert, J. Deep learning approaches to biomedical image segmentation. *Inform. Med. Unlocked* **2020**, *18*, 100297. [CrossRef]
8. Singh, M.K.; Singh, K.K. A review of publicly available automatic brain segmentation methodologies, machine learning models, recent advancements, and their comparison. *Ann. Neurosci.* **2021**, *28*, 82–93. [CrossRef]
9. Despotović, I.; Goossens, B.; Philips, W. MRI segmentation of the human brain: Challenges, methods, and applications. *Comput. Math. Methods Med.* **2015**, *2015*, 450341. [CrossRef]
10. Tomassini, S.; Falcionelli, N.; Sernani, P.; Burattini, L.; Dragoni, A.F. Lung nodule diagnosis and cancer histology classification from computed tomography data by convolutional neural networks: A survey. *Comput. Biol. Med.* **2022**, *146*, 105691. [CrossRef]
11. Fawzi, A.; Achuthan, A.; Belaton, B. Brain image segmentation in recent years: A narrative review. *Brain Sci.* **2021**, *11*, 1055. [CrossRef]
12. Minaee, S.; Boykov, Y.; Porikli, F.; Plaza, A.; Kehtarnavaz, N.; Terzopoulos, D. Image segmentation using deep learning: A survey. *IEEE Trans. Pattern Anal. Mach. Intell.* **2021**, *44*, 3523–3542. [CrossRef] [PubMed]
13. Krithika alias AnbuDevi, M.; Suganthi, K. Review of semantic segmentation of medical images using modified architectures of U-Net. *Diagnostics* **2022**, *12*, 3064. [CrossRef]
14. Sun, L.; Ma, W.; Ding, X.; Huang, Y.; Liang, D.; Paisley, J. A 3D spatially weighted network for segmentation of brain tissue from MRI. *IEEE Trans. Med Imaging* **2019**, *39*, 898–909. [CrossRef] [PubMed]

15. Wang, L.; Xie, C.; Zeng, N. RP-Net: A 3D convolutional neural network for brain segmentation from magnetic resonance imaging. *IEEE Access* **2019**, *7*, 39670–39679. [CrossRef]
16. Bontempi, D.; Benini, S.; Signoroni, A.; Svanera, M.; Muckli, L. CEREBRUM: A fast and fully-volumetric Convolutional Encoder-decodeR for weakly-supervised sEgmentation of BRain strUctures from out-of-the-scanner MRI. *Med. Image Anal.* **2020**, *62*, 101688. [CrossRef]
17. Ramzan, F.; Khan, M.U.G.; Iqbal, S.; Saba, T.; Rehman, A. Volumetric segmentation of brain regions from MRI scans using 3D convolutional neural networks. *IEEE Access* **2020**, *8*, 103697–103709. [CrossRef]
18. Laiton-Bonadiez, C.; Sanchez-Torres, G.; Branch-Bedoya, J. Deep 3D neural network for brain structures segmentation using self-attention modules in MRI images. *Sensors* **2022**, *22*, 2559. [CrossRef]
19. Svanera, M.; Benini, S.; Bontempi, D.; Muckli, L. CEREBRUM-7T: Fast and fully volumetric brain segmentation of 7 Tesla MR volumes. *Hum. Brain Mapp.* **2021**, *42*, 5563–5580. [CrossRef]
20. Cox, R.W. AFNI: Software for analysis and visualization of functional magnetic resonance neuroimages. *Comput. Biomed. Res.* **1996**, *29*, 162–173. [CrossRef]
21. Fracasso, A.; van Veluw, S.J.; Visser, F.; Luijten, P.R.; Spliet, W.; Zwanenburg, J.J.; Dumoulin, S.O.; Petridou, N. Lines of Baillarger in vivo and ex vivo: Myelin contrast across lamina at 7 T MRI and histology. *NeuroImage* **2016**, *133*, 163–175. [CrossRef]
22. Fischl, B. FreeSurfer. *NeuroImage* **2012**, *62*, 774–781. [CrossRef] [PubMed]
23. O'Brien, K.R.; Kober, T.; Hagmann, P.; Maeder, P.; Marques, J.; Lazeyras, F.; Krueger, G.; Roche, A. Robust T1-weighted structural brain imaging and morphometry at 7T using MP2RAGE. *PLoS ONE* **2014**, *9*, e99676. [CrossRef] [PubMed]
24. Yushkevich, P.A.; Piven, J.; Hazlett, H.C.; Smith, R.G.; Ho, S.; Gee, J.C.; Gerig, G. User-guided 3D active contour segmentation of anatomical structures: Significantly improved efficiency and reliability. *NeuroImage* **2006**, *31*, 1116–1128. [CrossRef]
25. Ronneberger, O.; Fischer, P.; Brox, T. U-Net: Convolutional networks for biomedical image segmentation. In Proceedings of the Medical Image Computing and Computer-Assisted Intervention–MICCAI 2015: 18th International Conference, Proceedings, Part III 18, Munich, Germany, 5–9 October 2015; Springer: Berlin/Heidelberg, Germany, 2015; pp. 234–241.
26. Ioffe, S.; Szegedy, C. Batch normalization: Accelerating deep network training by reducing internal covariate shift. In Proceedings of the International Conference on Machine Learning, Lille, France, 7–9 July 2015; pp. 448–456.
27. Tomassini, S.; Sbrollini, A.; Covella, G.; Sernani, P.; Falcionelli, N.; Müller, H.; Morettini, M.; Burattini, L.; Dragoni, A.F. Brain-on-Cloud for automatic diagnosis of Alzheimer's disease from 3D structural magnetic resonance whole-brain scans. *Comput. Methods Programs Biomed.* **2022**, *227*, 107191. [CrossRef] [PubMed]
28. Sugino, T.; Kawase, T.; Onogi, S.; Kin, T.; Saito, N.; Nakajima, Y. Loss weightings for improving imbalanced brain structure segmentation using fully convolutional networks. *Healthcare* **2021**, *9*, 938. [CrossRef] [PubMed]
29. Taha, A.A.; Hanbury, A. Metrics for evaluating 3D medical image segmentation: Analysis, selection, and tool. *BMC Med. Imaging* **2015**, *15*, 1–28. [CrossRef]
30. Le Bihan, D. How MRI makes the brain visible. In *Make Life Visible*; Springer: Singapore, 2020; pp. 201–212.
31. Ashburner, J.; Barnes, G.; Chen, C.C.; Daunizeau, J.; Flandin, G.; Friston, K.; Kiebel, S.; Kilner, J.; Litvak, V.; Moran, R.; et al. *SPM12 Manual*; Wellcome Trust Cent. Neuroimaging: London, UK, 2014; Volume 2464.
32. Jenkinson, M.; Beckmann, C.F.; Behrens, T.E.; Woolrich, M.W.; Smith, S.M. FSL. *NeuroImage* **2012**, *62*, 782–790. [CrossRef]
33. Zhang, F.; Breger, A.; Cho, K.I.K.; Ning, L.; Westin, C.F.; O'Donnell, L.J.; Pasternak, O. Deep learning based segmentation of brain tissue from diffusion MRI. *NeuroImage* **2021**, *233*, 117934. [CrossRef]
34. Kagadis, G.C.; Kloukinas, C.; Moore, K.; Philbin, J.; Papadimitroulas, P.; Alexakos, C.; Nagy, P.G.; Visvikis, D.; Hendee, W.R. Cloud computing in medical imaging. *Med. Phys.* **2013**, *40*, 070901. [CrossRef]
35. Erfannia, L.; Alipour, J. How does cloud computing improve cancer information management? A systematic review. *Inform. Med. Unlocked* **2022**, *33*, 101095. [CrossRef]

 *information*

Article

Improving Semantic Information Retrieval Using Multinomial Naive Bayes Classifier and Bayesian Networks

Wiem Chebil [1,*], Mohammad Wedyan [2], Moutaz Alazab [2,*], Ryan Alturki [3] and Omar Elshaweesh [4]

[1] Department of Computer Science, Higher Institute of Computer Science of Mahdia, University of Monastir, Monastir 5000, Tunisia
[2] Faculty of Artificial Intelligence, Al-Balqa Applied University, Al-Salt 19117, Jordan
[3] Department of Information Science, College of Computer and Information Systems, Umm Al-Qura University, P.O. Box 715, Makkah 21961, Saudi Arabia
[4] Department of Software Engineering, Information Technology College, Al-Hussein Bin Talal University, Ma'an 71111, Jordan
* Correspondence: wiem.chebil@isima.u-monastir.tn (W.C.); m.alazab@bau.edu.jo (M.A.)

Abstract: This research proposes a new approach to improve information retrieval systems based on a multinomial naive Bayes classifier (MNBC), Bayesian networks (BNs), and a multi-terminology which includes MeSH thesaurus (Medical Subject Headings) and SNOMED CT (Systematized Nomenclature of Medicine of Clinical Terms). Our approach, which is entitled improving semantic information retrieval (IMSIR), extracts and disambiguates concepts and retrieves documents. Relevant concepts of ambiguous terms were selected using probability measures and biomedical terminologies. Concepts are also extracted using an MNBC. The UMLS (Unified Medical Language System) thesaurus was then used to filter and rank concepts. Finally, we exploited a Bayesian network to match documents and queries using a conceptual representation. Our main contribution in this paper is to combine a supervised method (MNBC) and an unsupervised method (BN) to extract concepts from documents and queries. We also propose filtering the extracted concepts in order to keep relevant ones. Experiments of IMSIR using the two corpora, the OHSUMED corpus and the Clinical Trial (CT) corpus, were interesting because their results outperformed those of the baseline: the P@50 improvement rate was +36.5% over the baseline when the CT corpus was used.

Keywords: information retrieval; biomedical terminologies; multinomial naive Bayesian classifier; Bayesian networks

Citation: Chebil, W.; Wedyan, M.; Alazab, M.; Alturki, R.; Elshaweesh, O. Improving Semantic Information Retrieval Using Multinomial Naive Bayes Classifier and Bayesian Networks. *Information* **2023**, *14*, 272. https://doi.org/10.3390/info14050272

Academic Editors: Ognjen Arandjelović and Francesco Fontanella

Received: 12 January 2023
Revised: 23 April 2023
Accepted: 27 April 2023
Published: 3 May 2023

1. Introduction

The amount of data and information on the web is permanently increasing. Indeed, the web represents the most important source of knowledge and information that is quick and easy to access. Several information retrieval systems (IRSs) are available to users. An IRS's task is to identify the information most relevant to a user's query. This information can be a document, an image, a video, etc. In this paper, we focus especially on retrieving documents. A query is a set of words that represents a user's need for information. The two main tasks that characterize an IRS are the indexing task (documents and the query) and the matching task between the index documents and the index of the query. Indexing consists of extracting the most representative terms of the document (or of a query) that allows for an IRS to select a set of documents to respond to the users' queries. Term disambiguation is an essential step to improve the performance of an IRS, given the use of multi-terminologies for indexing. In multi-terminologies, a term may be related to more than one concept, so it is ambiguous. For example, "implantation procedure" and "implantation in uterus" are two different SNOMED-CT concepts that have the same term: "implantation, nos". This study proposes a new approach to improve information retrieval (IR) called "improving semantic information retrieval" (IMSIR). Our main contribution is to combine an unsupervised method (BN) and a supervised method (MNBC) to extract concepts and then filter the results using semantic information provided by the Unified Medical Language System (UMLS). The role of the filtering step is to retain the relevant concepts. We also exploited multi-terminologies instead of one terminology. The use of multi-terminologies has had good results in indexing biomedical documents [1], allowing IRSs to extract more concepts that are relevant to the user's query. Our approach exploits the structure of biomedical terminologies and the semantic information that these terminologies provide. In addition, IMSIR is based on the mechanism of inference which characterizes a Bayesian network (BN) to disambiguate terms and extract concepts and to match documents and queries. A BN is a graph that exploits a robust inference process for reasoning under uncertainty. The BN exploited by IMSIR performs a partial match that allows it to extract concepts that occur in the documents as well as concepts that partially occur in the documents. Moreover, IMSIR uses a multinomial naive Bayes classifier (MNBC) to extract concepts. The MNBC allows the IMSIR to enrich the index with new concepts whose terms do not occur in the documents [2]. For example, the concept "Bronchodilator Agents" belongs to the index of the document having the PMID = 11115306, although this concept does not occur in the document. In fact, experts can judge that concepts are relevant even when they do not occur in the document or in the query because they correspond to the context of the document or the query. Due to the fact that MNBC exploits features, these concepts can be extracted using MNBC. Machine learning is an efficient method for classification and its exploitation has led to good results [3], especially naive Bayes, which has been exploited in different works [4–7].

2. Related Work

In this section, we highlight the main approaches that have been proposed an IRS. The proposed approaches for IR that we cite are divided into unsupervised approaches and supervised approaches. We can cite some unsupervised approaches. In the work of Salton et al. [8], a similarity is computed using a vector space model (VSM) between the indexing terms of the query and the indexing terms of the documents. Based on a mathematical model, the authors of [9] proposed a probabilistic model that computes the likelihood of a document's relevance for a query [10]. These two IR models [8,9] do not use semantic resources, which leads to less precision. The possibility and necessity measures are used to map the query to the documents [2]. For document ranking, Ref. [11] suggested a generalized ensemble model (gEnM) that linearly merges numerous rankers. The authors in [12] proposed the matching of concepts and queries with a possibilistic network (PN) that is also used to match concepts and documents and to retrieve and rank documents. To retrieve documents, Ensan and Bagheri [13] presented a cross-language information retrieval approach using a language different from the one used by the user when writing the query. The work reported in [14] performed a new approach that exploited the proximity and co-occurrence of query terms in the document. Moreover, VSM is used to retrieve documents. The work in [15] proposed an unsupervised neural vector space model (NVSM) that defined representations of documents. NVSM learns document and word representations and rank documents based on their similarity to query representations. To improve IRS, the authors in [16] propose enriching the query by combining domain-specific and global ontologies. The authors computed weights for both semantic relationships and the occurrence of each concept. To evaluate the query expansion process, this was integrated into current search engines. The results showed an improvement of 10% in terms of precision. A user's profile and the context of their web history were exploited in [17] to improve IR.

We can site also some proposed supervised approaches for IR. The work in [18] defined relevance between a keyword style query and a document using a new deep learning model. The next and final step was a deep convolution stage, where, in order to compute the relevance, a deep feed-forward network is defined. The major limitation of this work is the small amount of training data used. The work described in [19] proposed the use of multinomial naive Bayes to improve IRS. The authors enriched the user's query using the following process: after retrieving documents for a user's query, the multinomial naive Bayes is exploited to extract relevant terms from retrieved documents. The document corpus is then processed and indexed. A limitation of this approach is that it depends on text and does not use semantic knowledge, which leads to low accuracy. The authors in [20] presented a neural semi-supervised framework to improve information retrieval. The framework is composed of two neural networks: an unsupervised network, which is a self-attention convolutional encoder–decoder network, and a supervised and sentence-level attention scientific literature retrieval network. The aim of combining the two networks is to detect the semantic information and learn the semantic representations in scientific literature datasets. Experiments using two datasets have shown encouraging results. The work of Prasath, Sarkar, and O'Reilly [21] proposed a supervised method to improve users' queries and ranking candidates' terms for indexing the query. The proposed framework is composed of two steps: the training stage and the testing stage. Pseudo-relevance feedback is used to have a set of candidates' terms. These are illustrated as a feature vector. These vectors contain the extracted context-based feature and the extracted resource-based features. A supervised method is exploited to refine and rank terms.

According to their theoretical methods and also when analyzing the index of documents and queries, we can conclude that the proposed unsupervised methods for IR ignore relevant concepts that do not occur in the documents [2]. In fact, these approaches extract only concepts that occur or partially occur in the document. The missed concepts can be extracted using supervised methods. The proposed supervised methods for IR suffer from low performance in indexing biomedical documents due to the lack of efficient features and a training corpus. To deal with the limitations of both supervised and unsupervised methods, we propose combining both using a BN that shows a good performance in indexing biomedical documents [22] and a MNBC that allows the extraction of new relevant concepts. The results are then filtered using UMLS [23].

To further improve an IRS, especially when using multi-terminologies, it is essential to include a word sense disambiguation (WSD) step. We can classify the WSD approaches as either supervised, external resource-based approaches [24] or free-knowledge and unsupervised approaches. We now describe some knowledge-based approaches. The work reported in [25] proposed implementing a supervised WSD using two deep learning-based models. The first model is dependent on a bi-directional long short-term memory (BiLSTM) network. The second is a neural network model with an appropriate top-layer structure. The authors in [26] developed an approach called deepBioWSD. It takes advantage of current deep learning and UMLS breakthroughs to build a model that exploits one single BiLSTM network. The proposed model produces a logical prediction for any ambiguous phrase. These embeddings were used to initialize a network to be trained. According to the experiments, WSD approaches based on supervised methods outperform other approaches. However, developing a distinct classifier for each ambiguous phrase necessitates a large amount of training data, which may not be available. The work described in [27] builds concept embeddings using recent approaches in neural word embeddings. Cosine similarity combined with the embeddings and an external-based method is exploited to find the correct meaning of a word, leading to high accuracy. The probability measure used by the naive Bayes was exploited in work [28], which evaluated the context of an ambiguous word. The relevant concept with the highest score was kept to represent the sense of the polysemic word. A similarity was computed in [29] between the description of the candidates' concepts and the context of the ambiguous word. [30] maps the documents to WordNet synsets. Definitions of UMLS [1] concepts were combined with word representations created on large corpora [31] to create a conceptual representation. The description of ambiguous terms' context was compared to the conceptual representation. However, a large training set is needed to test the method. Machine learning is an efficient approach exploited for classification in different fields

3. Materials and Method

The process of our information retrieval system IMSIR is composed of the following steps, as illustrated in Figure 1:

(1) Document, query and term pretreatment [12,32]
(2) Concept extraction using a multinomial naive Bayes classifier (MNBC)
(3) Term and concept extraction and disambiguation using a Bayesian network
(4) Filtering concepts
(5) Final indexes
(6) Matching queries and documents

Let us consider a document denoted d_j, a concept denoted c_f, and a term denoted t_j. d_i is a document that belongs to the corpus of documents that will be indexed, a $d_j \in \{d_1...d_U\}$, and U is the number of documents in the corpus. A $c_f \in \{c_1...c_M\}$ with M as the number of concepts in UMLS that correspond to MeSH descriptors and SNOMED-CT concepts. A concept is composed of a set of terms, for example, "Abortion, induced" is a concept and its terms are, respectively, "Abortion, induced", "Abortion, Rivanol", "Fertility Control, Post conception", "Abortion Failure", and "Adverse effects" [2]. A $t_i \in \{t_1...t_P\}$ with P is the number of terms that belong to all the concepts. A term can be composed of one or more than one word. A word $w_k \in \{w_1...w_L\}$, L is the number of words that belong to terms and documents. A query is denoted q_h and $q_h \in \{q_1...q_A\}$, A being the number of queries. First of all, documents, queries, and terms are pretreated. The pretreatment step consists of removing punctuation, pruning stop words, stemming the text, and dividing phrases into words. Then, the concepts are extracted using MNBC. The outputs of this step are concepts (classes) mapped to documents. In the next step, terms are extracted using BN, and the concepts are assigned and disambiguated. The output of this step is the indexes of the concepts. The two indexes of each document are merged and filtered. Thus, we obtain a final index for each query and document. Finally, documents are retrieved for each query by matching a query to each index of document and documents are ranked according to the score Equation (17).

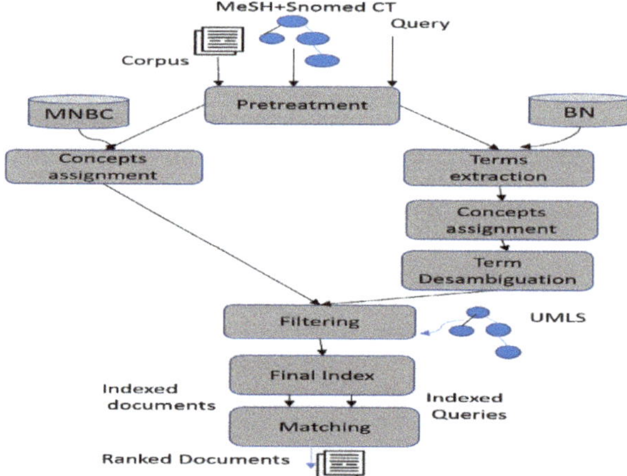

Figure 1. The process of IMSIR.

3.1. Concept Extraction Using a Multinomial Naïve Bayes Classifier

In this step, we use the MNBC to map concepts with documents with the aim of obtaining an index of concepts that represent the document. An MNBC is exploited for document classification [33], which consists of mapping classes and documents, using the statistical analysis of their contents. The classification is performed based on the documents that have already been classified. MNBC assumes the independence of variables and exploits probabilistic measures. It is characterized by the fact that the occurrence of one feature does not affect the probability of the occurrence of the other feature that characterized that category. A main advantage of MNBC is that it considers the Goss frequency, which is the frequency of the word and not the binary occurrence (whether the word occurs or not). The process of concept extraction using the MNBC is composed of the following steps (Figure 2): the training step and the classification step. The inputs of the training step are the already indexed documents and a set of classes $C = \{c_1, c_2, \ldots, c_M\}$ that corresponds to the set of MeSH and SNOMED CT concepts. The documents of the training set $(d_1, \ldots, d_v)$ (v is the number of documents) were indexed manually by experts with concepts that represent the classes of the document. The outputs of the training step are the probabilities $P(d_j|c_f)$. The probabilities, a test corpus, and the set of classes are the inputs of the classification step. A set of classified documents is the output of the last step (Equation (3)). The concepts (classes) are assigned to documents by computing the probabilities of documents knowing concepts $P(d_j|c_f)$, which is based on the probability that a word belongs to a given class (concept), also called likelihood. $P(d_j|c_f)$ is calculated as follows (Equations (1) and (2)) [19]:

$$P(d_j|c_f) = \prod_{t=1}^{L} p(w_t \mid c_f) \tag{1}$$

$p(w_t \mid c_f)$ is the probability of a word w_t that occurs in a class c_f in the training documents.

$nb(w_t, c_f)$ is the number of occurrences of w_t in the class c_f. $nb(c_f)$ is the total number of words in the class c_f.

$L = |T|$ is the length of the vocabulary,

$$p(w_t \mid c_f) = \frac{1 + nb(w_t, c_f)}{L + nb(c_f)} \tag{2}$$

The concept that will index a query or a document is selected using the maximizing function (Equation (3)):

$$c^*(d_j) = argmax_{c_f} P(c_f) \prod_{k=1}^{L} p(w_t \mid c_f) \tag{3}$$

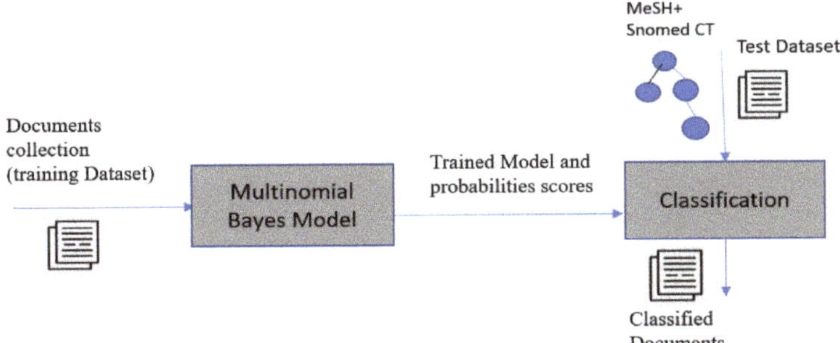

Figure 2. The process of concept extraction using MNBC.

3.2. Concept Extraction Using BN

To extract concepts, we employ a three-layer Bayesian network [22] (Figure 3). The network represents the following nodes: (i) the document to be indexed d_j (ii) a word of the document and of the term w_k, (iii) the term t_i and (iv) the dependency relationships that exist between the nodes. A document d_i belongs to the set of documents that will be indexed using our approach $\{d_1, d_2, \ldots, d_U\}$ (U is the number of documents that will be indexed). A term t_i belongs to the set of terms of MeSH and Snomed CT $\{t_1, t_2, \ldots, t_P\}$ (P is the number of terms).

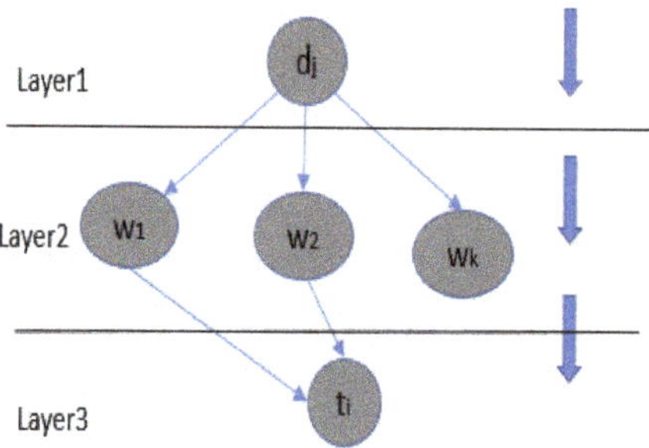

Figure 3. The BN for term extraction.

3.2.1. Evaluation of a Term

A term t_i is evaluated through the propagation of information given by the indexing term in the network once it is instantiated. Edges are activated by instantiating the term to the document. For each node, the conditional and marginal posterior probability are calculated given the conditional and marginal prior probability calculated according to Equations (4) and (5). According to the topology of the graph [22], we have:

$$P(t_i|d_j) = \sum_{\theta^r \in \theta^r} P(t_i|\theta^r)P(\theta^r|d_j) \times a \tag{4}$$

$$P(\theta^r|d_j) = \Pi_{w_k \in w(t) \wedge w(d)}(P(w_k|d_j)) \tag{5}$$

θ^W represents the set of possible configurations of the parents of the instantiated term t_i. θ^w is a possible configuration in θ^W.

$\theta^W = \{w_1, w_2\}, \{w_1, \neg w_2\}, \{\neg w_1, w_2\}, \{\neg w_1, \neg w_2\}$ are the possible configurations of the words $\{w_1, w_2\} of a term$ (the parents of t).

a is a coefficient whose values are included in the interval $[0, 1]$ with $a < 1$ if the words of a term are not in the same sentence. In the case where the words of a term are not in the same sentence, the coefficient a was tuned. $W(d)$: is the set of words of the document D. $W(t)$: is the set of words of the term T.

3.2.2. Computing the Weight of the Arc $P(w_k|d_j)$

To weigh the arc that links the nodes words to the document that will be indexed, we used the word frequency-inverse document frequency (wf/idf) measure. Thus, (Equations (6)–(8)) :

$$P(w_k|d_j) = wf_{kj} \times idf_k \tag{6}$$

$$wf_{kj} = \frac{freq_{kj}}{max_{r:1 \to p}(freq_{rj})} \tag{7}$$

$$idf = log\frac{Nu}{nd_k} \tag{8}$$

Nu is the number of documents in the corpus test. nd_k is the number of documents in which the word k appears. In addition, m denotes the total number of words in the document. Finally, $freq_{kj}$ is the number of times the word k appears in the document d_j. p is the number of words in the document that will be indexed. $freq_{kj}$ is the frequency of the word k in the document d_j .

3.2.3. Aggregation of Words of Terms $P(t_i|\theta^r)$

In our model, we adopted the five canonical forms proposed by Turtle in their Bayesian network Information Retrieval (IR) model for each type of search [34]. In fact, we replaced the query by an indexing term. Thus, an indexing term can be aggregated by a probabilistic sum or a Boolean operator (OR, AND, NOT) or one of its variations, the weighted sum. The aggregations are defined in Equations (9)–(13) to evaluate the conditional probabilities $P(T \mid \theta)$ (θ is all the set of parents of T) of a node T having n parents (n words) $\theta_1, \ldots, \theta_n$ and $P(\theta_1 = w_1) = p_1, \ldots, P(\theta_n = w_n) = p_n$

$$P_{or}(T \mid \theta^t) = 1 - (1 - p_1) - \ldots - (1 - p_n) \tag{9}$$

$$P_{and}(T \mid \theta^t) = p_1 \times \ldots \times p_n \tag{10}$$

$$P_{Not}(T \mid \theta_1^t) = 1 - p_1 \tag{11}$$

$$P_{Sum}(T \mid \theta^t) = \frac{p_1 + \ldots + p_n}{n} \tag{12}$$

$$P_{Weightedsum}(T \mid \theta) = \frac{(l_1 p_1 + \ldots + l_n p_n) l_t}{l_1 + \ldots + l_n} \tag{13}$$

The weight of the term and the word are denoted by l_t, l_n, respectively. A partial match between documents and terms is performed using our method. As a result, we used the disjunction to solve $P(t_e|\theta^r)$. If we consider a term t_e as a disjunctive Boolean query, candidate terms are those that have at least one word in the document d_j. However, $t_e = w_1 \vee w_2 \vee \ldots \vee w_p$ is the formula for a phrase t_e with p words.

3.2.4. Concept Assignment and Terms Disambiguation

To assign concepts to the terms, we compute the following Equation (Equation (14))

$$Sim(d_j, c_f) = Sim(d_j, t_i) = max_{t_i \in t(c_f)}(P(t_i|d_j)) \tag{14}$$

With $T(c_f)$ as a set of terms of a concept c_f.
The score of the sense of an ambiguous term T_j is computed as follows (Equation (15))

$$C_f^* = argmax_{C_s \in C(t_i)}(Sim(d_j, C_s)) \tag{15}$$

3.3. Filtering Based on UMLS

We merge the two indexes that are composed of concepts from both methods (BN and MNBC), putting the concepts with the highest scores in the first ranks, and we delete the duplicated concepts. Then, the UMLS is exploited to filter the concepts extracted in the previous step while keeping the relevant ones. Both the MNBC and BN methods can produce irrelevant concepts that contain a part of the words of their terms or all the words of their terms (in the case of using MNBC) and do not occur in the document. To deal with this limitation, we divide the set of concepts into two indexes: the secondary index (SI) and main index (MI). The MI is a set of concepts that have at least one term that has all of its words occurring in the document. The SI is a set of concepts where the words of all of their terms do not occur in the document.

MI = $\{MC_1, \ldots, MC_p, \ldots MC_v\}$, MC_p is a main concept. v is the number of MC. SI = $\{SC_1, \ldots, SC_f, \ldots SC_k\}$, SC_f is a secondary concept. K is the number of SC.

The SCs are then ranked according to the score computed in Equation (16). We hypothesize that if an SC is co-occurring and has semantic links (according to the UMLS) with the MI's L-initial MCs, it is more likely to be relevant. Finally, the n concepts with the highest scores are kept for indexing documents (n is tuned).

For example, the MeSH concepts "imaging, Three-Dimensional" and "coronary artery disease" are linked with the semantic relation "diagnoses", and the MeSH concept "Endocarditis, Bacterial" co-occurs 100 times with the MeSH concept "Penicillins" in MEDLINE.

The number of semantic relations is expressed by NR, and the frequency of co-occurrence is CF. z is the total number of co-occurrences between all MC and all SC. s is the total number of semantic relations between all MC and all SC.

3.4. Computing a Similarity between Queries and Documents $sim(q_h, d_j)$

We computed the similarity between a query and a document using a Bayesian network.

$$Sim(q_h, d_j) = P(q_h/d_j) \tag{16}$$

Thus, we computed $P(q_h/d_j)$ using Equation (4) by replacing a term with a query. To compute $p(q_h/d_j)$ we used $P(d_i/c_q)$, which is computed using $P(c_q/d_j)$ and the Bayes rule as follows:

$$P(c_q/d_j) = \frac{P(d_j/c_q)P(c_q)}{p(d_j)} \tag{17}$$

4. Results

Two corpora were used to evaluate our IR approach:

(1) OHSUMED (https://trec.nist.gov/ (Hersh et al., 1994) accessed on 23 April 2023), is a document collection that was used for the TREC-9 filtering track. This corpus is the same as that used in [12]. Details on this corpus are presented in [12].

(2) The Clinical Trial corpus 2021, which is composed of topics (descriptions of the user needs), clinical documents, and relevance judgments evaluated by experts. The topics correspond to the queries. This is the link to the corpus: http://www.trec-cds.org accessed on 12 May 2022.

We chose these two corpora because the first one is characterized by short queries and the second is characterized by long queries, which allowed us to test the performance of IMSIR using the two types of queries. Below is an example of a topic (query) in Clinical Trial corpus :

```
<topics task="2021 TREC Clinical Trials">
<topic number="-1">
A 2-year-old boy is brought to the emergency department by
their parents for 5 days of high fever
and irritability...
< /topic>
< /topics>
```

To test our approach, we indexed queries and documents using IMSIR and we computed the score (Equation (16)) between each query and document. The documents were then retrieved and ranked according to the score (Equation (16)) as a response to the query.

To evaluate our proposed information retrieval approach, we opted for the mean precision (MAP) (Equation (18)). We also computed the precision at ranks 5, 20, and 50. We compared the performance of our approach that exploits MNBC with the performance of our approach using a support vector machine (SVM) or a random forest classifier (RFC) instead of MNBC (Table 1). In addition, we computed the improvement rate ($\triangle MAP$) (19)), which highlights the added value of our contributions compared to a baseline, which is the work of [35] (Tables 2 and 3). This is a recent approach that exploits supervised methods and terminologies to improve the IRS. We also compared our work that exploits BN to match queries and documents with our work that exploits BM25 or VSM (vector space model) instead of BN (Tables 2 and 3) and with the approach of Mingying et al. [20], which is a recent approach that exploits a semi-supervised method. We also tested CIRM [12]

(Tables 2 and 3). Moreover, we computed Students' *t*-tests between the ranks (P@10, P@20, P@50, and MAP) obtained by each method tested and the baseline.

$$MAP = \frac{1}{N} \sum_{i=1}^{n} P@i \times R(i) \tag{18}$$

The total number of documents is n. The number of relevant documents is N. In addition, P@i indicates the accuracy of document retrieval. Finally, if the document is not relevant, then $R(i)$ is equal to 0 and if it is relevant, then $R(i)$ is equal to 1.

$$\triangle MAP = \frac{MAP_{methode} - MAP_{baseline}}{MAP_{baseline}} \times 100 \tag{19}$$

Table 1. Evaluation of IMSIR using different supervised methods when the corpus OHSUMED is exploited.

Approach	MAP	P@5	P@10	P@20	P@50
IMSIR-SVM	0.63	0.72	0.63	0.61	0.57
IMSIR-RFC	0.59	0.71	0.63	0.58	0.52
IMSIR-MNBC	0.67	0.75	0.62	0.60	0.56

Table 2. Evaluation of IMSIR when the corpus OHSUMED is exploited.

Approach	MAP	P@5	P@10	P@20	P@50
CIRM [12]	0.63 (+43.18%)	0.72 (+33.33%)	0.63 (+28.57%)	0.61 (+28.57%)	+0.57 (32.55%)
Baseline [35]	0.44	0.54	0.49	0.45	0.43
Mingying et al. [20]	0.65 (+47.72%)	0.70 (+29.62%)	0.6 (+22.44%)1	0.59 (+31.11%)	0.53 (+23.25%)
IMSIR-VSM	0.59 (+34.09%)	0.71 (+31.48%)	0.63 (+28.57%)	0.58 (+28.57%)	0.52 (+20.93%)
IMSIR-BM25	0.62 (+40.90%)	0.71 (+31.48%)	0.62 (+26.53%)	0.54 (+28.57%)	0.51 (+18.60%)
IMSIR-BN	0.67 (+52.27%) *	0.75 (+38.88%) *	0.62 (+26.53%) *	0.60 (+33.33%)	0.56 (+30.23%) *

* a substantial difference at $p < 0.05$.

As shown in Tables 1–3, the performance of our information retrieval system (IMSIR) is better than the baseline and the approach of [20] in terms of MAP and precision in different ranks of documents. Moreover, our proposed approach shows comparable results with CIRM. Furthermore, compared to the baseline, IMSIR is statistically significant. These results highlight the interest in the similarities proposed in IMSIR, which exploits a statistical and semantic weight for ranking concepts and proves that the structure of RB and the information propagation mechanism are adequate for controlled indexing. In addition, MNBC brings new relevant concepts, especially those whose terms do not occur in the document or in the query. The combination of BNs, a BNC, and the use of UMLS for filtering contributes to the retaining of relevant concepts and improvement of extraction and ranking of concepts. Table 4 shows the interest in using the filtering step in the process of IMSIR. In fact, the performance of IMSIR becomes greater when applying the filtering step. Using the co-occurrences and semantic relations provided by UMLS allows for the deletion of irrelevant concepts, especially those where a part of their words do not occur in the document or all of their words do not occur in the document. It is also clear also that IMSIR performs better when using the Clinical Trial corpus (CTC) than when using the OHSUMED corpus

(Table 2). These results are explained by the fact that IMSIR exploits statistic measures that demonstrate good results when using long queries. Moreover, according to Table 1, our approach returns better results when using the supervised method MNBC than when using SVM or RFC. Tables 2 and 3 also highlight the use of BN to match queries, as the performance of IMSIR-BN outperforms those of IMSIR-VSM and IMSIR-BM25. Moreover, IMSIR-BN achieved better performance when NC = 5 (Table 5 and Figure 4).

Table 3. Evaluation of IMSIR when the Clinical Trial corpus is exploited.

Approach	MAP	P@5	P@10	P@20	P@50
CIRM [12]	0.61 (+35.55%)	0.74 (+32.14%)	0.65 (+25%)	0.62 (+%)	0.60 (+33.33%)
Baseline [35]	0.45	0.56	0.50	0.47	0.45
Mingying et al. [20]	0.65 (44.44%)	0.76 (35.71%)	0.61 (22.00%)	0.60 (26.65%)	0.56 (24.44%)
IMSIR-VSM	0.62 (+37.77%)	0.74 (+32.14%)	0.64 (+23.07%)	0.60 (+21.66%)	0.55 (+22.22%)
IMSIR-BM25	0.65 (+44.44%)	0.73 (+30.35%)	0.63 (+21.15%)	0.59 (+20.33%)	0.54 (+20%)
IMSIR-BN	0.69 (+53.33%) *	0.78 (+39.28%) *	0.65 (+25%) *	0.63 (+36.50%)	0.59 (+31.11%) *

* a substantial difference at $p < 0.05$. IMSIR-VSM: VSM was used to perform IMSIR when matching queries and documents. IMSIR-BM25: BM25 was used to perform IMSIR when matching queries and documents. IMSIR-BN: BN was used to perform IMSIR when matching queries and documents.

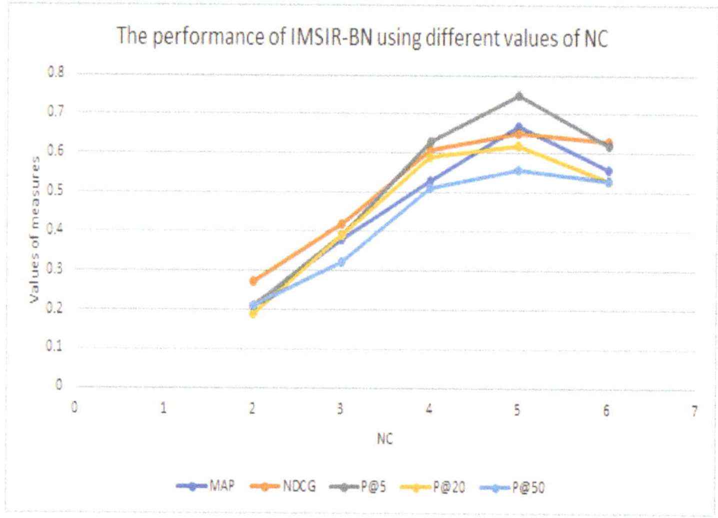

Figure 4. Tuning the value of NC. C is a concept, T_i is a term, W_k is a word belonging to a document or to a term, and D_j is a document.

Table 4. Evaluation of IMSIR with and without the step of filtering.

Approach	MAP	P@5	P@10	P@50
IMSIR-BN *	0.49	0.46	0.32	0.27
IMSIR-BN	0.69	0.78	0.65	0.59

IMSIR-BN *: is IMSIR-BN without the step of filtering. The performance of IMSIR-BN was tested with a different number of concepts (NC) in the indexes of the queries (NC) (Table 3) in order to keep the right NC.

Table 5. The performance of IMSIR-BN using different values of NC.

Rank	NC = 2	NC = 3	NC = 4	NC = 5	NC = 6
MAP	0.21	0.38	0.53	0.67	0.56
NDCG	0.27	0.42	0.61	0.65	0.63
P@5	0.21	0.39	0.63	0.75	0.62
P@20	0.19	0.39	0.59	0.62	0.53
P@50	0.21	0.32	0.51	0.56	0.53

5. Conclusions

This study developed a novel IRS called IMSIR that allows the improvement of the process of indexing documents and queries by adding new relevant concepts to the indexes. In fact, our approach combines a BN with three layers, MNBC, and terminologies to extract, disambiguate, and rank concepts. The BN allows extraction of concepts that occur and partially occur in the documents, and MNBC allows for enrichment of the index with relevant concepts that do not occur in the document. A semantic method is also exploited in IMSIR by using the terminologies; in fact, concepts are extracted when their terms occur in the documents or queries. An added value of our approach is the filtering step after the extraction of concepts using the supervised and the unsupervised methods. These methods do not perform an exact match; thus, irrelevant concepts may be extracted, and a filtering step is required in order to keep relevant concepts. This step exploits the properties of UMLS which are semantic relations and co-occurrences. In addition, IMSIR aims to enhance the ranking of retrieved unstructured documents in an IRS by using an efficient score to rank documents. Moreover, the experiments with IMSIR using the Clinical Trial corpus highlighted the added value of combining the inference mechanism of BN, MNBC, and the biomedical terminologies' structure and their semantics to extract, disambiguate, and rank concepts and documents. Furthermore, the experiments allowed us to determine how many concepts were used to index the queries. In the future, we aim to use the same methods described in this paper to enhance IRS through query expansion. In addition, we intend to employ more terminologies, as it will obtain a performance increase over the use of one terminology alone. Moreover, we aim to improve the ranking of concepts step after filtering.

Author Contributions: Conceptualization,W.C.; methodology, W.C.; software, W.C. and M.W.; validation, W.C., M.W., R.A., M.A. and O.E.; formal analysis and investigation, W.C. and M.W.; resources, W.C. and M.W.; data curation, W.C.; writing—original draft preparation, W.C.; writing—review and editing, R.A., M.A. and O.E.; visualization, all the authors; supervision, all the authors; project administration, all the authors; funding acquisition, all the authors. All authors have read and agreed to the published version of the manuscript.

Funding: This research received no external funding.

Data Availability Statement: The data exploited in our experiments are available at the following links: https://trec.nist.gov/, accessed on 1 August 2022 and http://www.cs.cmu.edu/~rafa/ir/ohsumed.html, accessed on 12 September 2022.

Conflicts of Interest: The authors declare no conflict of interest.

References

1. Chebil, W.; Soualmia, L.F.; Omri, M.N.; Darmoni, S.J. Indexing biomedical documents with a possibilistic network. *J. Assoc. Inf. Sci. Technol.* **2016**, *67*, 928–941. [CrossRef]
2. Chebil, W.; Soualmia, L.F.; Dahamna, B.; Darmoni, S.J. Indexation automatique de documents en santé: Évaluation et analyse de sources d'erreurs. *IRBM* **2012**, *33*, 316–329. [CrossRef]
3. Alazab, M. Automated malware detection in mobile app stores based on robust feature generation. *Electronics* **2020**, *9*, 435. [CrossRef]

4.   De Stefano, C.; Fontanella, F.; Marrocco, C.; di Freca, A.S.A. Hybrid Evolutionary Algorithm for Bayesian Networks Learning: An Application to Classifier Combination. In *Applications of Evolutionary Computation. EvoApplications 2010*; Lecture Notes in Computer Science; Springer: Berlin/Heidelberg, Germany, 2010; Volume 6024.

5.   Shen, Y.; Zhang, L.; Zhang, J.; Yang, M.; Tang, B.; Li, Y.; Lei, K. CBN: Constructing a Clinical Bayesian Network based on Data from the Electronic Medical Record. *J. Biomed. Inform.* **2018**, *88*, 1–10. [CrossRef] [PubMed]

6.   Malviya, S.; Tiwary, U.S. Knowledge-Based Summarization and Document Generation using Bayesian Network. *Procedia Comput. Sci.* **2018**, *89*, 333–340. [CrossRef]

7.   de Campos, C.P.; Zeng, Z.; Ji, Q. Structure Learning of Bayesian Networks Using Constraints. In Proceedings of the 26th Annual International Conference on Machine Learning, Montreal, QC, Canada, 14–18 June 2009; Association for Computing Machinery: New York, NY, USA.

8.   Salton, G.; Wong, A.; Yang, C. A vector space model for automatic indexing. *Commun. ACM* **1975**, *18*, 438–446. 613–620. [CrossRef]

9.   Robertson, S.; Jones, K.S. Relevance weighting of search terms. *J. Am. Soc. Inf. Sci.* **1976**, *27*, 129–146. [CrossRef]

10.   Salton, G.; Fox, E. A.; Wu, H. Extended boolean information retrieval. *Commun. ACM* **1983** , *26*, 1022–1036. [CrossRef]

11.   Wang, Y.; Choi, I.; Liu, H. Generalized ensemble model for document ranking in information retrieval. *arXiv* **2015**, arXiv:1507.08586.

12.   Chebil, W.; Soualmia, L.F.; Omri, M.; Darmoni, S.J. Possibilistic Information Retrieval Model Based on a Multi-Terminology. In Proceedings of the ADMA Advanced Data Mining and Applications, Nanjing, China, 18 November 2018.

13.   Ensan, F.; Bagheri, E. Retrieval model through semantic linking. In Proceedings of the 10th ACM International Conference on Websearch and Data Mining, Cambridge, UK, 6–10 February 2017; pp. 181–190.

14.   Sneiders, E. Text retrieval by term cooccurrences in a query based vector space. In Proceedings of the COLING 2016 the 26th International Conference on Computational Linguistics, Osaka, Japan, 11–16 December 2016; Technical Papers; pp. 2356–2365.

15.   Kenter, T.; Borisov, A.; Van Gysel, C.; Dehghani, M.; de Rijke, M.; Mitra, B. Neural networks for information retrieval. In Proceedings of the 40th International ACM SIGIR Conference on Research and Development in Information Retrieval, Tokyo, Japan, 7–11 August 2017; pp. 1403–1406.

16.   Jaina, S.; Seejab, K.; Jindal, R. A fuzzy ontology framework in information retrieval using semantic query expansion. *J. Inf. Manag. Data Insights* **2021**, *1*, 300–307. [CrossRef]

17.   Chebil, W.; Wedyan M.O.; Lu, H.; Elshaweesh, O.G. Context-Aware Personalized Web Search Using Navigation History. *Int. J. Semant. Web Inf. Syst.* **2020**, *16*, 91–107. [CrossRef]

18.   Mohan, S.; Fiorini, N.; Kim, S.; Lu, Z. A fast deep learning model for textual relevance in biomedical information retrieval. In Proceedings of the 2018 World Wide Web Conference, Lyon, France, 23–27 April 2018; pp. 77–86.

19.   Silva, S.; Seara, V.A; Celard, P.; Iglesias, E.L.; Borrajo, L. A query expansion method using multinomial naive bayes. *Appl. Sci.* **2021**, *11*, 10284. [CrossRef]

20.   Xu, M.; Du, J.; Xue, Z.; Kou, F.; Xu, X. A semi-supervised semantic-enhanced framework for scientific literature retrieval. *Neurocomputing* **2021**, *461*, 450–461. [CrossRef]

21.   Prasath, R.; Sarkar, S.; OReilly, P. Improving cross language information retrieval using corpus based query suggestion approach. In Proceedings of the International Conference on Intelligent Text Processing and Computational Linguistics, Cairo, Egypt, 14–20 April 2015; pp. 448–457.

22.   Chebil, W.; Soualmia, L.F.; Omri, M.; Darmoni, S.J. Indexing biomedical documents with Bayesian networks and terminologies. In Proceedings of the 12th International Conference on Intelligent Systems and Knowledge Engineering (ISKE), Nanjing, China, 24–26 November 2017.

23.   Bodenreider, O. The unified medical language system umls integrating biomedical terminology. *Nucleic Acids Res.* **2004**, *32*, 267–270. [CrossRef] [PubMed]

24.   Wedyan, M.; Alhadidi, B.; Alrabea, A. The effect of using a thesaurus in Arabic information retrieval system. *Int. J. Comput. Sci.* **2012**, *9*, 431–435.

25.   Zhang, C.; Bis, D.; Liu, X. Biomedical word sense disambiguation with bidirectional long short-term memory and attention-based neural networks. *BMC Bioinform.* **2019**, *28*, 159–182. [CrossRef] [PubMed]

26.   Pesaranghader, A; Matwin, S.; Sokolova, M.; Pesaranghader, A. Deepbiowsd effective deep neural word sense disambiguation of biomedical text data. *J. Am. Med. Inform. Assoc.* **2020**, *26*, 438–446. [CrossRef] [PubMed]

27.   Sabbir, A.; Jimeno-Yepes, A.; Kavuluru, R. Knowledge-Based Biomedical Word Sense Disambiguation with Neural Concept Embeddings. In Proceedings of the 2017 IEEE 17th International Conference on Bioinformatics and Bioengineering (BIBE), Washington, DC, USA, 23–25 October 2017. [CrossRef]

28.   Yepes, A.; Berlanga, R. Knowledge based word concept model estimation and refinement for biomedical text mining. *J. Biomed. Inform.* **2015**, *53*, 300–307. [CrossRef] [PubMed]

29.   Lesk, M. Automatic sense disambiguation using machine readable dictionaries how to tell a pine cone from an ice cream cone. In *SIGDOC '86: Proceedings of the 5th Annual International Conference on Systems Documentation*; Association for Computing Machinery: Washington, DC, USA, 1986; pp. 24–26.

30.   Voorhees, E.M. Using Wordnet to Disambiguate Word Senses for Text Retrieval. In *SIGIR '93: Proceedings of the ACM SIGIR, Conference on Research and Development in Information Retrieval*; Association for Computing Machinery: Washington, DC, USA, 1993; pp. 171–180.

Information **2023**, *14*, 272

31. Tulkens, S.; Suster, S.; Daelemans, W. Using distributed representations to disambiguate biomedical and clinical concepts. In Proceedings of the 15th Workshop on Biomedical Natural Language Processing, Berlin, Germany, 12 August 2016; Volume 53, pp. 77–82.
32. Chebil, W.; Soualmia, L.F. Improving semantic information retrieval by combining possibilistic networks, vector space model and pseudo-relevance feedback. *J. Inf. Sci.* **2023**. [CrossRef]
33. Raschka, S. Naive Bayes and Text Classification I-Introduction and Theory. *arXiv* **2014**, arXiv:1410-5329.
34. Turtle, H.; Croft, W.B. Inference Networks for Document Retrieval. In Proceedings of the ACM SIGIR Conference on Research and Development in Information Retrieval, Association for Computing Machinery, Brussels Belgium, 5–7 September 1990; pp. 1–24.
35. Xu, B.; Lin, H.; Yang, L.; Xu, K.; Zhang, Y.; Zhang, D.; Yang, Z.; Wang, J.; Lin, Y.; Yin, F. A supervised term ranking model for diversity enhanced biomedical information retrieval. *BMC Bioinform.* **2019**, *20*, 590. [CrossRef] [PubMed]

Article

The Process of Identifying Automobile Joint Failures during the Operation Phase: Data Analytics Based on Association Rules

Polina Buyvol *, Irina Makarova, Aleksandr Voroshilov and Alla Krivonogova

Naberezhnye Chelny Institute, Kazan Federal University, Syuyumbike Prosp. 10a, 423812 Naberezhnye Chelny, Russia
* Correspondence: skyeyes@mail.ru

Abstract: The increasing complexity of vehicle design, the use of new engine types and fuels, and the increasing intelligence of automobiles are making it increasingly difficult to ensure trouble-free operation. Finding faulty parts quickly and accurately is becoming increasingly difficult, as the diagnostic process requires analyzing a great amount of information. Therefore, we propose an approach based on association rules, a machine learning technique, to simplify the defect detection process. To facilitate its use in a real repair company environment, we have developed a web service that allows a repairman to simultaneously identify nodes with a high probability of failure. We have described the structure and working principles of the developed web service, as well as the procedure for its application, which resulted in the discovery of several useful non-trivial rules. We have presented several rules resulting from the use of this interactive tool, which allow repairers to detect possible defects in the relevant components, during the diagnostic process, quickly and easily. These rules are also well supported and can be used by procurement departments to make tactical decisions when selecting the most promising suppliers and manufacturers. The methodology developed allows the evaluation of the effectiveness of changes in the design and technology for the manufacture and operation of individual vehicle components, analyzing the change in the composition of parts combinations over time.

Keywords: association rules; defect analysis; automobile repair; decision support; web service

Citation: Buyvol, P.; Makarova, I.; Voroshilov, A.; Krivonogova, A. The Process of Identifying Automobile Joint Failures during the Operation Phase: Data Analytics Based on Association Rules. *Information* 2023, 14, 257. https://doi.org/10.3390/info14050257

Academic Editors: Amar Ramdane-Cherif, Ravi Tomar and TP Singh

Received: 8 March 2023
Revised: 18 April 2023
Accepted: 21 April 2023
Published: 25 April 2023

1. Introduction

Today, the growing sophistication of vehicle construction, the introduction of modern engine types, the transition to alternative fuels, and the growing share of electrical and electronic equipment complicates the process ensuring consumers experience trouble-free operation of their vehicles. Diagnostic work includes an increasing proportion of repair and maintenance work and requires a certain level of expertise. Troubleshooting faults during diagnosis quickly and accurately is becoming increasingly difficult, due to the great volume of information to be processed, but it is necessary to restore vehicle performance quickly and accurately. Because a vehicle is a complex technical system consisting of some interacting elements, intelligent analysis methods and instruments are needed to identify complex relationships, determine the most critical components, and establish repair sequences.

At the same time, the likelihood of unmanned vehicles entering public roads around the world is increasing. In this regard, the service concept will change: if for traditional vehicles, on-demand repairs when a malfunction is detected and periodic technical maintenance are accepted, then for unmanned vehicles, daily pre-trip inspections are more appropriate. This change is due to the fact that in the vehicle–driver system for unmanned vehicles the role of the driver is excluded, but indirect signs (smell, sound, noise, vibration) can be used to identify an existing malfunction and predict a possible failure in the near future. Depending on the ability to perform its functions and on compliance with the requirements established by the normative-technical and design documentation, an object can be in good, fault, and inoperable states. When the limiting state is reached, the object is removed from operation (Figure 1). The complexity of vehicle troubleshooting also lies in the fact that not all malfunctions lead to a fault state in vehicles.

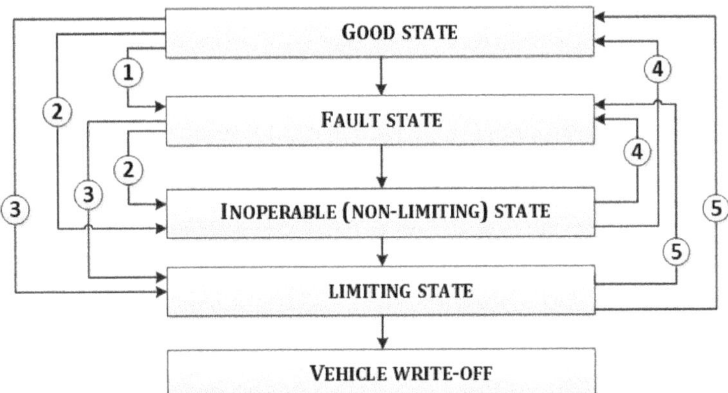

Figure 1. Scheme of vehicle states and their corresponding events: 1—damage; 2—failure; 3—marginal failure; 4—restoration; 5—overhaul.

In this case, defective parts affect other joint parts and can lead to a faulty state, resulting in the inoperable state of the entire vehicle. This type of failure is called dependent. This is a failure that occurs as a result of a change in the reliability characteristics of other elements. Thus, being operable at the time of the check, the vehicle may fail suddenly and, as a result, create a potential safety hazard for both the passengers in the vehicle and other road users. These situations can be avoided if all the defective parts are identified in time. Through external manifestation, one can single out explicit failures and implicit (hard to detect) failures, for the identification of which special means are used. In this regard, the urgency of the problem of identifying defective technical systems components increases.

A lot of research on investigating methods to analyze technical systems reliability has been conducted. Typical examples include fault tree analysis [1], logical probabilistic methods [2], failure and effects analysis [3], multivariate failure analysis [4], fuzzy logic [5], failure distribution function construction [6,7], and equipment diagnostic methods [8]. However, these methods have several limitations. Moreover, the first three methods require in-depth system knowledge of the system structure and the processes that occur. The others cannot take into account the interaction of the joint or cooperating components.

Machine learning methods have come to occupy an important place in artificial intelligence engineering research. It has become a powerful tool for developing efficient and accurate solutions to many decision-making and optimization problems, based on accumulated data [9,10]. In particular, decision trees have been used to diagnose malfunctions in traditional vehicles [11], for unmanned vehicles researchers have used the support vector machine [12], neural networks [13], the prediction of the unmanned vehicles' behavior has been carried out using the support vector machine [14], decision trees [15], and neural networks [16].

Machine learning does not have a high barrier to entry and the analyst does not need to be a highly skilled expert in automotive design.

We hypothesize that the use of machine learning methods for automotive repair post diagnostics will improve customer service by simultaneously identifying assemblies with a high likelihood of failure.

2. Application of Association Rules: A Literature Review

Association rules are one of the machine learning methods, which identify a set of frequent items from a large data set.

Association rules have been used successfully in the past to predict customer purchasing behavior [17,18], analyze bank customer deposit data, analyze school entry scores and three-year average grades [19], predict software failures [20], and identify factors affecting the probability and severity of traffic accidents [21]. The main advantage of association rules is that they can be easily understood and interpreted in programming languages [22].

Because of their ease of interpretation, association rules are widely used as a method of intelligent analysis in a variety of fields. Typical examples include the study of shopping patterns [19], the identification of factors affecting the outcome of accidents [23], disease outbreaks [24], the evaluation of suppliers [25] and software reliability [26], risk prediction for marine logistics services [27], construction of automobile insurance recommendation systems [28,29], and the formation of options combinations chosen when buying cars [30].

It has also been applied in studying the reliability of complex technical systems. Thus, in [31,32], the authors proposed to study the failure correlation coefficients between types of failures and their causes based on failure data of computer numerical control systems with lifetime data using association rules. In [33], rules were used in the detection of anomalous performance of process monitoring systems, such as fluid level monitoring systems; in [34], rules were used for evaluating the reliability of an electric power system. Moreover, ref. [35] was devoted to established fuzzy association rules built for explaining the relationship between the characteristics which can be measured and the failure of an industrial boiler. The authors of [36] used association rules and principal component analysis or support vector data description for finding anomalous performance in an unmanned aerial vehicle system. The authors of [37] proposed, for the first time, to use communication data mining to detect abnormal or malicious vehicle operation.

A structure for mining consistent samples of vehicle maintenance behavior from historical repair records under predefined support and thresholds was proposed in the study by [38]. A rule-based procedure was offered to predict the next maintenance interval and provide necessary information about spare parts. The proposed procedure can be extended for exploring repair actions and conducting root cause analysis to provide maintenance managers with more valuable recommendations on corrective actions to be taken to avoid further breakdown. The timestamp information can also be used to prioritize maintenance activities.

In [39], a correlation analysis was derived from the relationships among the default components, the time of manufacture, the place of operation, and the vehicle type. The researchers used SPSS Clementine with a minimum support value of 0.2% and a minimum confidence level of 20%.

In [40], a power system control device recorded a number of diesel engine thermal parameters, including the master cylinder exhaust temperature, the average exhaust temperature, the scavenging air temperature, the main bearing oil outlet temperature, and the cylinder coolant outlet temperature. Using MATLAB rules, the linking system failures (nozzle, air cooler, oil system, cooling system) and the thermal characteristics connected with engine defaults were created.

Warranty data is a crucial factor in vehicle improvement issues. For example, in [41] the authors used failure statistics accumulated over the last three years on a warranty for heavy excavator equipment and diesel engines to identify correlation patterns between the manufacturing process, the defaults that occurred during the guarantee period, and the subsequent failure series. They used SAS Enterprise Miner 6.2 software to find these patterns.

The algorithm in [42] used the concept of basis sets and database manipulation techniques to establish meaningful associations between vehicle characteristics and failure causes. The association rule represents these relationships, where the first part of the rule contains a set of attributes representing the data about vehicles (production date, repair date, mileage, transmission, engine type) and the second part of the rule contains the data representing the fault codes associated with failures. Because consistent exploration of warranty data can be very useful for product manufacturers, the authors went further and used association rules in their next study to look for patterns and correlations between follow-up requests to warranty service centers [43].

The future of the automotive industry is linked to the development of smart connected cars, with different reliability requirements than classic cars. As a result, more information in the form of fault codes can be stored in the vehicle's onboard systems. In [44], the authors used the Apriori intelligent association rule analysis algorithm to find and improve the reliability of the fault code rules that result in component failures. Additionally in [45], the authors used association rules to establish the relationship between the diagnostic trouble codes generated during the car's operation and stored in the memory bus, and the codes for repair actions that were taken to eliminate the malfunctions. Previously, they cleared the initial array of training data from anomalous records data on repairs either with a duration that far exceeded the standards, or that provoked repeated visits to service enterprises due to poor-quality troubleshooting, or was associated with unreasonable overspending of spare parts. They implemented their methodology in the prototype of the system, which used a distributed web client-server architecture. We observed the development of this work in [46]. It presented a system of ontology built on the basis of association rules that were applied not only to the database of encoded data on diagnostic faults codes and the date of their fixation, and the work performed, but also to a textual description of the repair, including complaints from the client, regarding used spare parts. The author built three groups of rules: combinations of failed parts, combinations of observed failures symptoms and detected defective parts, and combinations of failure signs for the parts and repair methods applied.

Thus, in many cases, the researchers aimed to find sequential patterns that were designed to identify the relationship between the production conditions and failures and to determine the failures series. However, insufficient consideration has been paid to the diagnostic process, which involves identifying the subcomponents that are likely to fail simultaneously, when examining the interaction of the joint components. Furthermore, when analyzing the works, the researcher only used statistical and analytical platforms to obtain and derive the dependencies, making it difficult to apply the obtained results to real service company situations.

In [47], the authors developed a desktop application in Visual Studio IDE that used, for the analytical process, data on the vehicle name, the vehicle part code, and the faulty part code. These association rules explained some extent the intrinsic correlation between the vehicle assembly components and the faulty parts, but they are not applicable to direct vehicle diagnosis. Therefore, the goal of this study is to develop an artificial intelligence-based methodology that can identify possible failures in the assembled parts, in a timely manner, for direct diagnosis by a repairman.

To achieve this goal, the following issues needed to be addressed:

- Select quality criteria and a rule generation algorithm, determine the requirements for the input data structure and prepare statistical data on the technical defects in the vehicles, including information on the defective parts;
- Design and develop a component reliability analysis web service that allows the generation of association rules interactively and generates a list of parts to be inspected;
- Generate association rules and analyze their quality.

3. Materials and Methods

3.1. The Essence of the Association Rules Method and Selected Quality Criteria

As applied to our subject area, associative rules are sets of vehicle parts and components that often turn out to be defective together. The simultaneity of the failure is fixed at the time the vehicle enters the service center and is registered with a specific document: a reclamation act if the vehicle is under warranty, and a work order if the vehicle is in the post-warranty period. The goal of an association rules is to find all frequent sets of parts above a user-defined support threshold and to generate all association rules above different confidence thresholds. The result is a set of related entities called condition M (antecedent) and result N (consequent), written M→N ("M follows N"). Thus, an association rule is expressed in the form "If a condition–default part, then the result–default part is". The quality of each rule is evaluated by its degree of support, confidence, and lift. This choice of criteria was due to the capabilities of the Accord.NET machine learning framework, which was used to develop the module for generating association rules.

The support is defined as the ratio of the number of transactions in which the condition (M) and the result (N) of the rule occur simultaneously to the total number of transactions in the database (W):

$$S(M{\rightarrow}N) = \text{Number } (M \cap N)/W \tag{1}$$

In our case, a transaction is understood as a document containing information about the identified defective parts (reclamation act or work order). The degree of support can take values from 0 to 1.0, which is equivalent to a range of 0–100%. If the support value is $S(M{\rightarrow}N) = X\%$, it means that $X\%$ of the documents in the database contain a combination of M and N. Thus, the higher the support value, the more often the rule occurs. However, it should be borne in mind that the database usually contains a large number of documents (from several thousand to millions). Therefore, even a small value of support (tenths or hundredths of a percent) is equivalent to a significant value of the absolute number of documents: tens, hundreds, and thousands of examples of the simultaneous appearance of a condition and a result.

The confidence is the ratio of the number of documents containing the condition and the result is the number of documents containing only the condition:

$$C(M{\rightarrow}N) = P\,(M \cap N)/P(M) \tag{2}$$

The confidence level can also take values from 0 to 1.0, which is equivalent to a range of 0–100%. If the support value is $C(M{\rightarrow}N) = Y\%$, this means that the document containing the condition M also contains the consequence N in $Y\%$ of cases.

The lift is the ratio of the product of the frequency with which a condition and result occur simultaneously and the frequency with which they occur separately:

$$L(M{\rightarrow}N) = P\,(M \cap N)/(P(M){\cdot}P(N)) \tag{3}$$

The lift over 1 is an indication that conditions and results are more likely to occur together in the document than independently. That means, the occurrence of the condition default part in a given claim is more likely to lead to the result default part, and vice versa. If the value of the lift is $L = Z$, this means that in the document containing the condition M, the result N will occur Z times more often than any other component. A rule with a lift value higher than 1 may be regarded as significant. The lift below 1 is an indication that conditions and results occur individually more often than they occur together in a transaction. In other words, in this case, there is an "anti-rule" where the occurrence of a condition has a negative impact on the occurrence of the outcome. Finally, a high value close to 1 indicates that the conditions and outcomes occur together with the same frequency as they occur separately in a transaction. This implies that the conditions and outcomes do not affect each other's occurrences.

3.2. Algorithm for Generating Association Rules

The general approach to generating association rules consists of two enlarged steps. During the first step, sets that occur with a frequency that belongs to the range ($support_{minimum}$, $support_{maximum}$) are formed. During the second step, those sets whose confidence and lift do not belong to the ranges ($confidence_{minimum}$, $confidence_{maximum}$) and [$lift_{minimum}$, $lift_{maximum}$) are deleted.

The ancestor of rule extraction algorithms is the Apriori algorithm [48]. A significant drawback is its low productivity. A complete enumeration of combinations requires a large amount of computational resources and does not allow the quick generation of rules that satisfy the given hyperparameters. Therefore, subsequently researchers have proposed various modifications to this algorithm (AprioriTID and AprioriHybrid) and other methods (FP growth [49], ECLAT [50]) to build efficient processes for extracting frequent sets of elements for association rules.

The work [51], presents the UniqAR algorithm, which allows the generation of unique classification association rules that allow one single class label to be generated and, thus, has 100% confidence. However, as noted in the book [52], "strong rules are not necessarily interesting". After the elimination of rules lower than the given level of support and confidence, we get the so-called strong rules. Indeed, at high levels of support and confidence the resulting rules are generally trivial, and the resulting combinations are well known to the road transport operating specialist. Therefore, the task of selecting the minimum and maximum levels of support, confidence and lift is, in fact, an additional research task in finding the optimal hyperparameters of the association rules algorithm.

Another hyperparameter that affects the speed of work and the composition of the resulting association rules is the power of the rule. This is the number of objects (in our case, parts) that are included in the rule. High values of power, support, confidence, lift, on the one hand, provide the generation of more rules and, on the other hand, require more computing resources, so it is necessary to find a balance between the values of the rule quality indicators and the speed. Moreover, as mentioned above, more rules do not mean they are better.

For our research, we decided to use the Apriori algorithm, which is built into the Accord.NET machine learning framework. On the one hand, it has simplicity, on the other hand, there is no reason to reject it at the beginning of the development of a method for generating combinations of vehicle parts that fail together.

3.3. Technique for Detecting Joint Defective Parts

To achieve the research goal, a methodology for detecting and using defective joint components based on association rules was developed. To generate association rules using the Apriori algorithm, it is necessary to prepare a data set for training. This data set should contain records in the form presented in Table 1.

Table 1. Structure of the initial data.

ID	DEFECTED_DETAIL
CLAIM _1	DEFECTED_DETAIL_1
CLAIM _1	DEFECTED_DETAIL_2
CLAIM _1	DEFECTED_DETAIL_3
. . .	. . .
CLAIM _J	DEFECTED_DETAIL_1
CLAIM _J	DEFECTED_DETAIL_2
CLAIM _J	. . .
CLAIM _J	DEFECTED_DETAIL_I

CLAIM is understood as the number of the document that includes the results of diagnosing a vehicle when it is contacted by a service center (reclamation report or work order). In fact, such a structure of initial data is easy to obtain from the information system database of a vehicle service enterprise, where information about customer requests and repair results is entered.

Further training is carried out, as a result of which an array of rules is obtained. The structure of the rules array with a power equal to two has the form presented in Table 2.

Table 2. Structure of the resulting association rules.

ANTECEDENT	CONSEQUENT	SUPPORT, pcs.	SUPPORT*, %	CONFIDENCE, %	LIFT
DEFECTED_DETAIL_1	DEFECTED_DETAIL_2	S_1	S^*_1	C_1	L_1
DEFECTED_DETAIL_2	DEFECTED_DETAIL_4	S_2	S^*_2	C_2	L_2
DEFECTED_DETAIL_3	DEFECTED_DETAIL_1	S_3	S^*_3	C_3	L_3
...	...	...	...	...	...

After the formation of such a rule base, it can be used as a recommendation system. When the client enters, the diagnostician starts the inspection and discovers the defective part. Then the rule base is searched for matches with the ANTECEDENT column and one or more rule consequences that include this part as an antecedent are returned. After evaluating the quality criteria, the service specialist can rank the order for the further checking of parts from this list according to descending quality indicators. Thus, the probability that all the defective parts will be detected increases, and the time of the defective process is also reduced.

4. Results

4.1. The Structure and Algorithms of the Recommendation Web Service

We have developed an AutoAnalytics web service for facilitating the validation of the proposed methodology. It uses a knowledge base of patterns obtained by applying the association rules method to vehicle failure statistics.

The following components were created for the web service:

- A rule generation component that runs in the background on a schedule;
- A component that implements the proposed recommendation mechanism in the form of a preliminary list of faults.

Association rule generation is computationally intensive and, therefore, runs in the background on a schedule.

To generate association rules, data from tables, such as part numbers, work orders, claim reports, etc., are needed. The CLAIM_NUMBER field is used as the transaction (event) number. The DETAIL field is used as data for analysis (Figure 2). The rules are generated according to the methodology described in Section 3.3.

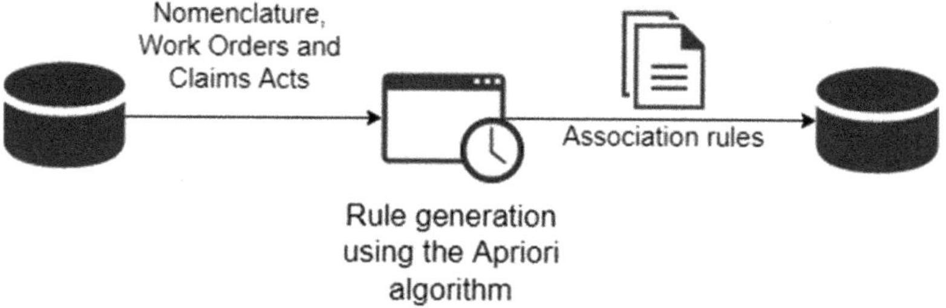

Figure 2. Block diagram of the knowledge base generation module.

To extract recommendations, we needed information about the claim and its defective parts, as well as data obtained from the database tables: nomenclature, association Rules (Figure 3). Users send the identifier of the found defective part in a request to the application server, which searches in the records of the association rules table filled in after the work of the previous module, where this part is included as a condition, and gives the client a list of selected parts that are included in the rules as consequences.

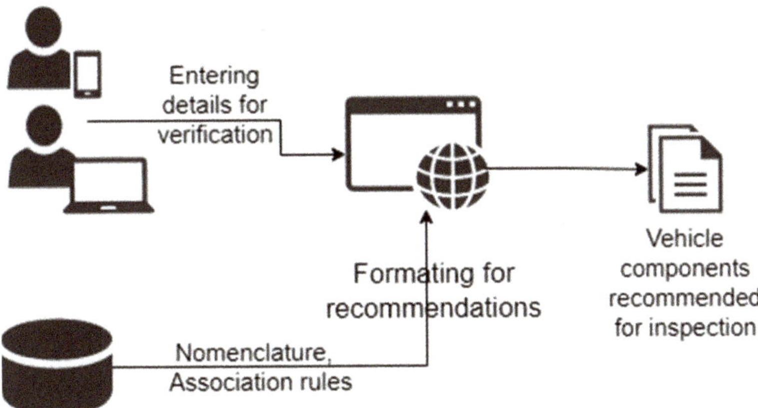

Figure 3. Block diagram of the generating recommendations module.

An appropriate database schema was developed (Figure 4) to store the generated association rules, for the purpose of analyzing the dynamics of changes in combinations of defective assemblies, taking into account the type of vehicle and operating zone. For this, the regions_id, model_id fields and the region and model tables are provided.

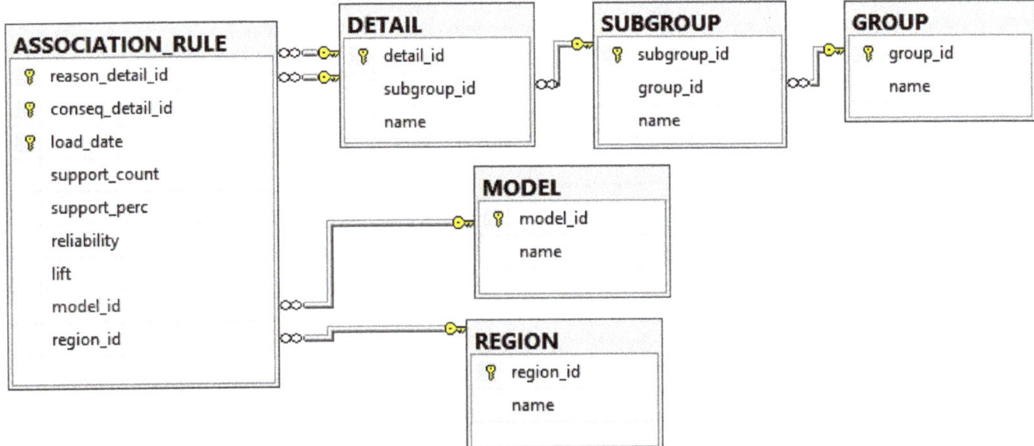

Figure 4. Database schema.

The solution was built using a three-tier architecture: database, server, and user interface (Figure 5). As a database management system, we chose PostgreSQL, which is open source and distributed free of charge. We also chose a combination of C# and ASP.NET frameworks because of their suitability for quickly building web applications. Entity Framework Core was used to speed up the development, and simplify and speed up the database operations. When implementing the front-end of the site, HTML and CSS were involved. In addition, JavaScript was used for the site dynamics. To implement association rules, we used the Accord.NET framework, a machine learning platform written entirely in C#. This library allowed us to integrate most of the business logic into the application itself, which made the service architecture easier and simpler.

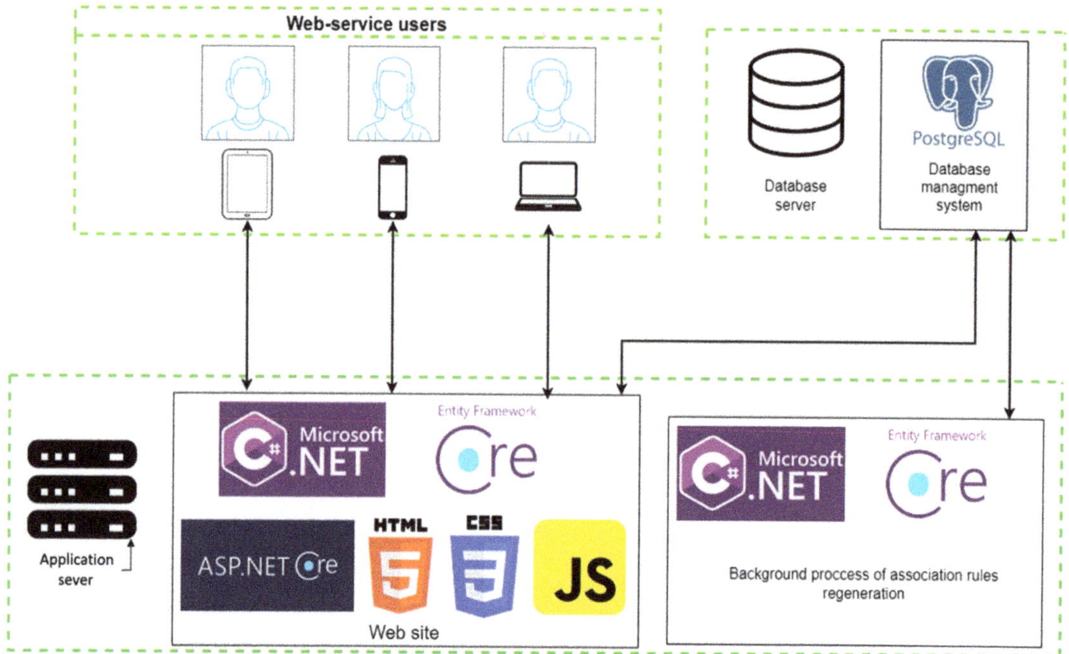

Figure 5. Web service architecture and technology stack.

When a fault is detected, the employee enters the corresponding node into the fault list and a list of nodes that are recommended to be checked will be available. For the convenience of the user, a three-stage input of already detected defective parts is implemented: first, the group to which the part belongs should be selected, then the subgroup, and only then the part itself should be selected. For example, if it was found that the pressure sensor is faulty, then, accordingly, we must first select the devices group and the pressure gauges subgroup. The list of parts recommended for inspection is displayed in a similar way. This mechanism is similar to the hierarchical way of organizing the nomenclature directory and is familiar to a warehouse or car service worker. Several assemblies can be added to the set of faults. A list of parts recommended for inspection is sorted according to descending rule validity, where the detected defective parts are listed as "antecedent" (Figure 6). The set of recommended parts is updated as they are added.

By implementing this tool as a web service adapted to mobile platforms, we enable maintenance professionals to use their mobile devices to obtain up-to-date information as they work.

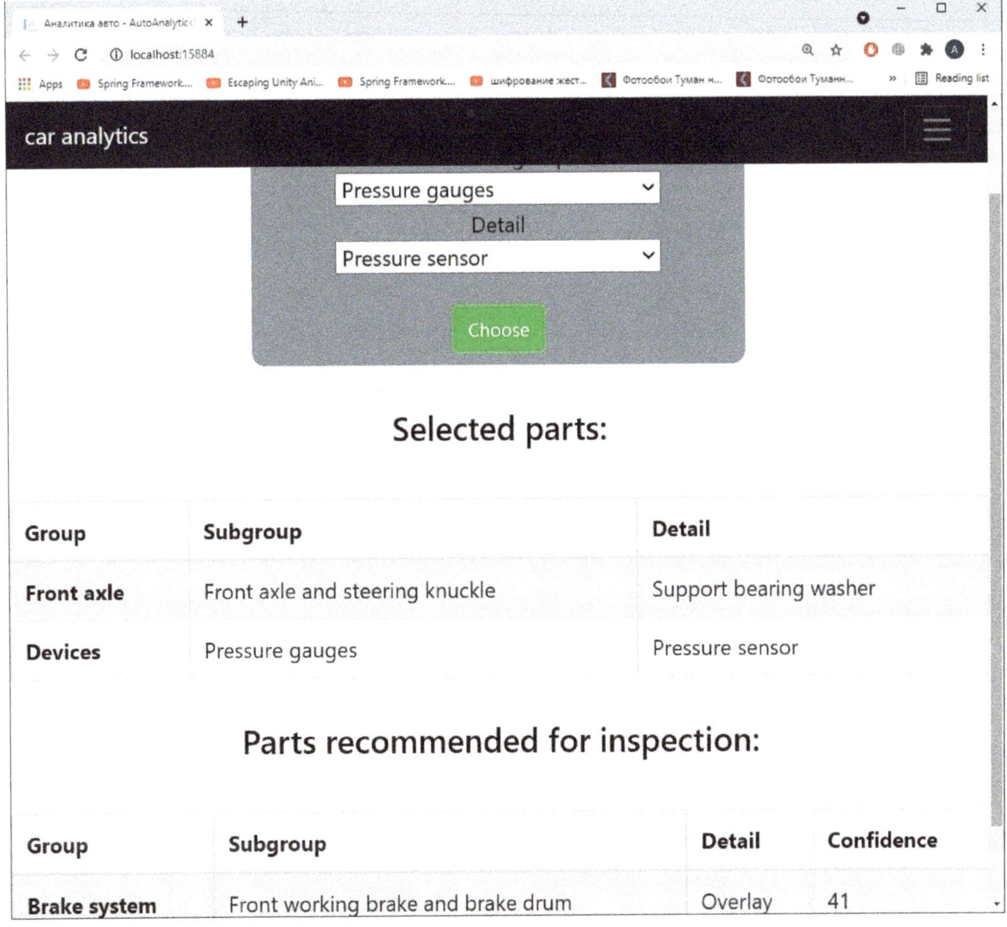

Figure 6. AutoAnalytics web service interface.

4.2. Testing a Web Recommendation Service: Derived Association Rules

To test this technique, we used real failure statistics for the years 2008–2009, containing 3126 records of failed parts in vehicles that were under warranty service and had a mileage of 1000 km to 10,000 km. This fleet of vehicles operated on the territory of the second category of operating conditions (R1-T4, R2-T1, T2, T3, T4, R3-T1, T2, T3), which means:

Road surfaces:

- R1—improved capital cement concrete, monolithic, reinforced concrete or reinforced prefabricated, asphalt concrete, paving stones and mosaics on a concrete base;
- R2—improved lightweight from crushed stone, gravel and sand treated with binders from cold asphalt concrete;
- R3—transitional crushed stone and gravel.

Terrain type (determined by height above sea level):

- T1—flat, up to 200 m;
- T2—slightly hilly, over 200 to 300 m;
- T3—hilly, over 300 to 1000 m;
- T4—mountainous, over 1000 to 2000 m.

As a result, we obtained a number of non-trivial useful rules. They included rules linking the components of the brake system and transmission system, the electrical system, the electrical system and brake system, and the engine cooling system (Table 3). The confidence level of these rules was high and reached 77%. Although some rules have a 20% confidence level, the high level of lift, which is significantly higher than 1, allowed us to talk about the usefulness of these rules.

Table 3. Obtained meaningful association rules.

Rule Number	Antecedent, Group	Antecedent, Subgroup	Antecedent, Detail	Support, %	Confidence	Lift
	Consequent, Group	Consequent, Subgroup	Consequent, Detail			
1	Brake system	Front working brake and brake drum	Cover	0.61	76.9	52.5
	Front axle	Front axle and steering knuckle	Support bearing disc			
2	Front axle	Front axle and steering knuckle	Support bearing disc	0.61	41.7	52.5
	Brake system	Front working brake and brake drum	Cover			
3	Brake system	Bypass brake valve	Dual-magister valve	0.26	26	9.8
	Devices	Oil pressure gauge	Gauge			
4	Devices	Tyre pressure gauge	Pressure gauge	0.14	21.2	8
	Devices	Pressure gauge	Gauge			
5	Electrical equipment	Generator	Relay regulator	0.10	41.7	28.8
	Devices	VK403B	Reversing light switch			
6	Cooling system	Fan and its drive	Electromagnetic clutch engagement sensor	0.08	26.6	17.2
	Cooling system	Thermostat	Thermostat			

The wear of the brake system components leads to the appearance of vibrations during braking. In this case, there is accelerated wear of the mating elements of the suspension and steering of the vehicle, namely, as follows from rule 1, the support bearing disc. Rule 2 is a mirror of rule 1, which confirms the relationship between the brake system and the running gear.

The analysis of rule 4 did not reveal the relationship between the elements, even indirectly. During troubleshooting, it was revealed that the air pressure drop indicator in the third circuit was on. Inspection of the electrical circuit indicated a failure in the emergency air drop sensor. At the same time, the emergency oil pressure alarm in the engine was constantly on due to a failure in the emergency oil pressure sensor. In both cases, the sensors were found to be replaceable.

According to rule 3, there should have been a connection between the failure of the dual-magister valve, which was detected in the slow release of the parking brake system, and the failure of the emergency oil pressure gauge, which manifested itself in the constant burning of the emergency oil pressure alarm in the engine. However, the analysis of the design does not provide for the identification of patterns between them.

A malfunction in the relay regulator in the generator often contributes to an increase in the voltage of the onboard electrical network, which in turn leads to accelerated wear of its electrical components. This explains the high reliability of rule 5, which in this case was 41.7%, this is the proportion of cases when the relay regulator fails, the reversing light switch is found to be defective.

If the electromagnetic clutch engagement sensor fails, the temperature regime of the engine increases, and the likelihood of it overheating increases. This can explain the accelerated degradation of the thermostat element and rule 6.

Thus, the selected composition of transactions does not allow us to explain the patterns found in all cases. This can be explained either by the absence of such a relationship or by the need to include additional information about both the failures themselves and the vehicles, and the characteristics of the components.

The described method is designed primarily for staff in vehicle repair and maintenance posts. Its use can accelerate the vehicle diagnostic process by providing employees with recommended vehicle parts to check.

It also allows purchasing departments to use well supported rules to select the most promising suppliers and manufacturers, and to develop tactical solutions to eliminate less reliable part suppliers. In addition, the structure of the association rules themselves can be dynamically analyzed to assess the impact of changes made to the design of individual vehicle model parts, as well as the design and manufacture of technology.

5. Conclusions

Manufacturers of modern intelligent vehicles are forced to look for new tools and techniques for ensuring the smooth operation of their customers' automobiles. The diagnostic decision support methodology described, based on the relationships obtained, can be used by workers with little experience or qualifications. During the next visit by the vehicle to the service center for maintenance or repair, the employee enters default parts and receives a set of parts made by the application for checking. The aforementioned methodology and the web application developed will improve the quality of and speed up the diagnostics. In addition, designers and engineers can use association rules to analyze possible causes of defective parts based on defects that commonly occur. Automotive manufacturers' purchasing departments can also use such a rules base to select parts suppliers.

Future research will focus on developing the proposed methodology to update the rules base in real-time after each repair and to provide information on the results of defect inspection and the process of repairing the vehicle. Obviously, it needs to investigate the speed of different algorithms for generating association rules.

Author Contributions: Conceptualization, P.B.; methodology, P.B.; formal analysis, I.M., A.V. and A.K.; investigation, P.B., I.M., A.V. and A.K.; resources, P.B. and A.V.; software, A.V. and A.K.; writing—original draft preparation, P.B. and A.V.; supervision, P.B.; writing—review and editing, P.B., I.M., A.V. and A.K.; visualization, P.B. and A.V. All authors have read and agreed to the published version of the manuscript.

Funding: This research received no external funding.

Data Availability Statement: Data acquisition can be discussed with the corresponding author.

Conflicts of Interest: The authors declare no conflict of interest.

References

1. James, A.T.; Gandhi, O.P.; Deshmukh, S.G. Fault diagnosis of automobile systems using fault tree based on digraph modeling. *Int. J. Syst. Assur. Eng. Manag.* **2018**, *9*, 494–508. [CrossRef]
2. Ryabinin, I.A.; Strukov, A.V. Quantitative examples of safety assessment using logical-probabilistic methods. *Int. J. Risk Assess. Manag.* **2018**, *21*, 4–20. [CrossRef]
3. Makarova, I.; Mukhametdinov, E.; Mavrin, V. Unified information environment role to improve the vehicle reliability at life cycle stages during the transition to industry 4.0. In Proceedings of the 2019 12th International Conference on Developments in eSystems Engineering (DeSE), Kazan, Russia, 7–10 October 2019; pp. 800–805.
4. Khabibullin, R.G.; Makarova, I.V.; Belyaev, E.I.; Suleimanov, I.F.; Pernebekov, S.S.; Ussipbayev, U.A.; Junusbekov, A.S.; Balabekov, Z.A. The study and management of reliability parameters for automotive equipment using simulation modeling. *Life Sci. J.* **2013**, *10*, 828–831.
5. Yadav, O.P.; Singh, N.; Goel, P.S.; Itabashi-Campbell, R. A Framework for Reliability Prediction During Product Development Process Incorporating Engineering Judgments. *Qual. Eng.* **2003**, *15*, 649–662. [CrossRef]
6. Makarova, I.; Shubenkova, K.; Buyvol, P.; Shepelev, V.; Gritsenko, A. The Role of Reverse Logistics in the Transition to a Circular Economy: Case Study of Automotive Spare Parts Logistics. *FME Trans.* **2021**, *49*, 173–185. [CrossRef]
7. Makarova, I.; Buyvol, P.; Mukhametdinov, E.; Pashkevich, A. Risk analysis in the appointment of the trucks' warranty period operation. In *Advances in Intelligent Systems and Computing, Proceedings of the 39th International Conference on Information Systems Architecture and Technology–ISAT 2018: Part III, Nysa, Poland, 16–18 September 2018*; Springer International Publishing: Cham, Switzerland, 2019; Volume 854, pp. 293–302.
8. Gritsenko, A.; Shepelev, V.; Zadorozhnaya, E.; Shubenkova, K. Test diagnostics of engine systems in passenger cars. *FME Trans.* **2020**, *48*, 46–52. [CrossRef]

9.   Drakaki, M.; Karnavas, Y.L.; Tzionas, P.; Chasiotis, I.D. Recent Developments Towards Industry 4.0 Oriented Predictive Maintenance in Induction Motors. *Procedia Comput. Sci.* **2021**, *180*, 943–949. [CrossRef]

10.   Alpaydin, E. *Introduction to Machine Learning*; MIT Press: Cambridge, MA, USA, 2020; 712p.

11.   Tian, J.; Wang, D.; Chen, L.; Zhu, Z.; Shen, C. A stable adaptive adversarial network with exponential adversarial strategy for bearing fault diagnosis. *IEEE Sens. J.* **2022**, *22*, 9754–9762. [CrossRef]

12.   Shi, Q.; Zhang, H. Fault Diagnosis of an Autonomous Vehicle With an Improved SVM Algorithm Subject to Unbalanced Datasets. *IEEE Trans. Ind. Electron.* **2021**, *68*, 6248–6256. [CrossRef]

13.   Jing, R.; Green, M.; Huang, X. Chapter 8—From traditional to deep learning: Fault diagnosis for autonomous vehicles. In *Learning Control: Applications in Robotics and Complex Dynamical Systems*; Elsevier: Amsterdam, The Netherlands, 2021; pp. 205–219.

14.   Kumar, P.; Perrollaz, M.; Lefevre, S.; Laugier, C. Learning-based approach for online lane change intention prediction. In Proceedings of the 2013 IEEE Intelligent Vehicles Symposium (IV), Gold Coast, QLD, Australia, 23–26 June 2013; pp. 797–802.

15.   Yi, H.; Edara, P.; Sun, C. Modeling mandatory lane changing using Bayes classifier and decision trees. *IEEE Trans. Intell. Transp. Syst.* **2013**, *15*, 647–655. [CrossRef]

16.   Shi, Q.; Zhang, H. An improved learning-based LSTM approach for lane change intention prediction subject to imbalanced data. *Transp. Res. Part C Emerg. Technol.* **2021**, *133*, 103414. [CrossRef]

17.   Ngai, E.W.T.; Xiu, L.; Chau, D.C.K. Application of data mining techniques in customer relationship management: A literature review and classification. *Expert Syst. Appl.* **2009**, *36*, 2592–2602. [CrossRef]

18.   Ali Alan, M.; Ince, A.R. Use of Association Rule Mining within the Framework of a Customer-Oriented Approach. *Eur. Sci. J.* **2016**, *12*, 81–99. [CrossRef]

19.   Lin, R.-H.; Chuang, W.W.; Chuang, C.L.; Chang, W.S. Applied Big Data Analysis to Build Customer Product Recommendation Model. *Sustainability* **2021**, *13*, 4985. [CrossRef]

20.   Yusupbekov, N.R.; Gulyamov, S.M.; Usmanova, N.B.; Mirzaev, D.A. Estimation of software reliability based on association rules. *Math. Methods Eng. Technol.* **2017**, *1*, 134–138.

21.   Yakupova, G.A.; Makarova, I.V.; Buyvol, P.A.; Mukhametdinov, E.M. Method of association rules in the analysis of road traffic accidents. *Transp. Sci. Technol. Manag.* **2020**, *11*, 40–44. (In Russian) [CrossRef]

22.   Introduction to the Analysis of Association Rules. Available online: https://loginom.ru/blog/associative-rules (accessed on 1 February 2023).

23.   Makarova, I.; Yakupova, G.; Buyvol, P.; Mukhametdinov, E.; Pashkevich, A. Association rules to identify factors affecting risk and severity of road accidents. In Proceedings of the 6th International Conference on Vehicle Technology and Intelligent Transport Systems (VEHITS), Online, 2–4 May 2020; pp. 614–621.

24.   Chen, Z.; Ordonez, C.; Zhao, K. Comparing Reliability of Association Rules and OLAP Statistical Tests. In Proceedings of the 2008 IEEE International Conference on Data Mining Workshops, Pisa, Italy, 15–19 December 2008; pp. 8–17.

25.   Jiao, M.; Tang, J.; Xu, J. Evaluation of supplier reliability based on the association rule and AHP method. In Proceedings of the 2008 Chinese Control and Decision Conference, Yantai, China, 2–4 July 2008; pp. 2266–2270.

26.   Tjortjis, C.; Layzell, P.J. Using data mining to assess software reliability. In Proceedings of the 12th International Symposium on Software Reliability Engineering, Hong Kong, China, 27–30 November 2001; pp. 221–223.

27.   Jia, X.; Zhang, D. Prediction of maritime logistics service risks applying soft set based association rule: An early warning model. *Reliab. Eng. Syst. Saf.* **2021**, *207*, 107339. [CrossRef]

28.   Lesage, L.; Deaconu, M.; Lejay, A. A recommendation system for car insurance. *Eur. Actuar. J.* **2020**, *10*, 377–398. [CrossRef]

29.   Jeong, H.; Gan, G.; Valdez, E.A. Association Rules for Understanding Policyholder Lapses. *Risks* **2018**, *6*, 69. [CrossRef]

30.   Katba, C. Automobile Options Association Rule Mining (Data Mining). Available online: https://katba-caroline.com/automobile-options-association-rule-mining-data-mining (accessed on 12 June 2022).

31.   Liu, G.; Peng, C. Research on Reliability Modeling of CNC System Based on Association Rule Mining. *Procedia Manuf.* **2017**, *11*, 1162–1169. [CrossRef]

32.   Zhang, R.; Jia, Y.; Sun, D. Application of Data Mining in CNC Equipment Reliability Analysis. *J. Jilin Univ.* **2007**, *6*.

33.   Jie, X.; Wang, H.; Fei, M.; Du, D.; Sun, Q.; Yang, T.C. Anomaly behavior detection and reliability assessment of control systems based on association rules. *Int. J. Crit. Infrastruct. Prot.* **2018**, *22*, 90–99. [CrossRef]

34.   Chen, B.; Qin, H.; Li, X. A Data Mining Method for Extracting Key Factors of Distribution Network Reliability. In Proceedings of the 2018 2nd IEEE Conference on Energy Internet and Energy System Integration (EI2), Beijing, China, 20–22 October 2018; pp. 1–5.

35.   Hui, Z.; Bi-bo, J.; Zhuo-qun, Z. Fault diagnosis of industrial boiler based on competitive agglomeration and fuzzy association rules. In Proceedings of the 2010 International Conference on Computer, Mechatronics, Control and Electronic Engineering, Changchun, China, 24–26 August 2010; pp. 64–67.

36.   Pan, D. Hybrid data-driven anomaly detection method to improve UAV operating reliability. In Proceedings of the 2017 Prognostics and System Health Management Conference (PHM-Harbin), Harbin, China, 9–12 July 2017; pp. 1–4.

37.   Rezgui, J.; Cherkaoui, S. Detecting faulty and malicious vehicles using rule-based communications data mining. In Proceedings of the 2011 IEEE 36th Conference on Local Computer Networks, Bonn, Germany, 4–7 October 2011; pp. 827–834.

38.   Moharana, U.C.; Sarmah, S.P.; Rathore, P.K. Application of data mining for spare parts information in maintenance schedule: A case study. *J. Manuf. Technol. Manag.* **2019**, *30*, 1055–1072. [CrossRef]

39. Li, Y.; Du, X.; Yang, B. Application of Association Analysis and Visualization Methods in Car Parts Repair. In *Proceedings of the Joint International Mechanical, Electronic and Information Technology Conference*; Atlantis Press: Amsterdam, The Netherlands, 2015; pp. 1136–1141.

40. Cao, M.; Guo, C. Research on the Improvement of Association Rule Algorithm for Power Monitoring Data Mining. In Proceedings of the 2017 10th International Symposium on Computational Intelligence and Design (ISCID), Hangzhou, China, 9–10 December 2017; pp. 112–115.

41. Jeon, J.; Sohn, S.Y. Product failure pattern analysis from warranty data using association rule and Weibull regression analysis: A case study. *Reliab. Eng. Syst. Saf.* **2015**, *133*, 176–183. [CrossRef]

42. Buddhakulsomsiri, J.; Siradeghyan, Y.; Zakarian, A.; Li, X. Association rule-generation algorithm for mining automotive warranty data. *Int. J. Prod. Res.* **2006**, *44*, 2749–2770. [CrossRef]

43. Mokhtari, K.; Ren, J.; Roberts, C.; Wang, J. Application of a generic bow-tie based risk analysis framework on risk management of sea ports and offshore terminals. *J. Hazard. Mater.* **2011**, *192*, 465–475. [CrossRef]

44. Singh, K.; Shroff, G.; Agarwal, P. Predictive reliability mining for early warnings in populations of connected machines. In Proceedings of the 2015 IEEE International Conference on Data Science and Advanced Analytics (DSAA), Paris, France, 19–21 October 2015; pp. 1–10.

45. Chougule, R.; Rajpathak, D.; Bandyopadhyay, P. An integrated framework for effective service and repair in the automotive domain: An application of association mining and case-based-reasoning. *Comput. Ind.* **2011**, *62*, 742–754. [CrossRef]

46. Rajpathak, D.G. An ontology based text mining system for knowledge discovery from the diagnosis data in the automotive domain. *Comput. Ind.* **2013**, *64*, 565–580. [CrossRef]

47. Lei, Z.; Zi-dong, Z.; Xiao-dong, W.; Bin, S. The Applied Research of Association Rules Mining in Automobile Industry. In Proceedings of the 2009 WRI World Congress on Computer Science and Information Engineering, Los Angeles, CA, USA, 31 March–2 April 2009; pp. 241–245.

48. Agrawal, R.; Imieli´nski, T.; Swami, A. Mining association rules between sets of items in large databases. In Proceedings of the 1993 ACM SIGMOD International Conference on Management of Data, Washington, DC, USA, 25–28 May 1993; Volume 22, pp. 207–216.

49. Han, J.; Pei, J.; Yin, Y.; Mao, R. Mining Frequent Patterns without Candidate Generation: A Frequent-Pattern Tree Approach. *Data Min. Knowl. Discov.* **2004**, *8*, 53–87. [CrossRef]

50. Zaki, M.J.; Parthasarathy, S.; Ogihara, M.; Li, W. Parallel Algorithms for Discovery of Association Rules. *Data Min. Knowl. Discov.* **1997**, *1*, 343–373. [CrossRef]

51. Nasr, M.; Hamdy, M.; Hegazy, D.; Bahnasy, K. An efficient algorithm for unique class association rule mining. *Expert Syst. Appl.* **2021**, *164*, 113978. [CrossRef]

52. Han, J.; Kamber, M.; Pei, J. *Data Mining: Concepts and Techniques*, 3rd ed.; A volume in The Morgan Kaufmann Series in Data Management System; Morgan Kaufmann Publishers: Burlington, MA, USA, 2012; 740p.

 *information*

Article

FedUA: An Uncertainty-Aware Distillation-Based Federated Learning Scheme for Image Classification

Shao-Ming Lee [1] and Ja-Ling Wu [1,2,*]

[1] Department of Computer Science and Information Engineering, National Taiwan University, Taipei 10617, Taiwan; gene840802@gamil.com
[2] Graduate Institute of Networking and Multimedia, National Taiwan University, Taipei 10617, Taiwan
* Correspondence: wjl@cmlab.csie.ntu.edu.tw

Abstract: Recently, federated learning (FL) has gradually become an important research topic in machine learning and information theory. FL emphasizes that clients jointly engage in solving learning tasks. In addition to data security issues, fundamental challenges in this type of learning include the imbalance and non-IID among clients' data and the unreliable connections between devices due to limited communication bandwidths. The above issues are intractable to FL. This study starts from the uncertainty analysis of deep neural networks (DNNs) to evaluate the effectiveness of FL, and proposes a new architecture for model aggregation. Our scheme improves FL's performance by applying knowledge distillation and the DNN's uncertainty quantification methods. A series of experiments on the image classification task confirms that our proposed model aggregation scheme can effectively solve the problem of non-IID data, especially when affordable transmission costs are limited.

Keywords: federated learning; model aggregation; knowledge distillation; uncertainty in deep neural networks

Citation: Lee, S.-M.; Wu, J.-L. FedUA: An Uncertainty-Aware Distillation-Based Federated Learning Scheme for Image Classification. *Information* **2023**, *14*, 234. https://doi.org/10.3390/info14040234

Academic Editors: Amar Ramdane-Cherif, Ravi Tomar and TP Singh

Received: 27 February 2023
Revised: 4 April 2023
Accepted: 6 April 2023
Published: 10 April 2023

1. Introduction

The concept of FL was proposed by McMahan et al. in 2016 [1]. Its goal is to complete the training of a global model when target datasets are distributed to different devices (or clients), and the sensitivity of each dataset is of grave concern. In addition, the authors [1] also proposed a federated averaging (FedAvg) algorithm to complete the task of global aggregation, so that each client can complete the model training and keep their data locally. FedAvg prevents users from transmitting sensitive data from their side to the server by uploading the client-side's model gradients to the server instead. Then, the server aggregates the uploaded gradients to build a new global model to protect the privacy and security of every client's local data.

Although FedAvg claims to be able to deal with non-IID data, many studies have pointed out that the accuracy of FedAvg seriously drops if the processed data are non-IID [2,3]. The main reason for the performance degradation is that the non-IID data will cause the weights of the local models to diverge. More precisely, since the loss function of a regular neural network (NN) is non-convex, if FedAvg obtains the global model by conducting the mean operations, it will continuously increase the gap between the obtained result and the ideal model obtained by training on ideal IID datasets, which in turn makes the ensemble unable to converge, and deteriorates the learning performances [4]. In addition, FedAvg cannot fully utilize all of the information provided by clients, such as the inter-client gradient variations.

Currently, the primary methods for dealing with the problem of non-IID data can be divided into three categories: data-based, system-based, and algorithm-based [5]. The data-based category solves the non-IID problem directly and effectively through data sharing [3,6] or data augmentation [7] techniques. However, such methods often violate the spirit of FL because there is a risk of data privacy leakage due to the inability to practice data decentralization securely. In contrast, system-based methods usually use clustering techniques to cluster users for the construction of multi-centric frameworks [8,9], and users in the same group will have similar training data. The adopted data similarity estimation methods can be further divided into two types: estimating the similarity of the loss values and estimating the similarity of the user-end model weights. The realization of algorithm-based methods are very diverse and include regularization [10,11], fine-tuning [12], and personalization layers [13], and these are introduced in user-end training. There are also some standard techniques in machine learning, such as multi-task learning [14], lifelong learning [15], and knowledge distillation [16–20].

Guha et al. proposed DOSFL [16] as a "one-shot" FL architecture. Unlike the model distillation method, this architecture uses the dataset distillation method, in which the client distills the local data and uploads the synthetic data and learning rate to the server. The server combines the synthetic data from the users to train a global model. Jeong et al. proposed the federated distillation (FD) architecture [17], in which users upload the per-label mean logit vectors for each label to speed up communications. In addition, for facing non-IID problems, a federated augmentation (FAug) [17] algorithm is proposed to deal with them. FAug will ask all users to inform the server of the samples they lack, the algorithm will let the server train a GAN, and then then allow the user to download the GAN to expand their local data into IID patterns. Compared with FedAvg, FD, and DOSFL, FAug can significantly reduce communication costs, but the associated accuracy performance is somewhat poor. The architecture of FedMD [18], proposed by Li et al., requires a public dataset. The user first uses the public dataset for general training and the local private data for customized training. In the communication stage, the user uploads the logarithmic probability calculated from the public dataset, and the server averages the logarithmic probability uploaded by all users before learning. Compared with FedMD, FedDF [19] proposed by Lin et al. uses unlabeled data for distillation and transfers the distillation task from the user to the server side. The results show that FedDF has a better robustness in selecting distillation datasets and is suitable for the context of FL. Figure 1 shows the block diagram and information flow of FedDF, which is chosen as the major benchmark for our newly proposed work. Each of the indicated functional modules of Figure 1 will be detailed in the next Section. Chen et al. proposed FedBE [20]. The architecture of FedBE is based on FedDF, and it introduces the Bayesian inference for sampling more models, and applies Bayesian ensembles to obtain better global models. FedBE has been proven effective for resolving non-IID problems, and is compatible with other architectures that normalize user-side models.

As mentioned above, our work was developed based on the architecture of FedDF, but with the following notable characteristics:

1. The server quantifies the network's uncertainty of the uploading client, which serves as the basis for building a more adaptable aggregation scheme to deal with the inhomogeneity of client side models;
2. The server introduces the sample's quality evaluation to effectively sieve through samples to suppress the influences of data uncertainty and improve learning efficiency;
3. As a knowledge distillation aggregation architecture, our work can effectively separate the information of uncertainty and inter-class relationships. This separation helps solve the non-IID data issue and provides a good learning performance while limiting the transmission costs.

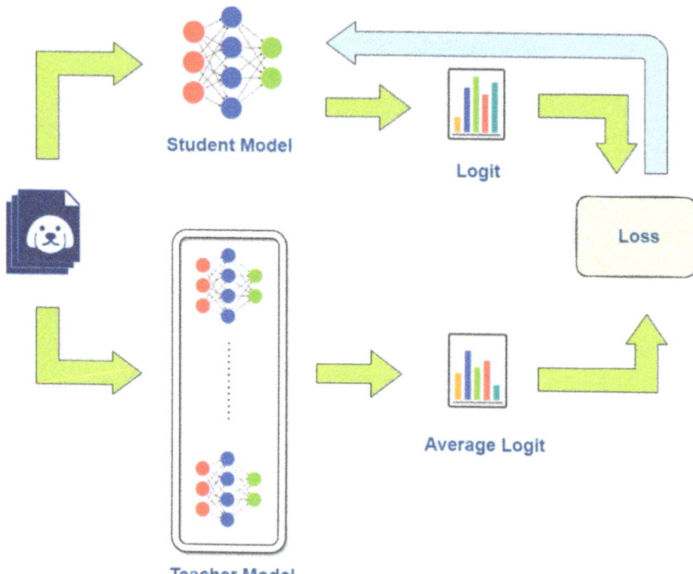

Figure 1. The schematic diagram and information flow of FedDF [19] (The green-color arrows indicates the forward training direction and the blue arrow the backward training one).

2. Preliminary Backgrounds

2.1. Knowledge Distillation

Initially, knowledge distillation was proposed to be applied for model compression [21], where the goal was to compress one or more large models (teacher models) into small models (student models). The resultant small models could effectively learn the so-called "important knowledge" from the pre-trained large models, allowing them to enhance a certain level of effectiveness associated with a specific requirement. Knowledge distillation is generally used to make small models have a better generalization ability. For example, as shown in Figure 2, a knowledge distillation-based classifier can effectively learn inter-class relations (a.k.a. dark knowledge) by regulating the distillation temperature in classification problems.

Knowledge distillation is a promising idea for federated learning. There are two practical reasons to support the above claims: First, it can alleviate over-fitting on the user side: In the context of FL, if users cannot perform model aggregation frequently due to the communication limitation, the differences in models among users will continue to accumulate, and the user model will learn too many useless local data features, compared to the general global IID data assumption in the server, those useless features behave like redundant noises, which in turn handicap the final aggregation result to approximate that of the ideal model. Applying knowledge distillation in the aggregation stage helps the global model sift through informative and valuable information for learning; therefore, it can alleviate the bias caused by incomplete or overtrained data that often occurs on the user side in FL.

As addressed above, knowledge distillation enables the global model to learn the inter-class relationship, which helps transfer the knowledge learned for a general multi-purposed model to a specific target-oriented model; this is the second reason for using knowledge distillation in FL. To dive into the reasoning of this claim in more detail: when the data are not independently and identically distributed, the inter-class relationship learned by the local model may be incomplete and inconsistent. An inappropriate aggregation scheme may not effectively transfer the genuine inter-class relationship to the global model.

Of course, effectively identifying the inter-class relationships suitable for FL becomes a research topic worthy of deeper investigation.

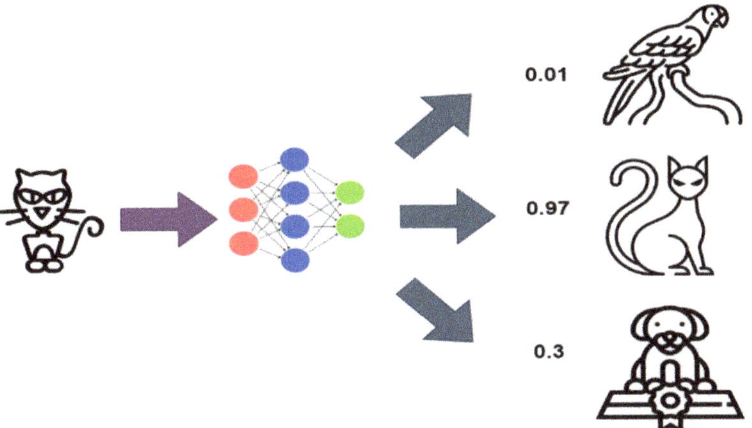

Figure 2. An illustrative example of learning dark knowledge through a knowledgeable teacher model in image classification problems. In this example, although we know that the input image falls in the cat category, we also know that the input has a higher probability of falling into the dog category than into the bird category. The pre-described classification order is the so-called "inter-class relationship".

However, the current distillation-based federated learning architectures have yet to effectively consider all of the advantages mentioned above. The uncertainty-aware distillation-based federated learning (shortened as FedUA) scheme proposed in this paper aims to provide a possible solution to improve the learning effect when both the non-IID data and the limited communication capacity occur at the same time.

2.2. Uncertainties in DNNs

General DNNs cannot express confidence levels; however, displaying confidence levels is increasingly essential for specific application domains, such as safety-critical tasks and medical applications. Therefore, studies on the uncertainty of NNs have also been investigated, including defining the sources of uncertainties in DNNs, quantifying uncertainties through various measures, and constructing correction networks, to name a few.

Generally speaking, the uncertainties of NNs can be divided into the following three types:

1. Data uncertainty—the uncertainty inherent in the data; even with a well-calibrated model, such an uncertainty still exists;
2. Model uncertainty—the model needs to be built with more knowledge. Generally speaking, this kind of uncertainty can be suppressed by improving the training process or calibrating the model;
3. Distributional uncertainty—the uncertainty of the distribution prediction itself. From another viewpoint, such an uncertainty can be an essential basis for out-of-distribution detection [22–25]. Figure 3 shows the classification of the uncertainties of NNs. We refer interested readers to find the detailed definition of each uncertainty class in [22].

Data uncertainty will manifest in the final forecast, such as estimating the outputs of a normalized exponential function (a.k.a. the Softmax output) for a classification task or the standard deviation of the predictions for a regression task. However, studies have found that NNs often suffer from overconfidence, and the normalized exponential function output is often poorly calibrated [26–28], resulting in imprecise uncertain estimates.

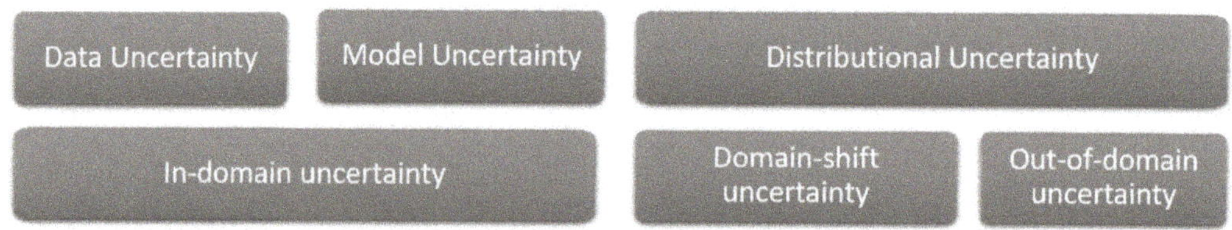

Figure 3. The Classification of the uncertainties of neural networks.

For FL, the adverse effects are more pronounced and trickier when the client data causes distributional uncertainty. Therefore, quantifying uncertainty is undoubtedly a critical information basis for model aggregation. This work investigates an FL architecture based on knowledge distillation (cf. Figure 4). We use the logarithmic probability extracted by the teacher model as the primary basis (i.e., the confidence measure) for the training process of the student model. If the confidence of the current sample can be effectively calculated under the multi-teacher structure, a more flexible and efficient knowledge transfer can be made successfully. Our proposed FedUA considers the influences of the uncertainties mentioned above and adds uncertainty quantification steps to clarify the global model's training objectives in the aggregation stage.

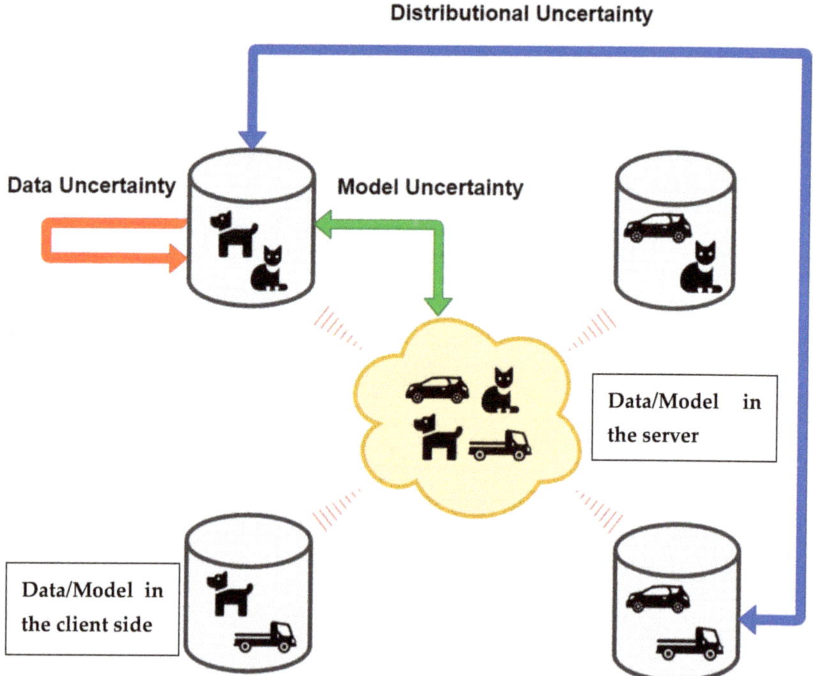

Figure 4. NN Uncertainties under the structure of federated learning.

3. The Proposed Method: FedUA

We designed our FL framework based on the so-called knowledge distillation-aware aggregation scheme to conquer the challenges of non-IID client data and the restricted communication between clients and the server. We add two core functional modules: the uncertainty measurement and the sample quality evaluator to enhance the overall system performance. The following subsections will describe these two modules' operations and architectures in detail.

3.1. Uncertainty Measurement

Due to the settings of FL, each client uses local data as the training dataset of the NN, resulting in a regional model; even under the same training parameters, the same data input may still produce highly inconsistent losses and predictions during the testing phase. The server side is decentralized in data, making it impossible to spy on or dig out which kind of data prediction the user model is good at. We bring the uncertainty measurement into the DNN as a vital basis for conducting the model aggregation process so that the server side can catch the confidence level of each participating user in the prediction generated by the specific input data to generate the subsequent integration results and strengthen the reliability of the global model.

Considering the knowledge distillation-based FL architectures, it is expected that in the aggregation stage, one can use referential information to approach the outcome of an ideal teacher model to perfect the knowledge inheritance. According to the model uncertainty, if an enormous amount of input data belongs to a specific object category in the model learning stage, the trained model should produce higher confidence concerning the output of this category in the inference stage. Regarding the distribution uncertainty, through practical measurements, all of the teacher models can participate in teaching the student models by "making use of their strengths and circumventing their weaknesses," which further enables the server side student models to have a more comprehensive classification ability.

To accommodate the variations of each client's data, we use the Gaussian discriminant analysis for each client to establish a Gaussian mixture model of its characteristic spatial density. Given a set of (X, Y), the establishment method is as follows:

for each class c with samples $X_c \subset X$ **do**

$$w_c \leftarrow \frac{|X_c|}{|X|} \tag{1a}$$

$$\mu_c \leftarrow \frac{1}{|X_c|} \sum_{X_c} f_w(X_c) \tag{1b}$$

$$\sigma_c \leftarrow \frac{1}{|X_c|} \left(f_w(X_c) - \mu_c\right) \left(f_w(X_c) - \mu_c\right)^T \tag{1c}$$

Prior to the model aggregation, a Gaussian mixture model is used to quantify the epistemic uncertainty of the current sample for a specific user-end model. The process is as follows:

$$z \leftarrow f_w(x) \qquad (f : a\ feature\ extractor) \tag{2a}$$

$$p(z) \leftarrow \sum_c w_c\, N(z; \mu_c; \sigma_c) \qquad (N : Gaussian\ model) \tag{2b}$$

For a given user-end model, we input sample x into feature extraction function f to obtain feature vector Z and its corresponding p(z). The feature space density probability of the server side samples associated with the current client side model can now be calculated.

At this time, the uncertainty measurement method we adopted is called the single deterministic model, that aims to reduce the computational burden of the model during training and testing. In addition, we used feature space as the quantization objective instead of the normalized exponential function (i.e., the SoftMax). The reason for this is because under the knowledge distillation-based FL architecture, the inter-class correlation of the data is beneficial to the aggregation model, and this relationship is reflected in the aleatoric uncertainty. Therefore, the aggregation process can exclude the influence of this factor to avoid the occurrence of an objective mismatch. Because of this consideration, we also made a comparative analysis in our experiment.

3.2. Sample Assessment

For typical knowledge distillation, the training data of the student and the teacher models are independently and identically distributed so that the two can achieve an efficient and stable knowledge inheritance. However, considering the situation of many teachers under the structure of FL, the teacher model uploaded by a client is prone to overfitting the local data. We hope that the server aggregation stage can effectively bring the global model towards a more generalized direction to eliminate this shortage.

To achieve the above purpose, we should carefully select the students' training data, so we include a sample evaluator in FedUA (cf. Figure 5) to be responsible for the sample evaluation task. At this stage, we followed the spirit of active learning and select samples with high epistemic uncertainty as the training data for the teacher model. We adopted the Bayesian active learning by disagreement (BALD) technique [29], that quantifies the uncertainty of the samples based on the Bayesian viewpoint, and mathematically it can be written as:

$$I(y; w | x, D) = H(y | x, D) - E_{p(w|D)}[H(y | x, w, D)], \tag{3}$$

where H (x | y) denotes the conditional entropy of x given y, and I (x, y | z) represents the conditional mutual information between x and y given z, respectively.

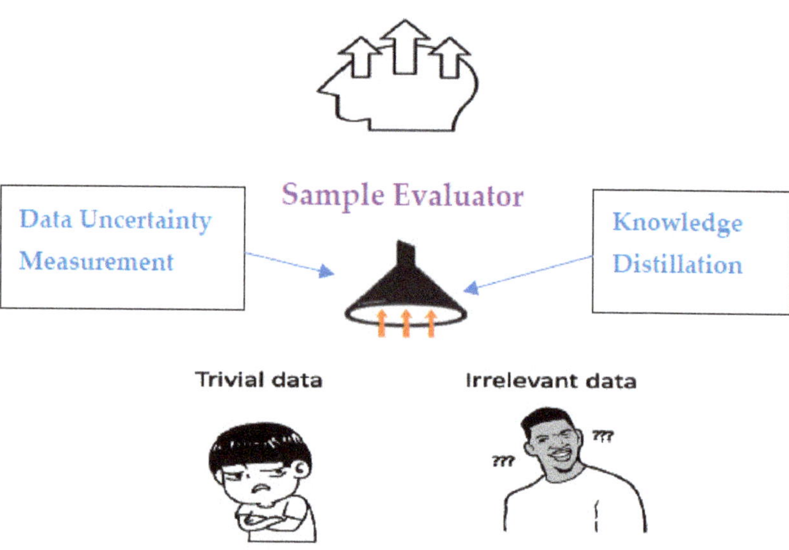

Figure 5. The effects of the sample evaluator (with the aid of knowledge distillation and measurement of data uncertainty).

In other words, we aim to find samples x, that maximize the mutual information between model output y and the model parameters {x, D}. From an information-theoretical point of view, the qualified samples should meet the following conditions: (1) Low confidence in average model output and (2) high confidence in a single sampling model output. Based on the above, for the samples with more prominent mutual information, it is harder to achieve consensus on the outputs between the models; therefore, they are what we are seeking.

In practice, the well-known Monte Carlo approximation can simplify the computation of the conditional mutual information in the above equation. That is,

$$I(y; w | x, D) \approx - \sum_c \left(\frac{1}{T} \sum_t p_c^t \right) \log \left(\frac{1}{T} \sum_t p_c^t \right) + \frac{1}{T} \sum_{c,t} p_c^t \log(p_c^t), \qquad (4)$$

where p_c^t denotes the output probability of class C for model T.

When applied to FL, as pre-described, we can comprehend the distillation process (cf. Equation (3)) as "samples that do not reach consensus among local models", should be taken with higher priority.

Samples with this characteristic will have a considerable divergence in the direction of model convergence during the training phase. Hence, they are more important for optimizing the global model on the server side. In our realization, the samples generated values computed from Equation (4) that are higher than a given threshold will be denoted as high-priority samples for optimization. Of course, the threshold value is accuracy-sensitive and is application dependent. In our experiments, this is empirically determined during the simulation iteration.

3.3. Overall Architecture

Under the mechanism of knowledge distillation, we hope that the student model can learn the inter-class relation of the ideal model well to suppress the adverse effects of data uncertainty. However, if the adopted uncertainty measurement is highly susceptible to data inhomogeneity, it will also be a disadvantage for the proposed FedUA. For example, suppose there is a sample with high data uncertainty from the viewpoint of the ideal model. For such a sample, the associated uncertainty measurement will output a low confidence. In contrast, from the perspective of class distinguishability, the more representative client (who can demonstrate a better interclass relation) will show a decrease in confidence value for this sample due to its native data uncertainty. This fact will degrade the overall performance of our FedUA. We found this problem when we tried to add the uncertainty measurement to the knowledge distillation-based FL, where the entropy of Softmax outputs of the NNs is applied to measure the data uncertainty. This finding explains why in our realization, we replace the Softmax entropy with its feature space density's counterpart (cf. Equation (2a)). Moreover, our experimental results, as illustrated in the next Section, will also justify that feature space density is less affected by the samples' native data uncertainty than that of the Softmax outputs of the NNs.

Figure 6 shows the schematic diagram of the overall architecture of our proposed FedUA. FedUA comprises two main boxes: the server box and the client box. As shown in the upper portion of Figure 6, the server box consists of five functional modules: the teacher evaluation, the sample assessment, the uncertainty measurement, the logits computation, and the student learning modules (note that the brown-colored arrows indicate the respective information flows of each functional module).

In each round, the server regards all user-end models uploaded in this round as the teacher model. When performing the teacher model evaluation, we capture the forward pass outputs of a specific NN layer and send them to the sample assessment and the uncertainty measurement modules for further analyses. The uncertainty measurement uses the selected features of the user-end model to represent the model outputs' weights. Instead of the original FedDF averaging operation in the logits combination module, we apply those weights to calculate the combined logarithmic probabilities (a.k.a. the ensemble logits). At the same time, the user-end model's prediction values are used for the sample quality evaluation, and the qualified samples (with prediction values more prominent than a pre-defined threshold) will be chosen as the training data of the teacher model. Then, after performing these preprocesses, the average parameters of the teacher models (computed through FedAvg) will be treated as the initial parameters of the student model. Then, we can perform the subsequent knowledge distillation (as indicated in the student learning module of Figure 6, we use the Kl-divergence to complete the corresponding calculation).

Following the end of the knowledge distillation, the trained student model will be sent back to the users as the global model for conducting the following local training (as indicated in the client box at the bottom of Figure 6).

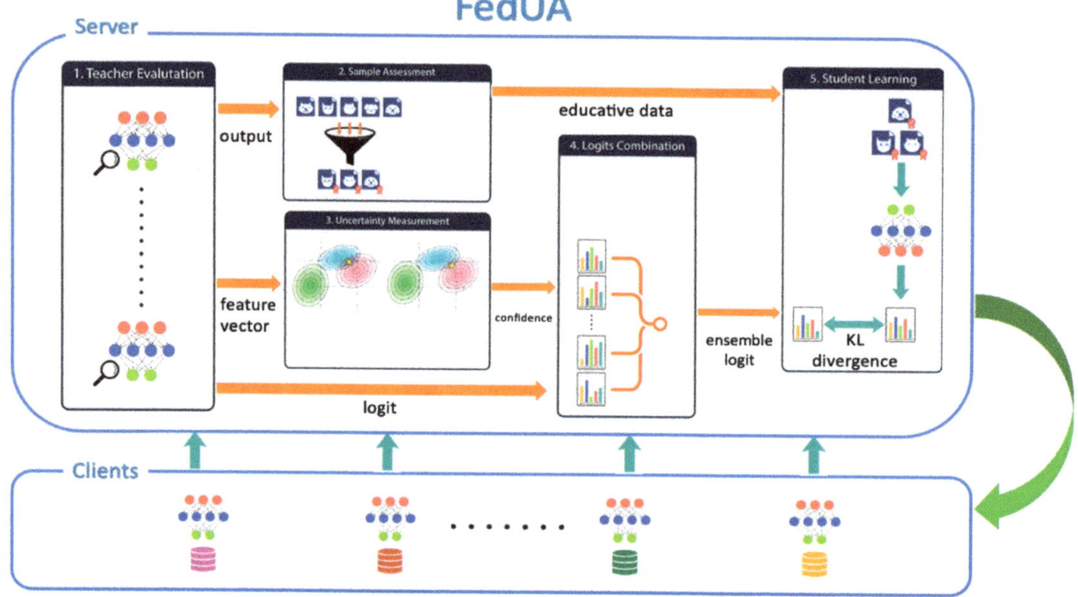

Figure 6. The schematic diagram and the information flow of our proposed FedUA.

4. Experiments

To verify the effectiveness of the adopted core methods, we conduct ablation analyses on the uncertainty measurement and sample evaluation, respectively. All experiments are repeated more than five times. The statistical average and variances will be reported as our experimental results.

4.1. Experimental Settings

(a)   Datasets and Network Models

We examined the proposed FedUA architecture in an image classification application. We selected ResNet-32 as the benchmarking neural network architecture and CIFAR-10 as the training dataset. We randomly picked 40,000 images from the training data as label data for local training on the client side. The remaining 10,000 images were used as unlabeled data for the server side distillation aggregation.

For the label data used for client training, we used the step method [20] as the baseline to achieve the goal of non-IID, and the Dirichlet to make different types of non-IID patterns. Under the step method, each client had many images of two specific categories and a few pictures of the remaining eight categories. The Dirichlet method uses a concentration parameter α (a.k.a. the concentration parameter), to regulate the Dirichlet distribution to produce data with different degrees of dispersion.

CIFAR-10 comprised ten categories of data composed of various vehicles and animals. The existence of inter-category relationships is beneficial for us to explore the correlations between the knowledge distillation, the uncertainty measurement, and the federated learning architecture. The obtained correlations helped to confirm the ability to learn the relationship between the classes and judge the effectiveness of the teacher model in the aggregation stage of federated learning.

(b) **Detailed Processes**

In order to facilitate comparisons and consider the parameter settings concerning related works, we set 40 rounds as the upper limit. The number of clients was assumed to be 10, and the reporting fraction was initialized to 1.0. The reporting fraction determines the number of randomly selected customers in each round, representing the proportion of models uploaded for subsequent aggregation.

Each round of local or server side training consisted of 20 epochs and applied the commonly adopted stochastic gradient descent method. We set the batch size to 32 on the local side and 128 on the server side. In addition, to adapt to knowledge distillation, relative entropy (i.e., the KL divergence) was used on the server side, and this is different from using cross entropy as the loss function on the local side.

4.2. Results and Analyses

4.2.1. Ablation Analysis

(a) **The Impact of the Sample Assessment**

In the ablation analysis of the sample assessment, we verified the effectiveness of Bayesian active learning by disagreement (BALD) first. Then, we considered the impact of the different sample ratios (SRs) on the unlabeled data.

We used random batch sampling as the benchmarking target for a fair comparison. Table 1 presents the relevant results of this examination, where we only depict the portions with a fixed unlabeled data sampling ratio because our experimental results demonstrate that BALD performs better under the condition associated with the same unlabeled data sampling ratio. Moreover, the results listed in Table 1 confirm that using BALD to screen out sample batches demonstrates a better meaning in learning, and therefore performance in accuracy, for the global model's optimization than using random batches traditionally.

Table 1. The performances under different settings in the sample assessment test. (Benchmark NN architecture: ResNet-32, training dataset: CIFAR-10, SR: sample ratio).

	SR = 0.2	**SR = 0.4**	**SR = 0.6**	**SR = 0.8**	**SR = 1.0**
Random	71.5 ± 0.61	71.3 ± 0.78	71.8 ± 0.66	**72.3 ± 0.66**	72.1 ± 0.55
BALD	72.0 ± 0.24	72.5 ± 0.36	**73.9 ± 0.46**	73.7 ± 0.33	73.2 ± 0.48

Interestingly, we also found that regardless of which filtering method was adopted, the best performance in some cases (other than iterative training) occurred with a complete dataset. For example, the best-performed SR setting for random and BLAD filtering is 0.8 and 0.6, respectively (cf. the boldfaced items in Table 1). A smaller sample ratio stands for less induced computational loads.

In conclusion, adding our proposed sample evaluation mechanism is beneficial, not only for the performance of distillation federated learning, but also helpful in reducing the computational burden on the server side.

(b) **The Impact of the Uncertainty Measurement**

In the rest of this subsection, we focus on the effectiveness of the uncertainty measurement. We calculate and compare the entropy of the feature space density outputs and the Softmax outputs in the inference mode when clients learn with non-IID data under the original learning settings.

CIFAR-10 has a high degree of data uncertainty because there is a specific correlation among the animal classes, and the same is valid for the vehicle data. Due to the assumptions of non-independence, we should pay attention to both model effects and distribution uncertainties. The former is because the data imbalance at the category level will affect the local model. The latter comes from the distribution uncertainty when the uploaded local model is compared with the ideal global model, that unavoidably has a distribution difference between the training and actual samples.

We take the example of a non-IID in Figure 7 for illustration. Client numbers 0, 2, 3, 5, 6, and 7 contain many images in two categories: vehicles and animals, so these clients should behave in a superior manner in the coarse-grained classification task. In contrast, clients numbers 1, 8, and 9 contain many images in two animal classes, and client number 4 contains many images in two vehicle classes. These clients tend to have specific behaviors in fine-grained classification associated with their highly correlated classes.

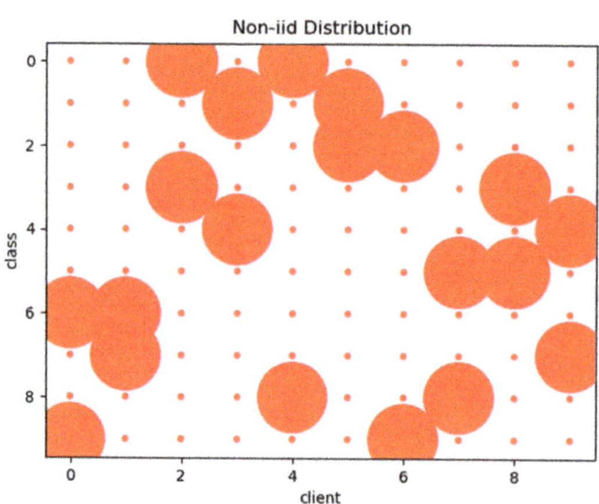

Class Number	0	1	2	3	4	5	6	7	8	9
Images	Airplane	Automobile	Bird	cat	Deer	Dog	Frog	Horse	Ship	Truck

Figure 7. An example of non-IID training data distribution (horizontal-axis: client number and vertical-axis: image class number).

For example, when the input sample belongs to the animal category, the No. 4 teacher model incorrectly classifies coarse-grained and fine-grained animal categories. Therefore, we should suppress the degree of the student model's referencing to the No. 4 teacher model. In addition, promoting the fine-grained ability of teacher models 1, 8, and 9 for animal classes is crucial in improving the student model's training effects for the later stages. Conversely, if the input sample belongs to the transportation category, we should lower the influences of No. 1, 8, and 9 teacher models. At the same time, the impact of the No. 4 teacher model should be increased in the aggregation stage.

Figures 8 and 9 show the distributions of the top-1 outputs' entropy of Softmax and feature space density, respectively. For ease of comprehension, we respectively illustrate the mean entropy values of Figures 8 and 9 in Figures 10 and 11. To emphasize the different behaviors of the two distributions in real applications, say out-of-distribution (OoD) detection as an example, let us explore the two-dimensional distribution patterns of the two in detail, as indicated in the red-colored rectangular boxes in Figure 12. Clearly, from Figure 12, the latter is a better choice than the former due to its higher sparsity in distribution.

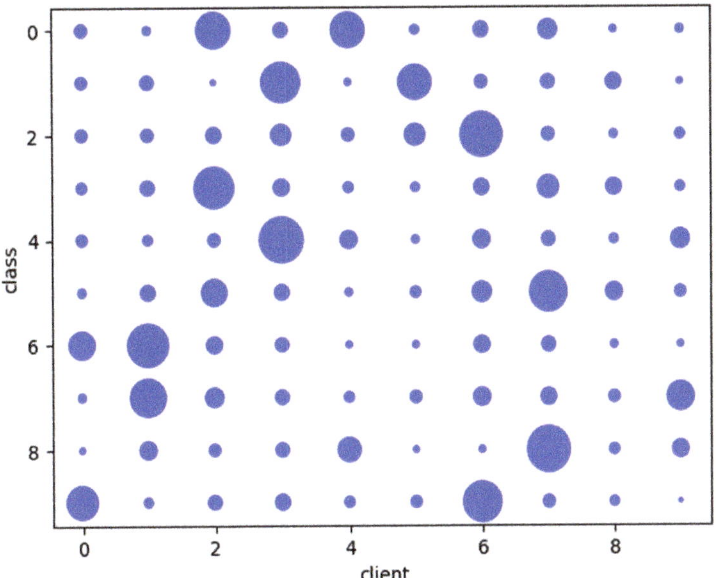

Figure 8. The statistical distribution of the top-1 SoftMax outputs' entropy.

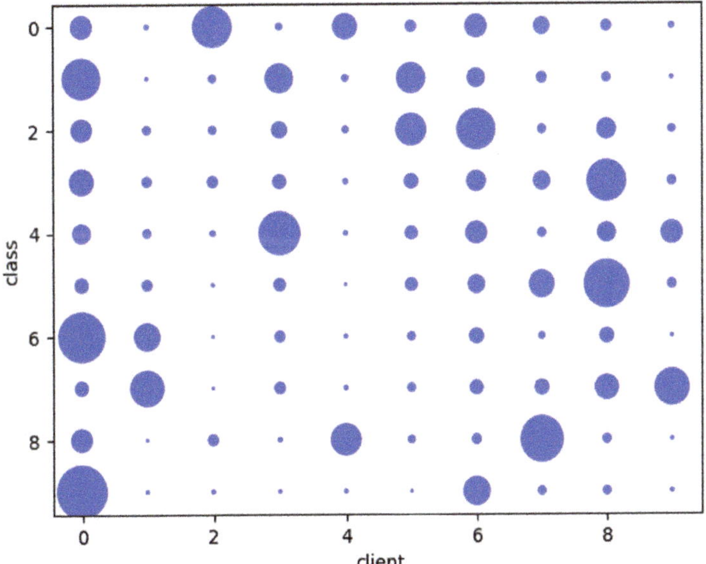

Figure 9. The statistical distribution of the top-1 feature space density's entropy.

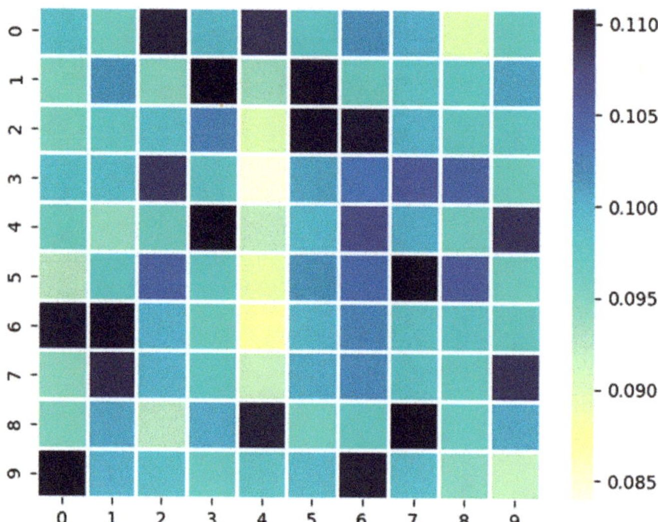

Figure 10. The mean of the entropy values obtained in Figure 8.

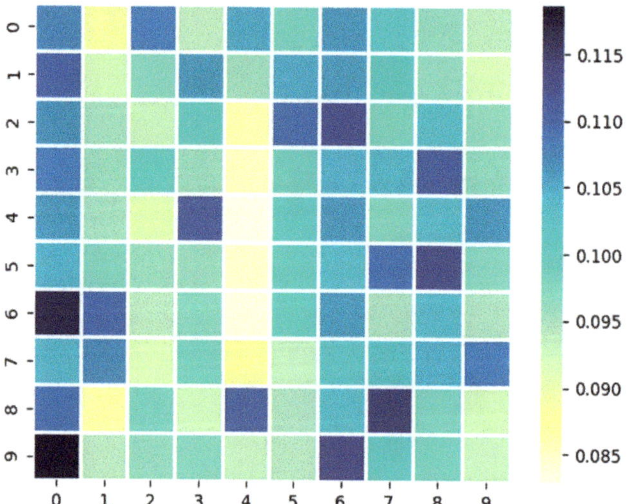

Figure 11. The mean of the entropy values obtained in Figure 9.

From the observations of the distributions and the mean values depicted in Figures 8–11, it is justified that both the proposed normalization function and the feature space density method enhance the classification performance under data and model uncertainties. Moreover, if we focus on the issue of distribution uncertainty in federated learning, the results associated with the feature space become more informative. That is, we can determine the correct classes from the darkness of the colors in Figure 11 as much more accessible than in Figure 10. This explains why our FedUA ultimately uses the feature space density method.

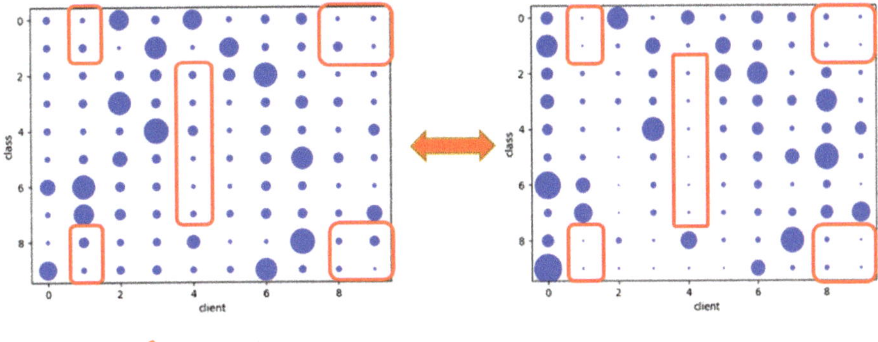

Figure 12. The sparsity comparison between the top-1 SoftMax outputs' entropy and the top-1 feature space density's entropy.

4.2.2. Performance Comparisons among the Benchmarked Works

(a)   **Learning Behaviors of the Different FL-schemes on Non-IID Data**

As addressed in Section 4.1. we implement FedUA based on the pre-described settings and use the step method [20] to find its effectiveness on non-IID data. Of course, we investigate the learning behaviors of competing aggregation approaches for comparison purposes. Figure 13 shows the learning curves of FedUA, FedDF, and FedAvg, where the vertical axis denotes the accuracy percentage and the horizontal axis stands for the round number.

Dataset	CIFAR-10
Labeled / Unlabeled	40,000/10,000
Non-IID Data	Step Method
Rounds	40
#Clients	10
Reporting Fraction	1.0
Network Model	ResNet-32
Epoch (client/server)	20/20
Batch Size (client/server)	32/128
Loss(client/server)	CE/KL

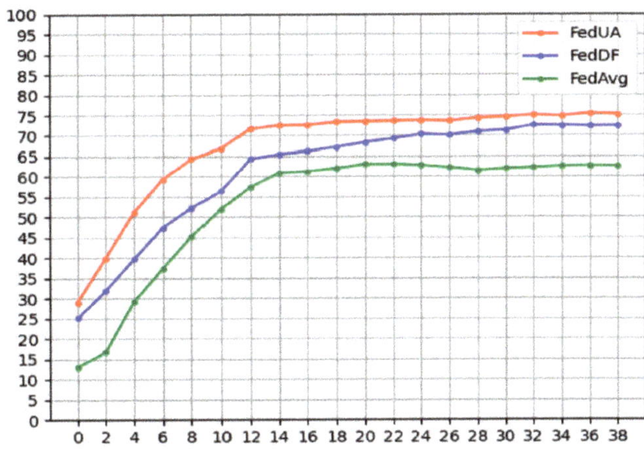

Figure 13. The learning curves of FedUA, FedDF, and FedAvg (horizontal-axis: the round number and vertical-axis: the accuracy percentage).

As shown in Figure 13, in the early stage, benefiting from knowledge distillation, the accuracy of the global models of FedUA and FedDF was significantly better than that of FedAvg. The performance-enhancing speed of the three is close, and it begins to slow down and converges to an upper limit after 14 rounds. In the end, both FedUA and FedDF outperform FedAvg, and the accuracy of FedUA is about 2–3% better than FedDF when the data distribution is non-IID.

(b) The Impact of Different Non-IID Data Partitions

To dive into the effects of non-IID data on various FL schemes, in this subsection, we examine the performances of FedUA, FedDF, and FedAvg under different non-IID settings. Figure 14, from left to right, shows the other non-IID data corresponding to the step method, the Dirichlet with $\alpha = 0.1$, and the Dirichlet with $\alpha = 0.5$. Table 2 compares the obtained classification accuracy among the benchmarked outcomes. The boldfaced items in Table 2 show that FedUA performs the best among the three.

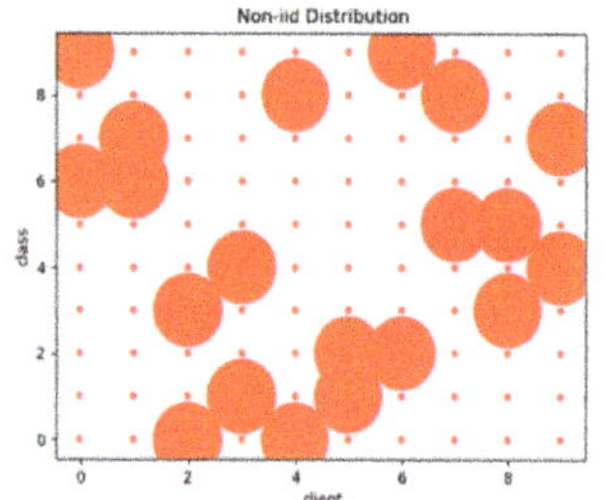
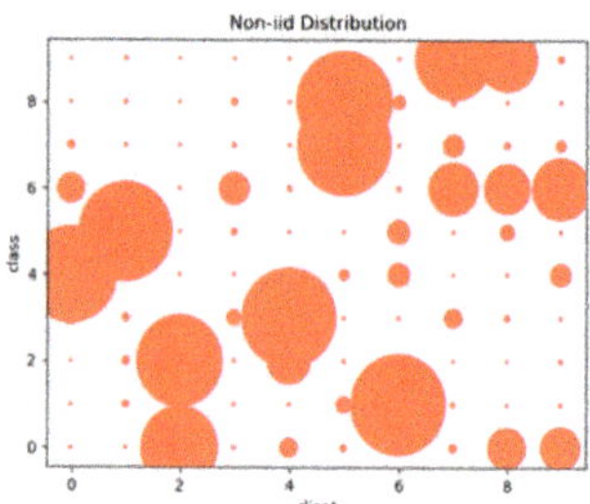
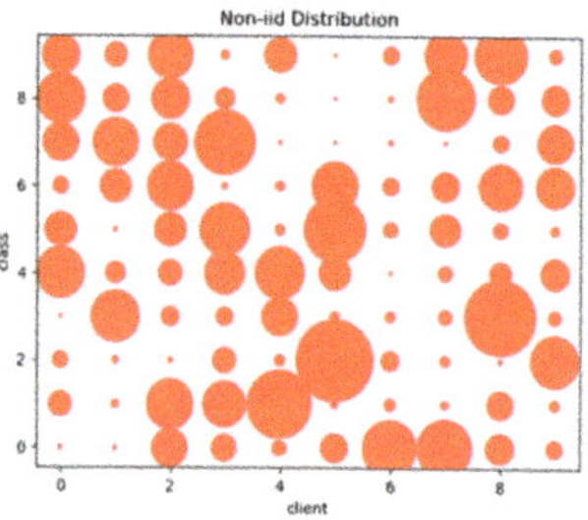

Figure 14. The distributions under different non-IID data settings.

Table 2. The performances of FedAvg, FedDF, and FedUA under different non-IID data. Settings illustrated in Figure 14. (Fed-α-Step and FedUA-FedAvg denote the Percentages of Accuracy Change vs. Parameter α when we take the Step method's and FedAvg's results as the reference, respectively.)

	Classification Accuracy vs. Parameter α		
	Step Method [20]	Dirichlet ($\alpha = 0.1$)	Dirichlet ($\alpha = 0.5$)
FedAvg	62.6 ± 0.23 1 1	59.6 ± 1.03 -4.7% 1	80.1 ± 0.45 $+28\%$ 1
FedDF FedDF-α-Step FedDF-FedAvg	72.3 ± 0.49 1 15.5 %	64.4 ± 0.93 -11% 8.0%	82.8 ± 0.47 $+14.5\%$ 3.4%
FedUA FedUA-α-Step FedUA-FedAvg	$\mathbf{74.8 \pm 0.45}$ 1 **19.5%**	$\mathbf{65.3 \pm 0.78}$ -12.7% 9.6%	$\mathbf{83.4 \pm 0.24}$ $+11.5\%$ 4.1%

More specifically, from Figure 14, we observed a more severe data imbalance and distribution uncertainty between clients under the Dirichlet ($\alpha = 0.1$) setting. The advantages of the FedUA core method are less prominent than the step method, which is only about 1.6% growth in accuracy compared with the counterpart of FedDF (cf. Table 2). Nevertheless, there is still a meaningful improvement in knowledge distillation compared to FedAvg. For the settings under Dirichlet ($\alpha = 0.5$), the client's data is closer to IID, and both FedUA and FedDF behaved normally and better than FedAvg. The reason is that under the data segmentation of Dirichlet, a more serious data imbalance and more complex feature space density patterns are derived, resulting in more difficulty in model uncertainty and distribution uncertainty estimation. Fortunately, considering the real-world usage of federated learning nowadays, the local data distribution between devices should tend to the step method, which embraces the adaptation of the proposed FedAU in federated machine learning.

(c) The Effects of Limited Allowable Communication Capacity

Finally, we consider the limited communication cost scenario faced by federated learning practices. Finding enough computational resources and large datasets to conduct accurate and concrete experiments is challenging in academia. To face this reality, we designed our simulations concerning the effects of limited allowable communication capacity by adjusting the clients' participation ratio (denoted by C) uploaded to the server for each round. We set different participation ratios and observed the corresponding results (cf. Table 3).

Table 3. The performances of FedAvg, FedDF, and FedUA under different participation ratios. (Accur-drop denotes the percentage of accuracy drop by taking C = 1.0 as the reference.)

	Classification Accuracy vs. Participation Ratio		
	C = 1.0	C = 0.7	C = 0.4
FedAvg	62.6 ± 0.23	61.3 ± 0.35	58.1 ± 0.26
Accur-drop	1	2.1%	7.2%
FedDF	72.3 ± 0.49	68.1 ± 0.61	63.8 ± 1.03
Accur-drop	1	5.9%	11.8
FedUA	74.8 ± 0.45	71.8 ± 0.56	68.3 ± 0.85
Accur-drop	1	4.0%	8.7%

From Table 3, our proposed FedUA performed the best concerning absolute classification accuracy. Moreover, Table 3 also indicates that the accuracy drop of FedAvg does not decline significantly if the participation ratio is lowered below the 0.7 settings, but that of FedDF declines the most of the three (ranges from 6% to 12%, approximately). While FedUA behaves in between with an accuracy drop ranging from 4% to about 9%.

When the participation ratio of the native distillation-based federated learning decreases, the function of the teacher model is insufficient. That is, knowledge deficiency has occurred, which may make the unlabeled data used in the server aggregation stage find no correct learning objectives. As a result, FedDF relies heavily on the client to participate in the aggregation stably. However, by introducing the sample evaluation and uncertainty measurement, FedUA somehow mitigates the impact of the above shortages and avoids the damage caused to the student model when meaningless or even erroneous learning mode scenarios occur. We can justify the above arguments from the experimental results obtained in Table 3.

5. Discussions and Conclusions
5.1. Current Progress in FL Dealing with Non-IID Data

Regarding the challenges faced in FL, people in different fields will have different perspectives. This paper focuses on the countermeasures we can take when the data distributions in FL are heterogeneous. One of the anonymous reviewers suggested lots of related literature [19,30–35] and asked us to make some focused summaries and comparisons among them. Therefore, before concluding our work, this section briefly summarizes various researchers' current efforts.

Smietanka et al. [30] briefly surveyed privacy-preserving techniques and applications concerning FL. Technique-wise, three kinds of data access-related security protection methods were discussed: differential privacy, secure multiparty computation, and homomorphic encryption. At the same time, FL-related applications in Google Gboard, Health, Retail, Finance, and Insurance were addressed as illustrative examples.

To combine the advantages of cloud-based and edge-based FL for speeding up the model training and improving the communication-computation trade-offs, ref. [31] proposed a hierarchical FL architecture using multiple edge servers to perform partial model aggregation before communicating with the cloud parameter server. Empirical experiments verified the analysis and demonstrated the benefits of this hierarchical architecture in different data distribution scenarios. In other words, introducing the intermediate edge servers can simultaneously reduce the end devices' model training time and energy consumption compared to cloud-based federated learning. However, ref. [31] ignored the effects of heterogeneous communication conditions and computing resources on different clients. Moreover, the performance-sensitive parameters need to be tuned empirically.

The authors of [32] pointed out that to train statistical models in a massive and heterogeneous network, naively minimizing an aggregate loss function may only benefit some involved devices. To face this shortage, ref. [32] proposed the so-called q-fair federated learning (q-FFL), that encourages a fairer (specifically, more uniform) accuracy distribution across devices in FL networks. Moreover, experimental results showed that with the aid of the newly devised aggregation mechanism q-FedAvg, q-FFL outperforms existing benchmarks regarding fairness, flexibility, and efficiency. Nevertheless, q-FFL increases the accuracy of poor-performing devices by sacrificing better-performing ones. This approach may not be suitable for performance-critical applications. Moreover, we need to determine the control parameter in advance again.

Tao et al. [19] proposed using ensemble distillation for model fusion, i.e., training the central classifier through unlabeled data on the outputs of the models from the clients. The authors claimed that the knowledge distillation technique would mitigate privacy risk and cost to the same extent as the baseline FL algorithms, but allowed flexible aggregation over heterogeneous client models that differed in size, numerical precision, or structure. They justified their claims through extensive empirical experiments on various CV/NLP datasets (CIFAR-10/100, ImageNet, AG News, SST2) and settings (heterogeneous models/data) by showing that the server model can be trained much faster, requiring fewer communication rounds than any existing FL technique known to them. Actually, ref. [19] inspired our work a lot.

Giovanni Paragliola and Antonio Coronato contributed a series of three papers [33–35] founded on the same kernel skills, targeting reducing communication costs in a federated healthcare environment. The inputs of the learning system were ECG waveforms of patients with various levels of risk associated with hypertension. The proposed FL framework comprised different learning strategies (varying in the numbers of cascaded dense layers and shared parameters).

To reduce the required communication costs in FL, ref. [33] presented an FL algorithm, TFedAvg, to train a time series (TS)-based model for the early identification of the level of risk associated with patients with hypertension in a federated healthcare environment. The primary framework of [33] consisted of two learning strategies, The FullNet Strategy and the PartialNet Strategy, *for which* TFedAVG exploits the whole model and a portion of the model to both guarantee the privacy and security of healthcare data, and reduce the communication costs between clients and aggregation server, respectively. Under three split local datasets conditions, ref. [33] presented two different settings concerning the types of data distribution across the regional nodes: (1) an IID setting where each node had 33% of the total samples, (2) a non-IID setting in which one of the nodes had 50% of the total samples while the other two nodes only had 25% each. Experimental results showed that the proposed approach improved from 3.01% to 11.09% in terms of classification accuracy and with a reduction of about 34% in terms of communication costs compared to the benchmarked works. Another contribution from [33] came from its summarization and comparative analyses of recent FL-related research statuses: Table 1 of [33] summarizes the studies on federated learning published between 2018 and 2022 regarding applications, adopted approaches, pros, and cons.

Yoshida et al. [6] continued and extended the discussion initiated in [33] concerning reducing the communication overhead with a further analysis, evaluating the trade-off between the performance and the communication costs. Such an analysis suggested a new learning strategy (LS) to reduce the total number of parameters shared during the FL process. The basic idea of [34] was to exploit subparts of the model in [33] by measuring the contribution of a subset of layers defining an ML model during the training process instead of the whole set of layers. To estimate the weight and the contribution of each layer, ref. [6] defined seven different learning strategies (LSs) aimed at selecting which parameters to transmit to the central server for the aggregation process, such that a trade-off between the requirement to bring down communication costs and the need to guarantee the highest classification performance could be reached. Compared with Google's FedAvg algorithm, experimental results show that the accuracy of the approach proposed in [34] ranges from 89.25% to 96.6%. In comparison, the improvements in reducing communication overheads range from 95.64% to 6%.

The catastrophic forgetting (CF) phenomenon occurs during an ML training process when the characteristics or distribution of new input instances differ significantly from previously observed ones. CF-induced new information may overwrite the previously learned knowledge of a neural network. A similar situation might occur in FL when the local data of each client cannot be considered representative of the overall data distribution due to class imbalance, distribution imbalance, and size imbalance, that causes the well-known non-IID data challenge to FL. By successfully transferring the problem of analyzing the occurrence of CF in FL as the analysis of the DNNs' training in a federated environment when dealing with non-stationary data, ref. [35] extended the use case scenario in [6] for evaluating the nature of CF events, and provided a quantification of when and how a CF event may happen during an FL process. Finally, the experimental results in [35] depicted an improvement in accuracy ranging from 2% to 28% among local clients affected by a CF event.

5.2. Conclusions and Possible Contributions of This Work

Federated learning, constrained by security and communication costs, has flourished recently. As the above section addresses, obtaining a standard solution for various and complex Non-IID types is still challenging and worthy of further exploration.

In practice in the past, the federated learning architecture that adopted knowledge distillation usually caused incomplete interclass relationships learned by the local model due to the imbalance of local training data, which in turn made the global model learning in the aggregation stage preliminary. Therefore, we started from the uncertainty analysis of DNNs, evaluated their effects on FL, and proposed a new architecture for model aggregation. The proposed scheme improves FL's performance by combining the knowledge distillation and the DNN's uncertainty quantification methods. A series of experiments on the image classification task confirms that our proposed model aggregation scheme can effectively solve the problem of non-IID data, especially when the affordable transmission cost is limited.

The possible contributions of our work can be summarized as follows:

1. We built an effective, adaptable aggregation scheme to deal with the inhomogeneity of client side models based on the proposed quantifiable network uncertainty of the uploading client;
2. Based on the evaluated sample quality, we introduced an effective sample sieve scheme to the server to suppress the influences of data uncertainty and improve the learning efficiency;
3. As a knowledge distillation aggregation architecture, our work can effectively separate the information of uncertainty and the inter-class relationships. This separation helps solve the non-IID data issue and provides a good learning performance while limiting the transmission cost;
4. Through a series of experiments on the image classification task, we confirmed that the proposed model aggregation scheme could effectively solve the problem of non-IID data, especially when the affordable transmission cost is limited.

In summary, in handling the problem of Non-IID, we hope that the uncertainty measurement and sample evaluation we propose can help consider real-world user data. They provide more information in the aggregation stage and make learning more effective. However, our current discussion only applies to the task of image classification. Moreover, lacking enough computing resources and practical use cases are difficult to find, we have yet to be able to experiment with more complex image datasets. Nevertheless, it may be possible to combine the advantages of the uncertainty method and distillation-based federated learning to create different collaboration models, that will be the main direction of our future efforts.

Author Contributions: Formal analysis, S.-M.L.; Funding acquisition, J.-L.W.; Investigation, S.-M.L. and J.-L.W.; Methodology, S.-M.L.; Project administration, J.-L.W.; Resources, J.-L.W.; Software, S.-M.L.; Supervision, J.-L.W.; Writing—original draft, S.-M.L.; Writing—review & editing, S.-M.L. and J.-L.W. All authors have read and agreed to the published version of the manuscript.

Funding: This research was funded by Minister of Science and Technology, Taiwan MOST 109-2218-E-002-015 and MOST 111-2221-E-002-134-MY3. And The APC was funded by MPDI.

Data Availability Statement: Not applicable.

Conflicts of Interest: The authors declare no conflict of interest.

References

1. McMahan, B.; Moore, E.; Ramage, D.; Hampson, S.; Arcas, B. Communication-Efficient Learning of Deep Networks from Decentralized Data. In *Artificial Intelligence and Statistics*; PMLR. 2017. Available online: https://proceedings.mlr.press/v54/mcmahan17a/mcmahan17a.pdf (accessed on 1 April 2023).
2. Kairouz, P.; McMahan, H.B.; Avent, B.; Bellet, A.; Bennis, M.; Bhagoji, A.N.; Bonawitz, K.; Charles, Z.; Cormode, G.; Cummings, R.; et al. Advances and Open Problems in Federated Learning. *Found. Trends Mach. Learn.* **2021**, *14*, 1–210. [CrossRef]
3. Zhao, Y.; Li, M.; Lai, L.; Suda, N.; Civin, D.; Chandra, V. Federated learning with non-IID data. *arXiv* **2018**, arXiv:1806.00582. [CrossRef]
4. Xiao, P.; Cheng, S.; Stankovic, V.; Vukobratovic, D. Averaging Is Probably Not the Optimum Way of Aggregating Parameters in Federated Learning. *Entropy* **2020**, *22*, 314. [CrossRef] [PubMed]
5. Zhu, H.; Xu, J.; Liu, S.; Jin, Y. Federated learning on non-IID data: A survey. *Neurocomputing* **2021**, *465*, 371–390. [CrossRef]
6. Yoshida, N.; Nishio, T.; Morikura, M.; Yamamoto, K.; Yonetani, R. Hybrid-FL for wireless networks: Cooperative learning mechanism using non-IID data. In Proceedings of the ICC 2020-2020 IEEE International Conference on Communications (ICC), Dublin, Ireland, 7–11 June 2020; pp. 7–11. [CrossRef]
7. Duan, M.; Liu, D.; Chen, X.; Tan, Y.; Ren, J.; Qiao, L.; Liang, L. Astraea: Self-Balancing Federated Learning for Improving Classification Accuracy of Mobile Deep Learning Applications. In Proceedings of the 2019 IEEE 37th International Conference on Computer Design (ICCD), Abu Dahbi, United Arab Emirates, 17–20 November 2019; pp. 246–254. [CrossRef]
8. Ghosh, A.; Hong, J.; Yin, D.; Ramchandran, K. Robust federated learning in a heterogeneous environment. *arXiv* **2019**, arXiv:1906.06629.
9. Ghosh, A.; Chung, J.; Yin, D.; Ramchandran, K. An Efficient Framework for Clustered Federated Learning. *IEEE Trans. Inf. Theory* **2022**, *68*, 8076–8091. [CrossRef]
10. Li, T.; Sahu, A.; Zaheer, M.; Sanjabi, M.; Talwalkar, A.; Smith, V. Federated optimization in heterogeneous networks. *Proc. Mach. Learn. Syst.* 2020; 2, 429–450.
11. Hsu, T.-M.H.; Qi, H.; Brown, M. Measuring the effects of non-identical data distribution for federated visual classification. *arXiv* **2019**, arXiv:1909.06335.
12. Wang, K.; Mathews, R.; Kiddon, C.; Eichner, H.; Beaufays, F.; Ramage, D. Federated evaluation of on-device personalization. *arXiv* **2019**, arXiv:1910.10252.
13. Arivazhagan, M.G.; Aggarwal, V.; Singh, A.; Choudhary, S. Federated learning with personalization layers. *arXiv* **2019**, arXiv:1912.00818.
14. Smith, V.; Chiang, C.-K.; Sanjabi, M.; Talwalkar, A. Federated multi-task learning. *Adv. Neural Inf. Process. Syst. NeurIPS* **2017**, *30*. Available online: https://papers.nips.cc/paper_files/paper/2017 (accessed on 1 April 2023).
15. Liu, B.; Wang, L.; Liu, M. Lifelong Federated Reinforcement Learning: A Learning Architecture for Navigation in Cloud Robotic Systems. *IEEE Robot. Autom. Lett.* **2019**, *4*, 4555–4562. [CrossRef]
16. Zhou, Y.; Pu, G.; Ma, X.; Li, X.; Wu, D. Distilled one-shot federated learning. *arXiv* **2020**, arXiv:2009.07999.
17. Jeong, E.; Oh, S.; Kim, H.; Park, J.; Bennis, M.; Kim, S.-L. Communication-efficient on-device machine learning: Federated distillation and augmentation under non-iid private data. *arXiv* **2018**, arXiv:1811.11479.
18. Li, D.; Wang, J. Fedmd: Heterogenous federated learning via model distillation. *arXiv* **2019**, arXiv:1910.03581.
19. Lin, T.; Kong, L.; Stich, S.; Jaggi, M. Ensemble Distillation for Robust Model Fusion in Federated Learning. *Adv. Neural Inf. Process. Syst.* **2020**, *33*, 2351–2363. [CrossRef]
20. Chen, H.-Y.; Chao, W.-L. Fedbe: Making bayesian model ensemble applicable to federated learning. *arXiv* **2020**, arXiv:2009.01974.
21. Hinton, G.; Vinyals, O.; Dean, J. Distilling the knowledge in a neural network. *arXiv* **2015**, arXiv:1503.02531.
22. Gawlikowski, J.; Vinyals, O.; Dean, J. A survey of uncertainty in deep neural networks. *arXiv* **2021**, arXiv:2107.03342.
23. Hendrycks, D.; Gimpel, K. A baseline for detecting misclassified and out-of-distribution examples in neural networks. *arXiv* **2016**, arXiv:1610.02136.

24. Gal, Y.; Ghahramani, Z. Dropout as a Bayesian approximation: Representing model uncertainty in deep learning. In Proceedings of the 33rd International Conference on International Conference on Machine Learning, New York, NY, USA, 19–24 June 2016; pp. 1050–1059.
25. Lakshminarayanan, B.; Pritzel, A.; Blundell, C. Simple and Scalable Predictive Uncertainty Estimation Using Deep Ensembles. *Adv. Neural Inf. Process. Syst.* **2017**, *30*. Available online: https://proceedings.neurips.cc/paper/2017/file/9ef2ed4b7fd2c81084 7ffa5fa85bce38-Paper.pdf (accessed on 1 April 2023).
26. Ovadia, Y.; Fertig, E.; Ren, J.; Nado, Z.; Sculley, D.; Nowozin, S.; Dillon, J.V.; Lakshminarayanan, B.; Snoek, J. Can You Trust Your Model's Uncertainty? Evaluating Predictive Uncertainty under Dataset Shift. *Adv. Neural Inf. Process. Syst.* **2019**, *32*. Available online: https://proceedings.neurips.cc/paper/2019/file/8558cb408c1d76621371888657d2eb1d-Paper.pdf (accessed on 1 April 2023).
27. Guo, C.; Pleiss, G.; Sun, Y.; Weinberger, K. On calibration of modern neural networks. In *International Conference on Machine Learning*; PMLR. 2017. Available online: https://proceedings.mlr.press/v70/guo17a/guo17a.pdf (accessed on 1 April 2023).
28. Mukhoti, J.; Kirsch, A.; van Amersfoort, J.; Torr, P.H.S.; Gal, Y. Deep Deterministic Uncertainty: A Simple Baseline. *arXiv* **2021**, arXiv:2102.11582.
29. Gal, Y.; Islam, R.; Ghahramani, Z. Deep bayesian active learning with image data. In *International Conference on Machine Learning*; PMLR. 2017. Available online: https://proceedings.mlr.press/v70/gal17a/gal17a.pdf (accessed on 1 April 2023).
30. Śmietanka, M.; Pithadia, H.; Treleaven, P. Federated Learning for Privacy-Preserving Data Access. 2020. Available online: https://papers.ssrn.com/sol3/papers.cfm?abstract_id=3696609 (accessed on 1 April 2023).
31. Lumin, L.; Zhang, J.; Song, S.; Letaief, K. Client-Edge-Cloud Hierarchical Federated Learning. In Proceedings of the IEEE International Conference on Communications (ICC, IEEE), Dublin, Ireland, 7–11 June 2020. [CrossRef]
32. Tian, L.; Sanjabi, M.; Beirami, A.; Smith, V. Fair Resource Allocation in Federated Learning. ICLR 2020. *arXiv* **2019**, arXiv:1905.10497.
33. Paragliola, G.; Coronato, A. Definition of a novel federated learning approach to reduce communication costs. *Expert Syst. Appl.* **2022**, *189*, 116109. [CrossRef]
34. Paragliola, G. Evaluation of the trade-off between performance and communication costs in federated learning scenario. *Futur. Gener. Comput. Syst.* **2022**, *136*, 282–293. [CrossRef]
35. Paragliola, G. A federated learning-based approach to recognize subjects at a high risk of hypertension in a non-stationary scenario. *Inf. Sci.* **2023**, *622*, 16–33. [CrossRef]

information

Article

Graph Neural Networks and Open-Government Data to Forecast Traffic Flow

Petros Brimos, Areti Karamanou, Evangelos Kalampokis * and Konstantinos Tarabanis

Information Systems Lab, Department of Business Administration, University of Macedonia, 54636 Thessaloniki, Greece
* Correspondence: ekal@uom.edu.gr

Abstract: Traffic forecasting has been an important area of research for several decades, with significant implications for urban traffic planning, management, and control. In recent years, deep-learning models, such as graph neural networks (GNN), have shown great promise in traffic forecasting due to their ability to capture complex spatio–temporal dependencies within traffic networks. Additionally, public authorities around the world have started providing real-time traffic data as open-government data (OGD). This large volume of dynamic and high-value data can open new avenues for creating innovative algorithms, services, and applications. In this paper, we investigate the use of traffic OGD with advanced deep-learning algorithms. Specifically, we deploy two GNN models—the Temporal Graph Convolutional Network and Diffusion Convolutional Recurrent Neural Network—to predict traffic flow based on real-time traffic OGD. Our evaluation of the forecasting models shows that both GNN models outperform the two baseline models—Historical Average and Autoregressive Integrated Moving Average—in terms of prediction performance. We anticipate that the exploitation of OGD in deep-learning scenarios will contribute to the development of more robust and reliable traffic-forecasting algorithms, as well as provide innovative and efficient public services for citizens and businesses.

Keywords: traffic flow forecasting; deep learning; graph neural networks; Artificial Intelligence; high-value data; open-government data

Citation: Brimos, P.; Karamanou, A.; Kalampokis, E.; Tarabanis, K. Graph Neural Networks and Open-Government Data to Forecast Traffic Flow. *Information* **2023**, *14*, 228. https://doi.org/10.3390/info14040228

Academic Editor: Michele Ottomanelli

Received: 28 February 2023
Revised: 17 March 2023
Accepted: 28 March 2023
Published: 7 April 2023

1. Introduction

Traffic forecasting, which is an important component of intelligent transportation systems, assists policymakers and public authorities to design and manage transportation systems that are efficient, safe, environmentally friendly, and cost-effective [1]. Traffic forecasts can be used to anticipate future needs and allocate resources accordingly, such as managing traffic lights [2,3], opening or closing lanes, estimating travel time [4], and mitigating traffic congestion [5–7]. The prediction of future traffic states based on historical or real-time traffic data can contribute to reducing the impact of negative effects on citizens such as health problems brought on by air pollution, and economic costs such as increased travel time spent, and wasted fuels, with a detrimental effect on both the environment and the quality of citizens' lives [8–11].

However, traffic forecasting is a challenging field due to the complex spatio–temporal dependencies that occur in the road network. Literature usually exploits three types of techniques to forecast future traffic conditions based on historical observations of traffic data; (i) traditional parametric methods including stochastic and temporal methods (e.g., Autoregressive Integrated Moving Average—ARIMA [12]); (ii) machine learning (e.g., Support Vector Machine [5]); and (iii) deep learning [11,13–15]. Artificial Intelligence approaches outperform parametric approaches due to their ability to deal with large quantities of data [16,17]. In addition, the parametric models fail to provide accurate results due to the stochastic and non-linear nature of traffic flow [18].

Recently, a significant number of research papers have explored the effectiveness of deep-learning algorithms on urban traffic forecasting, especially graph neural networks (GNNs) (e.g., [19–22]). These models manage to capture the complex topology of a road network, by extending the convolution operation from Euclidean to non-Euclidean space, while dynamic temporal dependencies are captured by the integration of recurrent units.

The rapid development of the Internet of Things (IoT) can significantly facilitate traffic forecasting by providing data sources (e.g., sensors), which generate large quantities of traffic data that can be analyzed to forecast the volume and density of traffic flow. Dynamic (or real-time) data with traffic-related information (e.g., counted number of vehicles, average speed) that are generated by sensors have only recently started being provided as open-government data (OGD) [23–25] available for free access and reuse. Dynamic OGD are suitable for creating value-added services, applications and high-quality and decent jobs, and are, hence, characterized as high-value data (HVD), with substantial societal, environmental, and financial advantages [26]. However, collecting and reusing this type of data requires addressing various challenges. For example, dynamic data are known for their high variability and quick obsolescence and should, hence, be immediately available and regularly updated to develop added-value services and applications. In addition, a recent work [27] showed that dynamic traffic data confront major quality challenges. These challenges are often caused by sensor malfunctions, e.g., brought on by bad climatic conditions [28].

This paper aims to investigate the exploitation of traffic OGD using state-of-the-art deep-learning algorithms. Specifically, we use two widely used and open-source GNN algorithms, namely Temporal Graph Convolutional Network (TGCN) and Diffusion Convolutional Recurrent Neural Network (DCRNN), and real-time traffic data from the Greek open-data portal to create models that accurately forecast traffic flow. The models forecast traffic flow in three time horizons, i.e., in the next 3 (short-term prediction), 6 (middle-term prediction), and 9 (long-term prediction) time steps (hours). The Greek data portal was selected for this work since it provides an Application Programming Interface (API) to access traffic data. We anticipate that the exploitation of OGD in deep-learning scenarios will contribute towards both (a) the development of more robust and reliable traffic-forecasting deep-learning algorithms and (b) the provision of innovative and efficient public services to citizens and businesses alike.

The rest of this paper is organized as follows. Section 2 presents related work describing traffic-forecasting, deep-learning approaches, and graph neural networks for traffic forecasting. Section 3 presents background knowledge regarding the two GNN algorithms employed in this work. Thereafter, Section 4 describes in detail the specific steps followed in this research. In addition, Sections 5 and 6 present the case study by first describing the collection of the traffic data from the Greek open-data portal (Section 5), and the pre-processing of the traffic data (Section 6), and then the creation and evaluation of the forecasting model (Section 7). Finally, Section 8 discusses the results of this paper and Section 9 concludes this paper.

2. Related Work

This section presents a review of the previous work in traffic forecasting, deep-learning approaches, and graph neural networks for traffic forecasting to facilitate readers' quick understanding of the key aspects of this work.

2.1. Traffic Forecasting

Intelligent Transportation Systems (ITS) aim to improve the operational efficiency of transportation networks by gathering, processing and analyzing massive amounts of traffic information [29]. This information is produced by sensors (e.g., loop detectors), traffic surveillance videos or Bluetooth devices that are in several control points of the transportation network such as roadways, highways, terminals, etc. In the rapid development of intelligent transportation systems, traffic forecasting has been considered a very important and developing area for both research and business applications, with a large range of published articles in the field [31]. Traffic forecasting is the process of estimating future traffic states given a continuum of historical traffic data. Moreover, it is one of the most challenging tasks among other time-series prediction problems because it involves huge amounts of data that have both complex spatial and temporal dependencies. In the context of traffic-forecasting problems, spatial dependencies of traffic time series, refer to the topological information of the transportation network and its effects on adjacent or distant traffic measurement points [32]. For instance, the traffic state in a particular location may (or not) be affected by traffic on nearby roads. Furthermore, complex temporal dynamics may include seasonality, periodicity, trend and other unexpected random events that may occur in a transportation network, such as accidents, construction sites and weather.

Data-driven traffic-forecasting modeling has been at the center of transportation research activity and efforts during the last three decades [1]. Most of the existing literature focuses on the prediction of three main traffic states, namely traffic flow (vehicles/time unit), average speed, and density (vehicles/distance). Many existing forecasting methods consider temporal dependence based on classic statistics such as Autoregressive Integrated Moving Average and its variants [33], and more complex machine-learning methods including Support Vector regression machine method [34], K-nearest neighbor models [35,36], and Bayesian networks [37]. Although both statistical and machine-learning models effectively consider the dynamic temporal features of past traffic conditions, they fail to extract spatial dependence.

2.2. Deep-Learning Approaches

Recently, the emerging development of deep-learning and neural networks has achieved significant success in traffic-forecasting tasks [13,14]. To model the temporal non-linear dependency, researchers proposed Recurrent Neural Networks (RNN) and their variants. RNNs are deep neural network models that manage to adapt sequential data but suffer from exploding and vanishing gradients during back propagation [38]. To mitigate these effects during training of such models, numerous research studies proposed alternative architectures based on RNNs, such as Long Short Memory networks (LSTM) and Gated Recurrent Units (GRU) [39–41]. These variants of RNNs use gated mechanisms to memorize long-term dependencies in sequence-based data including historical traffic information [42]. Their structure consists of various forget units that determine which information could be excluded from the prediction output, thus determining the optimal time windows [40]. GRUs are similar architectures to LSTMs, but they have a simpler structure being computationally efficient during training and their cells consist of two main gates, namely the reset gate and the update gate [40]. Similar to the traditional statistical and machine-learning models, the recurrent-based models ignore the spatial information that is hidden in traffic data, failing to adapt the road network topology. To this end, many research efforts focused on improving the prediction accuracy by considering temporal, as well as spatial features. An approach to capture the spatial relations among the traffic network is the use of convolutional neural networks (CNNs) combined with recurrent neural networks (RNN) for temporal modeling. For example, Ref. [43] proposed a 1-dimension CNN to capture spatial information of traffic flow combined with two LSTMs to adapt short-term variability and various periodicities of traffic flow. Another attempt to align spatial and temporal patterns is made by [44], proposing a convolutional LSTM that models a spatial region with a grid, extending the convolution operations applied on grid structures (e.g., images). Furthermore, Ref. [45] deployed deep convolutional networks to capture spatial relationships among traffic links. In this study the network topology is represented by a grid box, where near and far dependencies are captured at each convolutional operation. These spatial convolutions are then combined with LSTM networks that adapt temporal information.

In recent years, a significant amount of research papers have focused on embedding spatial information into traffic-forecasting models [14,21]. However, methods that use convolutional neural networks (CNNs) are limited to cases where the data have a Euclidean structure, thus they cannot fully capture the complex topological structure of transportation networks, such as subways or road networks. Graph Neural Networks (GNNs) have become the frontier of deep-learning research in graph representation learning, demonstrating state-of-the-art performance in a variety of applications [46]. They can effectively model spatial dependencies that are represented by non-Euclidean graph structures. To this end, they are ideally suited for traffic-forecasting problems because a road network can be naturally represented as a graph, with intersections modeled as graph nodes and roads as edges.

2.3. Graph Neural Networks for Traffic Forecasting

Deep-learning algorithms including convolutional neural networks (CNNs) and recurrent neural networks (RNNs) have significantly contributed to the progress of many machine-learning tasks such as object detection, speech recognition, and natural language processing [47–49]. These models can extract latent representations from Euclidean data such as images, audio, and text. For example, an image can be represented as a regular grid in Euclidean space, where a CNN can extract several meaningful features by identifying the topological structure of the image. Although neural networks effectively capture hidden patterns in Euclidean data, they cannot handle the arbitrary structure of graphs or networks [50]. There is an increasing number of applications where data are represented as graphs, such as social networks [51], networks in physics [52], and molecular graphs [53]. Traffic graphs are constructed based on traffic sensor data where each sensor is a node and edges are connections (roads) between sensors.

A traffic graph is defined as $G = (V, E, A)$, where V is the set of nodes that contains the historical traffic states for each sensor, E is the set of edges between nodes (sensors) and A the adjacency matrix. Recently, numerous graph neural network models have been developed in the traffic-forecasting domain, which effectively consider the spatial correlations between traffic sensors. They also integrate different sequence-based models, leveraging the RNN architecture to model the temporal dependency.

Graph Convolutional Networks (GCN) are variants of convolutional neural networks that operate directly on graphs [19,54] defining the representation of each node by aggregating features from the adjacent nodes. Graph convolutions play a central role in many GNN-based traffic-forecasting applications [20,55–57]. For example, Ref. [58] proposed the Spatio–Temporal Graph Convolutional Network model that introduces a graph convolution operator using spectral techniques by computing graph signals with Fourier transformations. In this study, researchers propose two spatio–temporal convolutional blocks by integrating graph convolutions and gated temporal convolutions to accurately predict traffic speed outperforming other baseline models. Moreover, Ref. [22] proposed the Temporal Graph Convolutional Network that also uses graph convolutions and spectral filters to acquire the spatial dependency while temporal dynamics are captured using the gated recurrent unit. In this study [54] authors proposed Graph WaveNet, a GNN model that captures spatial dependencies with graph convolutions and the construction of an adaptive adjacency matrix that learns spatial patterns directly from the data. Another technique that models spatial dependency initially proposed by [59], defines convolution operations as a diffusion process for each node in the input graph. Towards this direction, diffusion convolutional recurrent neural network (DCRNN) [60] manages to capture traffic spatial information using random walks in the traffic graph, while temporal dependency is modeled with an encoder-decoder RNN technique.

Besides the mentioned spatio–temporal convolutional models, there has been an increased interest in attention-based models in traffic-forecasting problems. Attention mechanisms are originally used in natural language processing, speech recognition, and computer vision tasks. They are also applied on graph-structured data as initially suggested by [57] as well as time-series problems [61]. The objective of the attention mechanism is to select the information that influences the prediction task most. In traffic forecasting, this information may be included in daily periodic or weekly periodic dependencies. For example, the authors in [62] deployed the Attention-Based Spatial–Temporal Graph Convolutional Network (ASTGCN) that simultaneously employs spatio–temporal attention mechanisms and spatial graph convolutions along with temporal convolutions for traffic flow forecasting. Towards this direction, there are many attention-based GNNs considered in the literature that accurately predict traffic states, being suitable for traffic datasets as they manage to assign larger weights to more important nodes of the graph [63–66].

3. Background

This section presents the background knowledge on the two types of graph neural network (GNN) algorithms that are employed in this work to forecast traffic flow, namely Temporal Graph Convolutional Networks (TGCN) and Diffusion Convolutional Recurrent Neural Networks (DCRNN). All notations and symbols utilized in this study are comprehensively listed in Table A1 of the Appendix A.

3.1. Temporal Graph Convolutional Network

In dynamic data produced by sensors, the Temporal Graph Convolutional Network (TGCN) algorithm [22] uses graph convolutions to capture the topological structure of the sensor network to acquire spatial embeddings of each node. Then the obtained time series with the spatial features are used as input into the Gated Recurrent Unit (GRU), which models the temporal features. The graph convolution encodes the topological structure of the sensor network and defines the spatial features of a target node also obtaining the attributes of the adjacent sensors. Following the spectral transformations of graph signals as proposed by [22,58,67,68], two graph convolution layers are defined as:

$$f_0(X, A) = Relu(W_0(D^{-1/2}\hat{A}D^{-1/2})X), \tag{1}$$

$$f_1(X, A) = \sigma(W_1(D^{-1/2}\hat{A}D^{-1/2})f_0), \tag{2}$$

where X is the feature matrix with the obtained traffic flows, A is the adjacency matrix, W_0, W_1 are the learnable weight matrices in the first and second layers, D is the degree matrix, $\hat{A} = A + I_N$ is the self-connection matrix, and $\sigma(), Relu()$ represent the non-linear activation functions.

The output of the spatial information is fed into a GRU network [22]. The gated unit captures temporal dependency by initially calculating the reset gate r_t and update gate u_t, which are then fed in a memory cell c_t. The final output of the unified spatio–temporal block h_t at time t takes as input the hidden traffic state at time $t - 1$ updating the current information with the previous time step, along with the current traffic information:

$$h_t = u_t * h_{t-1} + (1 - u_t) * c_t \tag{3}$$

3.2. Diffusion Convolutional Recurrent Neural Network

The Diffusion Convolutional Recurrent Neural Network (DCRNN) algorithm [60] uses a different graph convolution approach to model spatial dependencies. The model captures spatial information using a diffusion process by generating random walks on sensor graph G with a restart probability $a \in [0, 1]$. In summary, the diffusion convolution over a graph signal x as presented by the authors of DCRNN is defined as follows:

$$f * x = \sum_{k=0}^{K} (\theta_{k,1}(D_O^{-1}A)^k + \theta_{k,2}(D_I^{-1}A^\tau)^k)x \tag{4}$$

where k is the diffusion step, $D_O^{-1}A, D_I^{-1}A^\tau$ denote the transition and reverse matrices of the diffusion process, respectively, and θ_k are the parameters of the filter.

For the modeling of temporal dynamics, the framework also adapts the GRU architecture using an encoder-decoder method. Precisely, the historical traffic states are fed into the encoder and the decoder is responsible for the final prediction of the model. Both encoder and decoder combine the diffusion convolutions along with the GRUs, while the architecture of GRU is similar to the TGCN implementation.

4. Research Approach

The research approach of this work uses four steps, namely (1) data collection (Section 4.1); (2) data pre-processing (Section 4.2); (3) forecasting model creation (Section 4.3); and (4) forecasting model evaluation (Section 4.4). Python was used throughout the entire approach.

4.1. Data Collection

In this step available traffic data from data.gov.gr (accessed on 27 March 2023) are collected using the data.gov.gr (accessed on 27 March 2023) API. These data have been produced by sensors that are positioned in the Attica region in Greece. In addition, the position of the sensors is specified and mapped to latitude and longitude geographic coordinators.

4.2. Data Pre-Processing

This step aims to prepare the dataset for being used as an input for the creation of the two GNN models. Specifically, observations coming from unreliable sensors as well as anomalous observations are identified and removed. Towards this end, this step first explores the InterQuartile Range (IQR) of the vehicles measured by each sensor. IQR measures the spread of the middle half of the data by calculating the difference between the first and third quartiles and can help identify abnormal behaviors of sensors (e.g., sensors that repeatedly generate similar values). The first quartile (Q_1), is the value in the data set that holds 25% of the values below it, while the third quartile (Q_3), is the value in the data set that holds 25% of the values above it. IQR is then calculated as follows:

$$IQR = Q_3 - Q_1 \tag{5}$$

Thereafter, the missing observations of the dataset are identified. Missing observations are common in traffic data since they are dynamic data collected by sensors due to reasons such as failures of sensors, network faults, and other issues. Missing observations from the traffic data are identified and analyzed based on two dimensions; (i) the time dimension, where missing observations per day are calculated, and (ii) the sensors, where the total number of missing values per sensor is calculated. For the first case, the number of available observations is found and then subtracted from the number of observations that should be available for all sensors. For the second case, statistical analyses are employed to explore the distribution of the sensors' missing observations. Observations generated by sensors with large quantities of missing observations are removed. Finally, in this step, the anomalies per sensor are calculated. Specifically, we use the flow-speed correlation analysis to find anomalies in the measurements of the data. This kind of analysis relies on the fact that the number of vehicles counted by a sensor and their average speed are strongly correlated. Specifically, considering that each sensor measures data that pass from one or more lanes, the maximum number of vehicles that can pass in all lanes in one hour can be calculated as [69]:

$$number\_of\_vehicles = \frac{average\_speed * 1000}{average\_vehicle\_length + \dfrac{average\_speed}{3.6}} * number\_of\_lanes \tag{6}$$

where *average_speed* is the average speed provided by the sensors measured in km per hour and *average_vehicle_length* is the average length of the different types of vehicles, the fraction *average_speed*/3.6 represents the "safe driving distance" that should be kept between vehicles and is based on the vehicle speed, and *number_of_lanes* is the number of lanes in the road each sensor is positioned. The value of *average_vehicle_length* is set to 4. When the number of vehicles measured by a sensor in an hour is higher than this value, then the measurement is considered an anomaly and is removed from the dataset.

4.3. Forecasting Model Creation

This work uses two types of widely used and open-source GNN algorithms to forecast traffic flow, namely Temporal Graph Convolutional Network (TGCN) and Diffusion Convolutional Recurrent Neural Network (DCRNN). The two forecasting GNN models are created based on the preprocessed dataset of the previous step. Towards this end, the input traffic flow data are normalized to the interval [0, 1] using the min-max scaling technique. Moreover, the missing values are imputed using linear interpolation. In addition, 70% of the data are used for training, 20% for testing, and 10% for validation. The training and validation parts of the dataset were used to train and fine-tune the two GNN models, while the test part was to evaluate the created models. For each model, an adjacency matrix of the sensor graph is created based on the distances d_{ij} between the 406 sensors similar to most related studies using thresholded Gaussian kernel. $A_{ij} = 1$ if $\exp(-\frac{d_{ij}^2}{\sigma^2}) >= \epsilon$, otherwise $A_{ij} = 0$. σ^2, ϵ are thresholds that determine the distribution and sparsity of the matrix and are set to 10 and 0.5, respectively. The proposed algorithms use 12 past observations to forecast traffic flow in the next 3 (short-term prediction), 6 (middle-term prediction), and 9 (long-term prediction) time steps (hours). The created models were fine-tuned to determine the optimal values of the hyperparameters.

4.4. Forecasting Model Evaluation

The performance of both the created forecasting models was measured using three metrics, namely Root Mean Squared Error (RMSE), Mean Absolute Error (MAE), and Mean Absolute Percentage Error (MAPE). These metrics are computed as follows:

$$\text{RMSE} = \sqrt{\frac{1}{n} \sum_{i=1}^{n} (y_t - \hat{y}_t)^2} \tag{7}$$

$$\text{MAE} = \frac{1}{n} \sum_{i=1}^{n} |y_t - \hat{y}_t| \tag{8}$$

$$\text{MAPE} = \frac{1}{n} \sum_{i=1}^{n} \frac{|y_t - \hat{y}_t|}{y_t} * 100 \tag{9}$$

where y_t denotes the real traffic flow and $\hat{y}_t$ the corresponding predicted value.

Thereafter, the performance of the models was compared against the performance of two baseline models, i.e., Historical Average (HA) and Autoregressive Integrated Moving Average (ARIMA) that were created using the same dataset.

5. Data Collection

Data.gov.gr is the official Greek data portal for open-government data (OGD). The latest intorsion of the data portal was released in 2020 and provides access to 49 datasets published by the central government, local authorities, or other Greek public bodies classified in ten thematic areas including environment, economy, and transportation.

The major update and innovation of the latest version of the Greek OGD portal was the introduction of an Application Programming Interface (API) that enables accessing and retrieving the data through either a graphical interface or code. The API is freely provided and can be employed to develop various products and services including data intelligence applications. Acquiring a token is needed to use the API by completing a registration process. This process requires providing personal information (i.e., name, email, and organization) as well as the reason for using the API.

The introduction of the API enables the timely provision of dynamic data that are frequently updated. The API can be used, for example, to retrieve datasets describing data related to a variety of transportation systems (e.g., road traffic for the Attica region, ticket validation of Attica's Urban Rail Transport, and route information and passenger counts of Greek shipping companies). The frequency of data update varies.

The traffic data for the Attica region in Greece were collected from traffic sensors, which periodically transmit traffic information regarding the number of vehicles on specific roads of Attica along with their speed. The data are hourly aggregated to avoid raising privacy issues. Data are updated hourly with only one hour delay.

We used the API provided by data.gov.gr and collected 1,311,608 records for five months, specifically, from 2 August 2022 to 17 December 2022 (138 days). Figure 1 presents a snapshot of the traffic data. Each record includes (a) the unique identifier of the sensor ("deviceid") (e.g., "MS834"); (b) the road in which the sensor is located along with ("road_name"); (c) a detailed text description of its position ("road_info"); (d) the date and time of the measurement ("appprocesstime"); (e) the absolute number of the vehicles detected by the sensor during the hour of measurement ("countedcars"); and (f) their average speed in km per hour ("average_speed"). The exact position of the sensor is a text description in Greek language and usually provides details including whether the sensor is located on a main or side road, or on an exit or entrance ramp, the direction of the road (e.g., direction to center), and the distance to main roads (e.g., "200 m from Kifisias avenue").

deviceid	countedcars	appprocesstime	road_name	road_info	average_speed
MS394	432	2022-08-02T00:00:00Z	ΒΑΣ. ΣΟΦΙΑΣ	ΚΥΡΙΟΣ ΔΡΟΜΟΣ ΜΕ ΚΑΤΕΥΘΥΝΣΗ ΚΕΝΤΡΟ 100 Μ. ΠΡΙΝ...	48.853659
MS434	641	2022-08-02T00:00:00Z	Λ. ΜΕΣΟΓΕΙΩΝ	ΚΥΡΙΟΣ ΔΡΟΜΟΣ ΜΕ ΚΑΤΕΥΘΥΝΣΗ ΠΡΟΣ ΑΓ. ΠΑΡΑΣΚΕΥΗ...	44.975610
MS435	619	2022-08-02T00:00:00Z	Λ. ΜΕΣΟΓΕΙΩΝ	ΚΕΝΤΡΙΚΟΣ ΚΛΑΔΟΣ ΜΕ ΚΑΤΕΥΘΥΝΣΗ Λ. ΒΑΣ. ΣΟΦΙΑΣ ...	43.926829
MS436	803	2022-08-02T00:00:00Z	ΦΕΙΔΙΠΠΙΔΟΥ	ΚΥΡΙΟΣ ΔΡΟΜΟΣ ΜΕ ΚΑΤΕΥΘΥΝΣΗ ΑΝΑΤΟΛΙΚΑ 80 Μ. ΠΡ...	43.121951
MS437	429	2022-08-02T00:00:00Z	Λ. ΜΕΣΟΓΕΙΩΝ	ΚΥΡΙΟΣ ΔΡΟΜΟΣ ΜΕ ΚΑΤΕΥΘΥΝΣΗ ΚΕΝΤΡΟ (ΝΔ) ΠΡΙΝ Α...	54.585366
...	...	...	...	...	...
MS946	941	2022-12-30T06:00:00Z	Π. ΡΑΛΛΗ	ΚΥΡΙΟΣ ΔΡΟΜΟΣ ΜΕ ΚΑΤΕΥΘΥΝΣΗ ΑΘΗΝΑ 40 Μ. ΠΡΙΝ Α...	19.894737
MS945	178	2022-12-30T06:00:00Z	Λ. ΚΗΦΙΣΟΥ	ΡΑΜΠΑ ΕΞΟΔΟΥ ΠΡΟΣ Π. ΡΑΛΛΗ ΤΟΥ ΚΛΑΔΟΥ ΤΗΣ Λ. Κ...	24.052632
MS852	69	2022-12-30T06:00:00Z	ΕΘΝ. ΜΑΚΑΡΙΟΥ	ΚΥΡΙΟΣ ΔΡΟΜΟΣ ΜΕ ΚΑΤΕΥΘΥΝΣΗ ΠΡΟΣ ΑΛΙΜΟ, 150 Μ...	40.894737
MS962	568	2022-12-30T06:00:00Z	ΑΘΗΝΩΝ	ΡΑΜΠΑ ΕΞΟΔΟΥ ΠΡΟΣ Λ. ΚΗΦΙΣΟΥ ΤΟΥ ΚΛΑΔΟΥ ΤΗΣ Λ...	54.842105
MS989	612	2022-12-30T06:00:00Z	ΕΘΝ. ΜΑΚΑΡΙΟΥ	ΚΥΡΙΟΣ ΔΡΟΜΟΣ ΜΕ ΚΑΤΕΥΘΥΝΣΗ ΠΡΟΣ ΑΘΗΝΑ, 40 ΜΕΤ...	34.684211

Figure 1. A snapshot of the traffic data from data.gov.gr.

6. Data Pre-Processing

The traffic data that were retrieved by the Greek open-data portal were produced by 428 sensors. We manually translated the text description of the position of the sensors to latitude and longitude geographic coordinators to be able to present data in a map visualization. Specific position details are missing for one sensor (i.e., the sensor with identifier "MS339") making it impossible to find its exact coordinates. The sensors are positioned on 93 main roads of the region of Attica.

We then calculated the InterQuartile Range (IQR) of the counted vehicles measured by each sensor to understand the spread of the values. The IQR for each sensor ranges from 0 to 2985.5, while the mean IQR is 826.46. There are eight sensors with IQR equal to 0, which reveals an abnormal behavior since it means that the first and third percentiles are the same and that all the measurements of these sensors are very similar.

Thereafter, we searched for observations that are missing from the traffic data based on two dimensions; (i) the time, where we calculate the missing values per day; and (ii) the sensors, where the missing values per sensor are calculated. Given all the sensor measurements collected and that the traffic data are hourly aggregated, the total number of observations that would have been made by the 428 sensors over the course of the 138 days would be 1,417,536. However, 105,928 observations (or 7.47%) are missing. This number is significantly better than the 20.16% of missing values we discovered in our earlier work that analyzed traffic data collected from data.gov.gr from November of 2020 to June of 2022 [27]. Figure 2 presents the number of missing observations per day for the two different time periods.

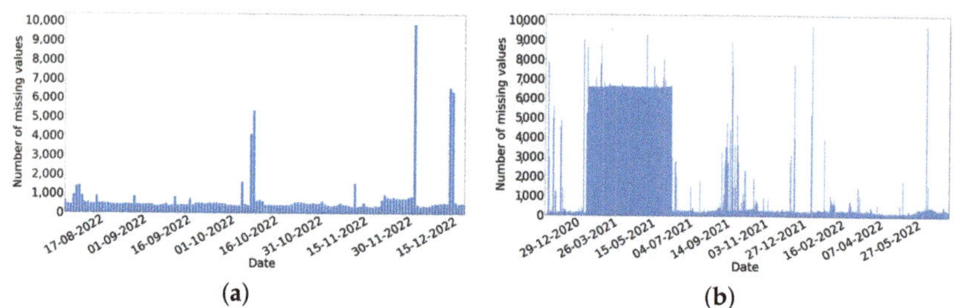

Figure 2. Number of observations that are missing per day for: (**a**) 2 August 2022 to 17 December 2022; and (**b**) 5 November 2020 to 31 June 2022.

We also calculate the percentage of missing observations for each sensor from 2 August 2022 to 17 December 2022. The median percent of missing observations is 33.1% meaning that half of the sensors have less than or equal percentages of missing observations to the median, and half of the sensors have greater than or equal percentages of missing observations to it. The 50% of the sensors have a percentage of missing observations in the range 3–5% (interquartile range box). In addition, according to the whiskers of the boxplot (bottom 25% and top 25% of the data values, excluding outliers), the percentage of missing observations of each sensor may be as low as 3% and as high as 7%. Furthermore, the absolute majority of the sensors (i.e., 85.5%) have less than 10% missing observations. Finally, only five sensors have more than 90% of missing observations and 22 sensors have more than 25% missing values. For the creation of the forecasting models we removed these 22 sensors.

Furthermore, we calculate the number and percentages of anomalies per sensor based on the flow-speed correlation analysis described in Section 3. In order to be able to calculate the number of vehicles that can pass in all lanes, we manually found the number of lanes that each sensor tracks and mapped them to the records. We discovered only 58 records generated by 26 sensors that count more vehicles than the number calculated by the flow-speed correlation analysis. This number is significantly lower compared to the 59.4% of anomalies found in our earlier work. These anomalous measurements relate to only 18 days, which is also a significant improvement related to the analysis of the traffic data from earlier dates when anomalies were generated almost throughout the entire time period. We also calculated the number of anomalies per sensor. This number ranges from 0 to 23 anomalies, while the mean number of detected anomalies per sensor is 0.13 anomalies. In order to create the forecasting model, anomalous observations were removed from the dataset.

7. Forecasting Traffic Flow

The creation and evaluation of the forecasting GNN models are based on the pre-processed dataset of the previous step. Specifically, 1,354,416 observations generated by 406 sensors were used to create the Temporal Graph Convolutional Network (TGCN) and Diffusion Convolutional Recurrent Neural Network (DCRNN) models and evaluate them against two baseline models, i.e., Historical Average (HA) and Autoregressive Integrated Moving Average (ARIMA).

Table 1 presents the results of the fine-tuning for both GNN models. More precisely, the learning rate is set to 0.001 for both GNN algorithms as well as a batch size of 50. The training process is deployed using the Adam optimizer for both algorithms. TGCN is trained for 100 epochs and DCRNN for 200 epochs. For the TGCN algorithm the graph convolution layer sizes are set to 64 and 10 units, respectively, while the two GRU layers consist of 256 units. Regarding DCRNN both the encoder and decoder consist of two recurrent layers with 64 units each. Following the paper definition, the maximum steps K of random walks on the graph for the diffusion process is set to 3.

Table 1. Optimal hyper-parameter values for the two forecasting models.

	TGCN	DCRNN
Learning rate	0.001	0.001
Batch size	50	50
Epochs	100	200
GCN layer sizes (1st/2nd layer)	64/10	-
GRU layer sizes (1st/2nd layer)	256/256	64/64
max steps of random walks	-	3

The performance of the GNN-based algorithms is compared with the performance of two baseline methods: Historical Average (HA) and Autoregressive Integrated Moving Average (ARIMA). Table 2 shows the comparison of the performance of different algorithms for three forecasting horizons. All the error metrics are calculated by computing the mean error of each sensor and then averaging it over all 406 sensors. Thus, the evaluation metrics presented in Table 2 represent the overall prediction performance of the proposed algorithms considering the three error metrics among the three forecasting horizons.

Table 2. Performance comparison for GNN and baseline models on the Greek OGD dataset.

Forecasting Horizon	Metric	HA	ARIMA	TGCN	DCRNN
3	RMSE	757.58	534.51	222.2	244.58
	MAE	556.35	466.47	125.12	151.10
	MAPE	7.06%	4.33%	3.98%	6.39%
6	RMSE	757.58	582.33	260.42	331.04
	MAE	556.35	501.13	146.73	212.52
	MAPE	7.06%	7.02%	3.96%	7.664%
9	RMSE	757.58	690.12	267.88	398.31
	MAE	556.35	589.98	156.06	263.54
	MAPE	7.06%	6.98%	4.01%	7.8%

Specifically, the two GNN algorithms that emphasize the modeling of spatial dependence perform better in terms of prediction precision compared to the baselines. The results show that the TGCN algorithm outperforms all the other methods regarding all the error metrics for all prediction horizons. For the 3 time steps forecasting horizon, the RMSE error of TGCN and DCRNN is decreased by 58.42% and 54.24% compared with ARIMA model, respectively, and by 70.66% and 67.77% compared with HA. Although the error metrics of all models are increased towards the 6 and 9 time steps horizons, GNN models maintain better prediction results compared with the baselines. To verify which GNN model captures more effectively the spatial–temporal dependencies of traffic flow, we compare the results of TGCN and DCRNN. According to Table 2, the TGCN model demonstrates the best prediction performance among all prediction steps, being able to capture not only short-term, but also long-term spatial–temporal dependencies of the traffic network. This indicates that, for the specific case study, where the traffic flow measurements come from an urban environment where sensors are located close to each other, the graph convolution operation of TGCN captures the complex topology of the sensor network better than the diffusion process of the DCRNN model. Since both models use a similar architecture to model the temporal traffic information with gated recurrent units (GRUs), TGCN effectively captures the spatial dependencies of traffic flows that are obtained from a dense, complex sensor network.

To diagnose the behavior of the proposed models, we created two learning curves (per model) that are calculated based on the metric by which the parameters of the model are optimized, in our case the loss function. To this end, a training learning curve, which is calculated from the loss of the training dataset, and a validation curve, which is calculated from the validation dataset were created. Figure 3 depicts the learning curves of the two GNN models with the number of training epochs in x-axis and scaled MAE in y-axis. TGCN algorithm achieves the lowest validation and training error suggesting superior performance during the training process. Moreover, both validation and training losses regarding the two models, decrease to a point of stability. In addition, Figures 4 and 5 show visualization results of the two GNN algorithms for the 3 h forecasting horizon, for sensors MS109 and MS985, between 16 December 2022 and 17 December 2022. It is observed that TGCN predicts traffic flow slightly better than DCRNN, specifically in high peaks of traffic flow. Therefore, TGCN is more likely to accurately predict abrupt changes in the traffic flow.

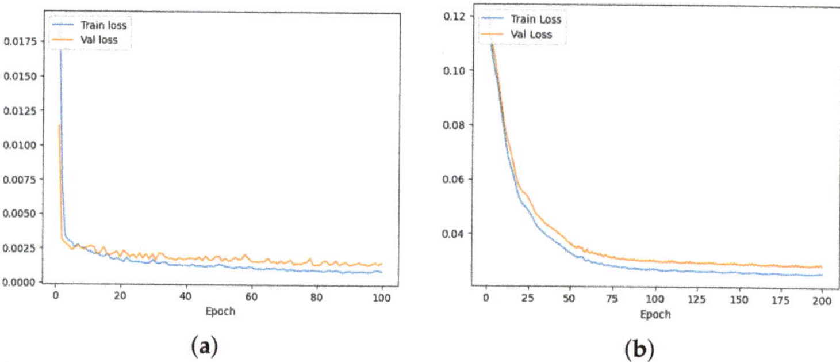

Figure 3. Learning curves with training and validation error for: (**a**) Temporal Graph Convolutional networks (TGCN), and (**b**) Diffusion Convolutional Recurrent Neural Network (DCRNN).

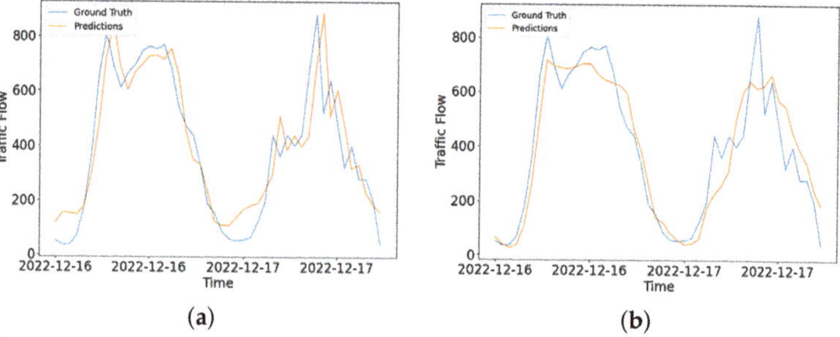

Figure 4. Visualizations of prediction results for the forecasting horizon of 3 h for sensor MS109: (**a**) Temporal Graph Convolutional networks (TGCN), and (**b**) Diffusion Convolutional Recurrent Neural Network (DCRNN).

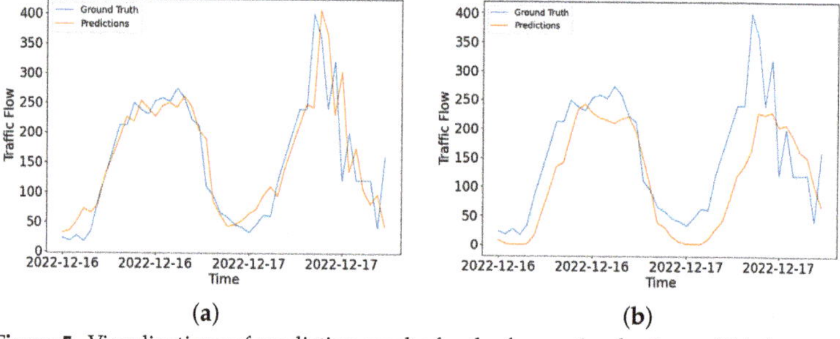

Figure 5. Visualizations of prediction results for the forecasting horizon of 3 h for sensor MS985: (**a**) Temporal Graph Convolutional networks (TGCN), and (**b**) Diffusion Convolutional Recurrent Neural Network (DCRNN).

8. Discussion

Traffic forecasting is a crucial component of modern intelligent transportation systems, which aim to improve traffic management and public safety [39,70,71]. However, it remains a challenging problem, as several traffic states are influenced by a multitude of complex factors, such as the spatial dependence of intricate urban road networks and complex temporal dynamics. In the literature, many studies have employed graph neural networks (GNNs) which have achieved state-of-the-art performance in traffic forecasting due to their powerful ability to extract spatial information from non-Euclidean structured data commonly encountered in the field of mobility data. The complex spatial dependency of traffic networks can be captured using graph convolutional aggregators on the input graph, while temporal dynamics can be extracted through the integration of recurrent sequential models.

Accessing historical traffic data is essential for deploying models in traffic forecasting. However, obtaining such data can be challenging due to privacy concerns, transmission, and storage restrictions. Most research studies on traffic forecasting using GNNs [19,20,22,58,60,62,65,66] have used open traffic data that are already cleansed and preprocessed, including METR-LA, a traffic speed dataset from the highway system of Los Angeles County containing data from 207 sensors during 4 months in 2012 preprocessed by [60], and Performance Measurement System (PeMS) Data (https://pems.dot.ca.gov/(accessed on 3 January 2023)) consisting of several subsets of sensor-generated data across metropolitan areas of California. Although these datasets are often used for benchmarking and comparing the prediction performance of various models, it is important to note that they may not reflect current traffic patterns. This is because traffic data are collected from a past period (e.g., in the case of METR-LA from 2012) and a specific geographical area, such as a highway system, rather than a densely populated urban environment, as typically found in cities.

In recent years, governments and the public sector have started to publish dynamic open-government data (e.g., traffic, environmental, satellite, and meteorological data) freely accessible and reusable on their portals [23–25]. This type of data can potentially facilitate the implementation of innovative machine-learning applications [72,73], including state-of-the-art algorithms in traffic forecasting. For instance, the Swiss OGD (https://opentransportdata.swiss/en/ (accessed on 1 November 2022)) portal provides real-time streaming traffic data that is updated every minute. The significance of dynamic open-government data for the deployment of graph neural networks (GNNs) in traffic forecasting cannot be overstated. First, these data sources are open and easily accessible through Application Programming Interfaces (APIs), enabling researchers to retrieve the necessary traffic information without undergoing procedures that include restricted authorization protocols. Furthermore, these data sources are often updated in real time, providing up-to-date traffic information for analysis and prediction. Second, the availability of such data allows for the evaluation and experimentation of relevant GNN models that are currently applied on commonly used benchmarking preprocessed datasets. Therefore, the use of dynamic open-government data has the potential to enhance the accuracy and efficiency of GNN-based traffic-forecasting models.

In this study, two well-known GNN variants, namely Temporal Graph Convolutional Networks (TGCN) and Diffusion Convolutional Recurrent Neural Networks (DCRNN), were used to forecast traffic flow. Specifically, the models were trained on the Greek OGD dataset, and following related literature, 12 past observations, equivalent to 12 past hours, were used to predict traffic flow in the next 3, 6, and 9 h. Before deploying the two models, the OGD traffic dataset underwent pre-processing to address missing observations and anomalies. As a result, sensors with more than 25% missing values and traffic observations detected as anomalies through flow-speed correlation analysis were excluded from the experiments. To model the network topology, a 406×406 adjacency matrix was created based on pairwise distances between traffic sensors.

Both GNN models achieved better prediction performance across all prediction horizons and among all error metrics (RMSE, MAE, MAPE) compared with the two baseline models. Overall, TGCN achieves the best prediction results compared with DCRNN and baselines. For this specific case study, TGCN captures spatial dependencies using graph convolutions from spectral theory, outperforms the DCRNN model that on the other hand, and captures spatial information using bidirectional random walks on the sensor graph with a diffusion process. In summary, both GNN-based models manage to efficiently capture the topological structure of the sensor graph, as well as complex temporal dynamics compared with traditional baselines that only handle time-related features (HA, ARIMA).

The traffic data used to forecast traffic flow in the region of Attica were retrieved by the Greek data portal using the provided API. The data include traffic measurements for the time period 2 August 2022 to 17 December 2022. The exploration of the data showed that the major quality difficulties, including a lot of missing observations as well as anomalous observations, found in the authors' earlier research [27] have been resolved to a large extent. As a result, these data can be used as a trusted source to make accurate predictions and, thereafter, take informed decisions.

9. Conclusions

The findings of this study demonstrate that open-government data (OGD) is an invaluable resource that can be leveraged by researchers to develop and train more advanced graph neural network (GNN) algorithms. However, the performance of GNNs is highly dependent on the quality and quantity of the data on which they are trained. OGD provides a unique opportunity for researchers to access vast amounts of data from various sources. These data can be used to train GNN models to generalize across a broad range of traffic datasets, resulting in more accurate predictions. Furthermore, the availability of OGD from multiple countries and governments can enable the development of more comprehensive models that can be used to forecast traffic patterns in different regions and under varying conditions. Finally, most studies in the field of traffic forecasting use historical traffic data on small intervals, typically 5 or 15 min. However, this study focuses on the case of the Greek OGD traffic data, which stores traffic data in one-hour intervals. To this end, we plan to conduct further research in dynamic data from other OGD portals that contain datasets of higher quality and smaller aggregated time intervals. In any case, we believe that by continuing to investigate the potential of OGD datasets, will advance the field of traffic forecasting and contribute to the development of more accurate and comprehensive models for predicting traffic patterns.

Author Contributions: Conceptualization, E.K. and A.K.; methodology, E.K.; software, A.K. and P.B.; data curation, A.K. and P.B.; writing—original draft preparation, A.K. and P.B.; writing—review and editing, A.K. and E.K.; supervision, E.K. and K.T.; project administration, E.K. and K.T. All authors have read and agreed to the published version of the manuscript.

Funding: This research received no external funding.

Institutional Review Board Statement: Not applicable.

Informed Consent Statement: Not applicable.

Data Availability Statement: Publicly available datasets were analyzed in this study. These data can be found here: https://data.gov.gr/datasets/road_traffic_attica/ (accessed on 1 September 2022).

Conflicts of Interest: The authors declare no conflict of interest.

Abbreviations

The following abbreviations are used in this manuscript:

ITS	Intelligent Transportation System
OGD	Open-Government Data
API	Application Programming Interface
JSON	JavaScript Object Notation
XML	eXtensible Markup Language
GNN	Graph Neural Networks

ARIMA   Autoregressive Integrated Moving Average
HA   Historical Average
SVR   Support Vector Regression
KNN   K-Nearest Neighbor
RNN   Recurrent Neural Network
GRU   Gated Recurrent Unit
LSTM   Long Short Memory
CNN   Convolutional Neural Network
GCN   Graph Convolutional Network
TGCN   Temporal Graph Convolutional Network
DCRNN   Diffusion Convolutional Recurrent Neural Network
ASTGCN   Attention-based Spatial–Temporal Graph Convolutional Network
IQR   InterQuartile Range
RMSE   Root Mean Squared Error
MAE   Mean Absolute Error
MAPE   Mean Absolute Percentage Error

Appendix A

Table A1. Table of notations and symbols used in this paper.

Notation	Description
G	A graph
V	The set of nodes of a graph $V = \{v_1, v_2, \ldots, v_n\}$
E	The set of edges of a graph $E = \{(v_i, v_j) | v_i, v_j \in V\}$
A	The adjacency matrix of a graph
$\hat{A}, I_N$	Self-connection adjacency matrix and Identity matrix
D	The degree matrix
X	The feature matrix consisting of historical traffic flows
f_0, f_1	First and second graph convolutional layers
W_0, W_1	Weight matrices of first and second layers
σ	A non-linear activation function
$ReLu$	The Rectified Linear Unit for an input x: $ReLu(x) = max(0, x)$
h_t, h_{t-1}	The output layer of a recurrent unit at time $t, t-1$
r_t, u_t	The reset and update gates of a GRU at time t
c_t	The memory cell of a GRU at time t
$f * x$	A diffusion convolution f over a graph signal x
θ_k	The parameters of a diffusion convolutional layer
D_I, D_O	Input and output degree matrices of the DCRNN model
IQR	the discrepancy between the 75th and 25th percentiles of the data $Q_3 - Q_1$

References

1.   Lana, I.; Del Ser, J.; Velez, M.; Vlahogianni, E.I. Road Traffic Forecasting: Recent Advances and New Challenges. *IEEE Intell. Transp. Syst. Mag.* **2018**, *10*, 93–109. [CrossRef]
2.   Varga, N.; Bokor, L.; Takács, A.; Kovács, J.; Virág, L. An architecture proposal for V2X communication-centric traffic light controller systems. In Proceedings of the 2017 15th International Conference on ITS Telecommunications (ITST), Warsaw, Poland, 29–31 May 2017; pp. 1–7.
3.   Navarro-Espinoza, A.; López-Bonilla, O.R.; García-Guerrero, E.E.; Tlelo-Cuautle, E.; López-Mancilla, D.; Hernández-Mejía, C.; Inzunza-González, E. Traffic Flow Prediction for Smart Traffic Lights Using Machine Learning Algorithms. *Technologies* **2022**, *10*, 5. [CrossRef]
4.   Ran, X.; Shan, Z.; Fang, Y.; Lin, C. An LSTM-Based Method with Attention Mechanism for Travel Time Prediction. *Sensors* **2019**, *19*, 861. [CrossRef] [PubMed]

5. Ata, A.; Khan, M.A.; Abbas, S.; Khan, M.S.; Ahmad, G. Adaptive IoT empowered smart road traffic congestion control system using supervised machine learning algorithm. *Comput. J.* **2021**, *64*, 1672–1679. [CrossRef]
6. Kashyap, A.A.; Raviraj, S.; Devarakonda, A.; Shamanth, R.N.K.; Santhosh, K.V.; Bhat, S.J.; Galatioto, F. Traffic flow prediction models—A review of deep learning techniques. *Cogent Eng.* **2022**, *9*, 2010510. [CrossRef]
7. Zahid, M.; Chen, Y.; Jamal, A.; Mamadou, C.Z. Freeway Short-Term Travel Speed Prediction Based on Data Collection Time-Horizons: A Fast Forest Quantile Regression Approach. *Sustainability* **2020**, *12*, 646. [CrossRef]
8. Cornago, E.; Dimitropoulos, A.; Oueslati, W. Evaluating the Impact of Urban Road Pricing on the Use of Green Transport Modes. *OECD Environ. Work. Pap.* **2019**.
9. Chin, A.T. Containing air pollution and traffic congestion: Transport policy and the environment in Singapore. *Atmos. Environ.* **1996**, *30*, 787–801. [CrossRef]
10. Rosenlund, M.; Forastiere, F.; Stafoggia, M.; Porta, D.; Perucci, M.; Ranzi, A.; Nussio, F.; Perucci, C.A. Comparison of regression models with land-use and emissions data to predict the spatial distribution of traffic-related air pollution in Rome. *J. Expo. Sci. Environ. Epidemiol.* **2008**, *18*, 192–199. [CrossRef]
11. Zhou, Q.; Chen, N.; Lin, S. FASTNN: A Deep Learning Approach for Traffic Flow Prediction Considering Spatiotemporal Features. *Sensors* **2022**, *22*, 6921. [CrossRef]
12. Kumar, P.B.; Hariharan, K. Time Series Traffic Flow Prediction with Hyper-Parameter Optimized ARIMA Models for Intelligent Transportation System. *J. Sci. Ind. Res.* **2022**, *81*, 408–415.
13. Yao, H.; Tang, X.; Wei, H.; Zheng, G.; Li, Z. Revisiting Spatial-Temporal Similarity: A Deep Learning Framework for Traffic Prediction. *Proc. AAAI Conf. Artif. Intell.* **2019**, *33*, 5668–5675. [CrossRef]
14. Yin, X.; Wu, G.; Wei, J.; Shen, Y.; Qi, H.; Yin, B. Deep Learning on Traffic Prediction: Methods, Analysis, and Future Directions. *IEEE Trans. Intell. Transp. Syst.* **2022**, *23*, 4927–4943. [CrossRef]
15. Qi, Y.; Cheng, Z. Research on Traffic Congestion Forecast Based on Deep Learning. *Information* **2023**, *14*, 108. [CrossRef]
16. George, S.; Santra, A.K. Traffic Prediction Using Multifaceted Techniques: A Survey. *Wirel. Pers. Commun.* **2020**, *115*, 1047–1106. [CrossRef]
17. Xie, P.; Li, T.; Liu, J.; Du, S.; Yang, X.; Zhang, J. Urban flow prediction from spatiotemporal data using machine learning: A survey. *Inf. Fusion* **2020**, *59*, 1–12. [CrossRef]
18. Chen, K.; Chen, F.; Lai, B.; Jin, Z.; Liu, Y.; Li, K.; Wei, L.; Wang, P.; Tang, Y.; Huang, J.; et al. Dynamic Spatio-Temporal Graph-Based CNNs for Traffic Flow Prediction. *IEEE Access* **2020**, *8*, 185136–185145. [CrossRef]
19. Bui, K.H.N.; Cho, J.; Yi, H. Spatial-temporal graph neural network for traffic forecasting: An overview and open research issues. *Appl. Intell.* **2022**, *52*, 2763–2774. [CrossRef]
20. Guo, K.; Hu, Y.; Qian, Z.; Sun, Y.; Gao, J.; Yin, B. Dynamic Graph Convolution Network for Traffic Forecasting Based on Latent Network of Laplace Matrix Estimation. *Trans. Intell. Transport. Syst.* **2022**, *23*, 1009–1018. [CrossRef]
21. Jiang, W.; Luo, J. Graph neural network for traffic forecasting: A survey. *Expert Syst. Appl.* **2022**, *207*, 117921. [CrossRef]
22. Zhao, L.; Song, Y.; Zhang, C.; Liu, Y.; Wang, P.; Lin, T.; Deng, M.; Li, H. T-GCN: A Temporal Graph Convolutional Network for Traffic Prediction. *IEEE Trans. Intell. Transp. Syst.* **2020**, *21*, 3848–3858. [CrossRef]
23. Kalampokis, E.; Tambouris, E.; Tarabanis, K. A classification scheme for open government data: Towards linking decentralised data. *Int. J. Web Eng. Technol.* **2011**, *6*, 266–285. [CrossRef]
24. Kalampokis, E.; Tambouris, E.; Tarabanis, K. Open government data: A stage model. In Proceedings of the Electronic Government: 10th IFIP WG 8.5 International Conference, EGOV 2011, Delft, The Netherlands, 28 August–2 September 2011; Springer: Berlin/Heidelberg, Germany, 2011; pp. 235–246.
25. Karamanou, A.; Kalampokis, E.; Tarabanis, K. Integrated statistical indicators from Scottish linked open government data. *Data Brief* **2023**, *46*, 108779. [CrossRef] [PubMed]
26. Parliament, E. Directive (EU) 2019/1024 of the European Parliament and of the Council of 20 June 2019 on open data and the re-use of public sector information (recast). *Off. J. Eur. Union* **2019**, *172*, 56–83.
27. Karamanou, A.; Brimos, P.; Kalampokis, E.; Tarabanis, K. Exploring the Quality of Dynamic Open Government Data Using Statistical and Machine Learning Methods. *Sensors* **2022**, *22*, 9684. [CrossRef] [PubMed]
28. Teh, H.Y.; Kempa-Liehr, A.W.; Wang, K.I.K. Sensor data quality: A systematic review. *J. Big Data* **2020**, *7*, 11. [CrossRef]
29. Mahrez, Z.; Sabir, E.; Badidi, E.; Saad, W.; Sadik, M. Smart Urban Mobility: When Mobility Systems Meet Smart Data. *IEEE Trans. Intell. Transp. Syst.* **2022**, *23*, 6222–6239. [CrossRef]
30. Vlahogianni, E.I.; Golias, J.C.; Karlaftis, M.G. Short-term traffic forecasting: Overview of objectives and methods. *Transp. Rev.* **2004**, *24*, 533–557. [CrossRef]
31. Vlahogianni, E.I.; Karlaftis, M.G.; Golias, J.C. Short-term traffic forecasting: Where we are and where we're going. *Transp. Res. Part C Emerg. Technol.* **2014**, *43*, 3–19. [CrossRef]
32. Ermagun, A.; Levinson, D. Spatiotemporal traffic forecasting: Review and proposed directions. *Transp. Rev.* **2018**, *38*, 786–814. [CrossRef]
33. Williams, B.M.; Hoel, L.A. Modeling and Forecasting Vehicular Traffic Flow as a Seasonal ARIMA Process: Theoretical Basis and Empirical Results. *J. Transp. Eng.* **2003**, *129*, 664–672. [CrossRef]
34. Yao, Z.; Shao, C.; Gao, Y. Research on methods of short-term traffic forecasting based on support vector regression. *J. Beijing Jiaotong Univ.* **2006**, *30*, 19–22.

35. Pang, X.; Wang, C.; Huang, G. A Short-Term Traffic Flow Forecasting Method Based on a Three-Layer K-Nearest Neighbor Non-Parametric Regression Algorithm. *J. Transp. Technol.* **2016**, *06*, 200–206. [CrossRef]
36. Zhang, X.L.; He, G.; Lu, H. Short-term traffic flow forecasting based on K-nearest neighbors non-parametric regression. *J. Syst. Eng.* **2009**, *24*, 178–183.
37. Sun, S.; Zhang, C.; Yu, G. A bayesian network approach to traffic flow forecasting. *IEEE Trans. Intell. Transp. Syst.* **2006**, *7*, 124–132. [CrossRef]
38. Jozefowicz, R.; Zaremba, W.; Sutskever, I. An Empirical Exploration of Recurrent Network Architectures. In Proceedings of the 32nd International Conference on Machine Learning, Lille, France, 6–11 July 2015; Volume 37, pp. 2342–2350.
39. Ma, X.; Tao, Z.; Wang, Y.; Yu, H.; Wang, Y. Long short-term memory neural network for traffic speed prediction using remote microwave sensor data. *Transp. Res. Part C Emerg. Technol.* **2015**, *54*, 187–197. [CrossRef]
40. Fu, R.; Zhang, Z.; Li, L. Using LSTM and GRU neural network methods for traffic flow prediction. In Proceedings of the 2016 31st Youth Academic Annual Conference of Chinese Association of Automation (YAC), Wuhan, China, 11–13 November 2016; pp. 324–328. [CrossRef]
41. Cui, Z.; Ke, R.; Pu, Z.; Wang, Y. Deep Bidirectional and Unidirectional LSTM Recurrent Neural Network for Network-wide Traffic Speed Prediction. *arXiv* **2018**. arXiv:1801.02143.
42. Yu, R.; Li, Y.; Shahabi, C.; Demiryurek, U.; Liu, Y. Deep learning: A generic approach for extreme condition traffic forecasting. In Proceedings of the 2017 SIAM international Conference on Data Mining, Houston, TX, USA, 27–29 April 2017; pp. 777–785. [CrossRef]
43. Wu, Y.; Tan, H. Short-term traffic flow forecasting with spatial-temporal correlation in a hybrid deep learning framework. *arXiv* **2016**. arXiv:1612.01022.
44. Shi, X.; Chen, Z.; Wang, H.; Yeung, D.Y.; Wong, W.k.; Woo, W.C. Convolutional LSTM Network: A Machine Learning Approach for Precipitation Nowcasting. In *Advances in Neural Information Processing Systems*; Cortes, C., Lawrence, N., Lee, D., Sugiyama, M., Garnett, R., Eds.; Curran Associates, Inc.: Red Hook, NY, USA, 2015; Volume 28.
45. Yu, H.; Wu, Z.; Wang, S.; Wang, Y.; Ma, X. Spatiotemporal Recurrent Convolutional Networks for Traffic Prediction in Transportation Networks. *Sensors* **2017**, *17*, 1501. [CrossRef]
46. Wu, Z.; Pan, S.; Chen, F.; Long, G.; Zhang, C.; Yu, P.S. A Comprehensive Survey on Graph Neural Networks. *IEEE Trans. Neural Netw. Learn. Syst.* **2021**, *32*, 4–24. [CrossRef]
47. Girshick, R. Fast R-CNN. In Proceedings of the IEEE International Conference on Computer Vision (ICCV), Santiago, Chile, 7–13 December 2015.
48. Li, J.; Chen, X.; Hovy, E.; Jurafsky, D. Visualizing and Understanding Neural Models in NLP. In Proceedings of the 2016 Conference of the North American Chapter of the Association for Computational Linguistics: Human Language Technologies, San Diego, CA, USA, 2–17 June 2016; pp. 681–691.
49. Abdel-Hamid, O.; Mohamed, A.R.; Jiang, H.; Deng, L.; Penn, G.; Yu, D. Convolutional Neural Networks for Speech Recognition. *IEEE/ACM Trans. Audio Speech Lang. Process.* **2014**, *22*, 1533–1545. [CrossRef]
50. Bronstein, M.M.; Bruna, J.; LeCun, Y.; Szlam, A.; Vandergheynst, P. Geometric Deep Learning: Going beyond Euclidean data. *IEEE Signal Process. Mag.* **2017**, *34*, 18–42. [CrossRef]
51. Qiu, J.; Tang, J.; Ma, H.; Dong, Y.; Wang, K.; Tang, J. DeepInf. In Proceedings of the 24th ACM SIGKDD International Conference on Knowledge Discovery & Data Mining, London, UK, 19–23 August 2018. [CrossRef]
52. Henrion, I.; Brehmer, J.; Bruna, J.; Cho, K.; Cranmer, K.; Louppe, G.; Rochette, G. Neural Message Passing for Jet Physics. In Proceedings of the Deep Learning for Physical Sciences Workshop, Long Beach, CA, USA, 4–9 December 2017.
53. Duvenaud, D.K.; Maclaurin, D.; Iparraguirre, J.; Bombarell, R.; Hirzel, T.; Aspuru-Guzik, A.; Adams, R.P. Convolutional Networks on Graphs for Learning Molecular Fingerprints. In *Advances in Neural Information Processing Systems*; Cortes, C., Lawrence, N., Lee, D., Sugiyama, M., Garnett, R., Eds.; Curran Associates, Inc.: Red Hook, NY, USA, 2015; Volume 28.
54. Wu, Z.; Pan, S.; Long, G.; Jiang, J.; Zhang, C. Graph wavenet for deep spatial-temporal graph modeling. In Proceedings of the 28th International Joint Conference on Artificial Intelligence, Macao, China, 10–16 August 2019; pp. 1907–1913.
55. Agafonov, A. Traffic Flow Prediction Using Graph Convolution Neural Networks. In Proceedings of the 2020 10th International Conference on Information Science and Technology (ICIST), Lecce, Italy, 4–5 June 2020; pp. 91–95. [CrossRef]
56. Zhang, Y.; Cheng, T.; Ren, Y.; Xie, K. A novel residual graph convolution deep learning model for short-term network-based traffic forecasting. *Int. J. Geogr. Inf. Sci.* **2020**, *34*, 969–995. [CrossRef]
57. Veličković, P.; Cucurull, G.; Casanova, A.; Romero, A.; Liò, P.; Bengio, Y. Graph Attention Networks. In Proceedings of the International Conference on Learning Representations, ICLR 2017, Toulon, France, 24–26 April 2017.
58. Yu, B.; Yin, H.; Zhu, Z. Spatio-Temporal Graph Convolutional Networks: A Deep Learning Framework for Traffic Forecasting. In Proceedings of the Twenty-Seventh International Joint Conference on Artificial Intelligence. International Joint Conferences on Artificial Intelligence Organization, Stockholm, Sweden, 13–19 July 2018. [CrossRef]
59. Atwood, J.; Towsley, D. Diffusion-Convolutional Neural Networks. In *Advances in Neural Information Processing Systems*; Lee, D., Sugiyama, M., Luxburg, U., Guyon, I., Garnett, R., Eds.; Curran Associates, Inc.: Red Hook, NY, USA, 2015; Volume 29.
60. Li, Y.; Yu, R.; Shahabi, C.; Liu, Y. Diffusion Convolutional Recurrent Neural Network: Data-Driven Traffic Forecasting. In Proceedings of the International Conference on Learning Representations, ICLR 2017, Toulon, France, 24–26 April 2017.

61. Liang, Y.; Ke, S.; Zhang, J.; Yi, X.; Zheng, Y. GeoMAN: Multi-level Attention Networks for Geo-sensory Time Series Prediction. In Proceedings of the Twenty-Seventh International Joint Conference on Artificial Intelligence, IJCAI-18, Stockholm, Sweden, 13–19 July 2018; pp. 3428–3434. [CrossRef]
62. Guo, S.; Lin, Y.; Feng, N.; Song, C.; Wan, H. Attention Based Spatial-Temporal Graph Convolutional Networks for Traffic Flow Forecasting. *Proc. AAAI Conf. Artif. Intell.* **2019**, *33*, 922–929. [CrossRef]
63. Do, L.N.; Vu, H.L.; Vo, B.Q.; Liu, Z.; Phung, D. An effective spatial-temporal attention based neural network for traffic flow prediction. *Transp. Res. Part C Emerg. Technol.* **2019**, *108*, 12–28. [CrossRef]
64. Yin, X.; Wu, G.; Wei, J.; Shen, Y.; Qi, H.; Yin, B. Multi-stage attention spatial-temporal graph networks for traffic prediction. *Neurocomputing* **2021**, *428*, 42–53. [CrossRef]
65. Zheng, C.; Fan, X.; Wang, C.; Qi, J. GMAN: A Graph Multi-Attention Network for Traffic Prediction. *Proc. AAAI Conf. Artif. Intell.* **2020**, *34*, 1234–1241. [CrossRef]
66. Bai, J.; Zhu, J.; Song, Y.; Zhao, L.; Hou, Z.; Du, R.; Li, H. A3T-GCN: Attention Temporal Graph Convolutional Network for Traffic Forecasting. *ISPRS Int. J. Geo-Inf.* **2021**, *10*, 485. [CrossRef]
67. Bruna, J.; Zaremba, W.; Szlam, A.; LeCun, Y. Spectral networks and deep locally connected networks on graphs. In Proceedings of the 2nd International Conference on Learning Representations, ICLR 2014, Banff, AB, Canada, 14–16 April 2014.
68. Kipf, T.N.; Welling, M. Semi-Supervised Classification with Graph Convolutional Networks. In Proceedings of the International Conference on Learning Representations, ICLR 2017, Toulon, France, 24–26 April 2017.
69. Bachechi, C.; Rollo, F.; Po, L. Detection and classification of sensor anomalies for simulating urban traffic scenarios. *Clust. Comput.* **2022**, *25*, 2793–2817. [CrossRef]
70. Wei, W.; Wu, H.; Ma, H. An autoencoder and LSTM-based traffic flow prediction method. *Sensors* **2019**, *19*, 2946. [CrossRef] [PubMed]
71. Kuang, L.; Yan, X.; Tan, X.; Li, S.; Yang, X. Predicting taxi demand based on 3D convolutional neural network and multi-task learning. *Remote Sens.* **2019**, *11*, 1265. [CrossRef]
72. Kalampokis, E.; Karacapilidis, N.; Tsakalidis, D.; Tarabanis, K. Artificial Intelligence and Blockchain Technologies in the Public Sector: A Research Projects Perspective. In Proceedings of the Electronic Government: 21st IFIP WG 8.5 International Conference, EGOV 2022, Linköping, Sweden, 6–8 September 2022; Springer: Berlin/Heidelberg, Germany, 2022; pp. 323–335.
73. Karamanou, A.; Kalampokis, E.; Tarabanis, K. Linked open government data to predict and explain house prices: The case of Scottish statistics portal. *Big Data Res.* **2022**, *30*, 100355. [CrossRef]

 *information*

Article

An Attention-Based Deep Convolutional Neural Network for Brain Tumor and Disorder Classification and Grading in Magnetic Resonance Imaging

Ioannis D. Apostolopoulos [1,*], Sokratis Aznaouridis [2] and Mpesi Tzani [3]

[1] Department of Medical Physics, School of Medicine, University of Patras, 26504 Rio, Greece
[2] Department of Computer Engineering and Informatics, University of Patras, 26504 Rio, Greece
[3] Department of Electrical and Computer Technology Engineering, University of Patras, 26504 Rio, Greece
* Correspondence: ece7216@upnet.gr

Abstract: This study proposes the integration of attention modules, feature-fusion blocks, and baseline convolutional neural networks for developing a robust multi-path network that leverages its multiple feature-extraction blocks for non-hierarchical mining of important medical image-related features. The network is evaluated using 10-fold cross-validation on large-scale magnetic resonance imaging datasets involving brain tumor classification, brain disorder classification, and dementia grading tasks. The Attention Feature Fusion VGG19 (AFF-VGG19) network demonstrates superiority against state-of-the-art networks and attains an accuracy of 0.9353 in distinguishing between three brain tumor classes, an accuracy of 0.9565 in distinguishing between Alzheimer's and Parkinson's diseases, and an accuracy of 0.9497 in grading cases of dementia.

Keywords: artificial intelligence; deep learning; attention module; feature fusion; magnetic resonance imaging

Citation: Apostolopoulos, I.D.; Aznaouridis, S.; Tzani, M. An Attention-Based Deep Convolutional Neural Network for Brain Tumor and Disorder Classification and Grading in Magnetic Resonance Imaging. *Information* 2023, 14, 174. https://doi.org/10.3390/info14030174

Academic Editors: Amar Ramdane-Cherif, Ravi Tomar and T P Singh

Received: 9 February 2023
Revised: 1 March 2023
Accepted: 7 March 2023
Published: 9 March 2023

1. Introduction

Deep learning (DL) [1] is a subset of artificial intelligence (A.I.) methods that has been gaining popularity in recent years, and it has been applied to various fields, including medical imaging, such as magnetic resonance imaging (MRI). MRI is a non-invasive imaging technique that uses a magnetic field and radio waves to produce detailed images of the body's internal structures [2]. The technique provides high-resolution images and can diagnose a wide range of conditions.

DL algorithms have been used to improve the diagnostic accuracy and efficiency of MRI by automating the analysis of images and identifying patterns that are not visible to the human eye [3]. The algorithms can be trained to recognize specific patterns, such as tumors or abnormal brain structures, which can aid in diagnosing diseases [4].

One of the main advantages of using DL in MRI is that it can analyze large amounts of data quickly and accurately [5]. Traditional image analysis methods require manual interpretation by radiologists, which can be time-consuming and subjective. DL algorithms can be trained to analyze images, automatically saving time and improving diagnostic accuracy.

One of the main applications of DL in MRI is analyzing brain images [6]. Algorithms have been developed to detect and diagnose brain tumors, identify brain structures, and predict the progression of diseases such as Alzheimer's [7]. This can aid in the early detection and treatment of brain diseases, leading to better patient outcomes. Another application of DL in MRI is the analysis of cardiac images [8]. Algorithms have been developed to detect and diagnose heart conditions, such as heart failure and arrhythmia. This can aid in the early detection and treatment of heart diseases.

DL algorithms have also been used to improve the diagnostic accuracy and efficiency of MRI in analyzing images of the prostate, liver, and pancreas; and to detect and diagnose prostate cancer, liver fibrosis, and pancreatic cancer [9].

One of the main challenges of using DL in MRI is the limited availability of labelled data. MRI images are often difficult to obtain and expensive to acquire, making it challenging to train DL algorithms. Additionally, the images can be of poor quality, making it difficult to detect specific patterns.

Another challenge is the variability in the MRI images. They can vary depending on the imaging protocol and the acquisition parameters. This can make it non-trivial to train generalized DL algorithms able to cope with variations in acquisition parameters.

Recent advances in the architecture and functions of convolutional neural networks (CNNs) [10] have allowed for a dramatic improvement in the performance of these models. Attention mechanisms [11] are a popular addition to CNNs, allowing for adaptive refinement of the feature maps to identify essential components of the image. Optimization strategies [12] such as batch normalization and dropout have enabled faster convergence and higher accuracy when applied to CNNs. In addition, feature fusion techniques such as sparse coding, autoencoders, and multi-resolution processing have been used to combine the strengths of multiple feature maps and have been shown to significantly improve the accuracy of CNNs.

The study proposes an innovative modification of a well-established CNN developed by the Virtual Geometry Group (VGG) [13] and named after it. We propose the integration of feature-fusion blocks and attention models to enrich the encapsulation of important image features that lead to more precise image classification. This network is employed to classify brain MRI images in brain tumor classification and brain disorder discrimination and grading.

The paper is structured as follows: In Section 2, we briefly describe the entities of DL, the attention modules, and feature-fusion blocks. The proposed network is described in detail. In addition, the employed datasets are presented. In the Section 3, the results of the study are presented. Discussion and concluding remarks take place in Sections 5 and 6, respectively.

2. Related Work

Sadad et al. [14] utilized the Unet architecture with ResNet50 as a backbone to perform segmentation on the Figshare dataset, achieving an impressive intersection over union (IoU) score of 0.9504. The researchers also employed preprocessing and data augmentation techniques to improve classification accuracy. They used evolutionary algorithms and reinforcement learning in transfer learning to perform multi-classification of brain tumors. The study compared the performance of different DL models, such as ResNet50, DenseNet201, MobileNet V2, and InceptionV3; and demonstrated that the proposed framework outperformed the state-of-the-art methods. The study also applied various CNN models to classify brain tumors, including MobileNet V2, Inception V3, ResNet50, DenseNet201, and NASNet, achieving accuracies of 91.8%, 92.8%, 92.9%, 93.1%, and 99.6%, respectively. The NASNet model showed the highest accuracy among all the models.

Allah et al. [15] investigated the effectiveness of a novel approach to classify brain tumor MRI images using a VGG19 feature extractor and one of three different types of classifiers. To address the shortage of images needed for deep learning, the study employed a progressive, growing generative adversarial network (PGGAN) augmentation model to generate "realistic" brain tumor MRI images. The findings demonstrated that the proposed framework outperformed previous studies in accurately classifying gliomas, meningiomas, and pituitary tumors, achieving an accuracy rate of 98.54%.

In [16], a novel hybrid CNN-based architecture was proposed to classify three types of brain tumors using MRI images. The approach involves utilizing two methods of hybrid deep learning classification based on CNN. The first method combines a pre-trained Google-Net model of the CNN algorithm for feature extraction with SVM for pattern classification, while the second method integrates a finely tuned Google-Net with a soft-max classifier. The performance of the proposed approach was evaluated on a dataset containing a total of 1426 glioma images, 708 meningioma images, 930 pituitary tumor images, and 396 normal brain images. The results revealed that the finely tuned Google-Net model achieved an accuracy of 93.1%. However, the accuracy was improved to 98.1% when the Google-Net was combined with an SVM classifier as a feature extractor.

Kang et al. [17] applied transfer learning and utilized pre-trained deep convolutional neural networks to extract deep features from MRI images of the brain. These extracted features were evaluated using various machine learning classifiers, and the top three performing features were selected and combined to form an ensemble of deep features. This ensemble was then used as input to several machine learning classifiers to predict the final output. We evaluated the effectiveness of different pre-trained models as deep feature extractors, various machine learning classifiers, and the impact of the ensemble of deep features for brain tumor classification using three openly accessible brain MRI datasets. The experimental results revealed that using an ensemble of deep features significantly improved performance, and SVM with radial basis function kernel outperformed other machine learning classifiers, particularly for large datasets.

Sivaranjini et al. [18] used a DL neural network to classify MRI images of healthy individuals and those with Parkinson's disease (PD). The researchers utilized the AlexNet convolutional neural network architecture to improve the accuracy of Parkinson's disease diagnosis. By training the network with MRI images and testing it, the system was able to achieve an accuracy rate of 88.9%. This demonstrates that deep learning models can aid clinicians in diagnosing PD more objectively and accurately in the future.

Bhan et al. [19] successfully diagnosed PD from MRI images. The LeNet-5 architecture and a dropout algorithm achieved 97.92% accuracy using batch normalization on a large dataset consisting of 10,548 images. This method has the potential to accurately diagnose various stages of PD.

Hussain et al. [20] presented a 12-layer CNN for binary classification and detection of Alzheimer's disease using brain MRI data. The proposed model's performance is evaluated and compared to existing CNN models based on accuracy, precision, recall, F1 score, and receiver operating characteristic (ROC) curve using the open access series of imaging studies (OASIS) dataset. The model attained an accuracy of 97.75%, higher than any previously published CNN models on this dataset.

Salehi et al. [21] used to detect and classify Alzheimer's Disease (AD) at an early stage by analyzing MRI images from the ADNI database. The dataset consisted of 1512 mild, 2633 normal, and 2480 AD images. The CNN model achieved a remarkable accuracy of 99%, surpassing the performance of several other studies.

3. Materials and Methods

3.1. Deep Learning

DL is a branch of machine learning that uses neural networks with many layers to learn patterns and features from data [10]. It is based on the idea that a neural network can learn to recognize data patterns like a human brain does [1].

DL algorithms comprise multiple layers of artificial neural networks, interconnected layers of nodes, or artificial neurons. These layers work together to extract features from the data and make predictions. The first layer of a DL network is typically responsible for recognizing simple features, such as edges or shapes, while the last layer makes the final prediction.

DL networks are trained using large amounts of data, fed into the network, and used to adjust the weights and biases of the artificial neurons. The goal is to adjust these weights and biases, so the network can correctly classify or predict the output for new input data. This process is known as supervised learning, where the network is trained on labelled data [22].

DL networks are also used in unsupervised learning, where the network is not provided with labelled data and has to find patterns and features on its own. This can be useful for image compression, anomaly detection, and generative models.

3.2. Attention Feature-Fusion VGG19

The study proposes an innovative expansion of the baseline VGG19 network. It circumvents the baseline's hierarchical feature extraction method using handcrafted feature fusion blocks and feature concatenation. The latter modification led to the creation of the Feature-Fusion VGG19 network [23]. Here, we propose an extension of this network that leverages the attention modules. The components of the Attention Feature-Fusion VGG19 (AFF-VGG19) network are analytically described below.

3.2.1. Virtual Geometry Group

VGG19 is a CNN architecture developed by the Visual Geometry Group (VGG) at the University of Oxford in 2014 [13]. It is a deeper version of the VGG16 architecture known for its excellent performance in image classification tasks.

The VGG19 network architecture comprises 19 layers, including 16 convolutional layers and three fully connected layers. The convolutional layers extract features from the input image, while the fully connected layers are used for classification. The architecture uses a small 3×3 convolutional kernel size, which allows for a deeper architecture with more layers.

One of the critical features of the VGG19 architecture is its use of tiny convolutional filters, which allows the network to learn more fine-grained features from the input image [13]. Additionally, the network uses many filters in each convolutional layer, which allows it to learn many features from the input image.

The VGG19 network was trained on the ImageNet dataset, a large dataset of images labelled with one of 1000 different classes. The network was trained using the stochastic gradient descent (SGD) optimization algorithm with a batch size of 128 and a learning rate of 0.001.

The VGG19 network achieved state-of-the-art performance on the ImageNet dataset [24], with an accuracy of 92.7% on the validation set. This made it one of the most accurate CNNs at its release, and it is still considered a very accurate network today.

Training VGG19 from scratch involves learning 143 million parameters, a colossal number for small-scale datasets. Supplying such a network with inadequate amounts of data results in underfitting. Therefore, we considered the fine-tuning option, which borrows the architecture and some pre-assigned conditions to reduce the number of trainable parameters. The initial network training defined the untrainable weights, which were defined using the ImageNet database [24]. Though the latter dataset consists of irrelevant images (non-medical), successful training helps the network learn how to extract low-level image features (e.g., edges, shapes), which are met in medical images also. Therefore, it is fair to transfer this knowledge to other domains [25–27].

We propose that the VGG19 network of the study is fine-tuned to extract approximately 5 million trainable parameters to circumvent underfitting. We selected to train the deep convolutional layers and freeze the first ones because abstract and high-level features are learned from the deeper convolutional layers of the network.

3.2.2. Feature Fusion Modification

Feature fusion is a modification to the traditional CNN architecture that allows for extracting features from an input image in a non-hierarchical way. This modification is used to improve the performance of CNNs in tasks, such as image classification, object detection, and semantic segmentation.

In traditional CNNs, features are extracted hierarchically. Lower-level features are learned in the earlier layers, and higher-level features are learned in the later layers. However, feature fusion CNNs extract features from multiple layers simultaneously and combine to form a single set of features. This allows for extracting low-level and high-level features in a non-hierarchical way.

There are several different ways to implement feature fusion in a CNN. One standard method combines different layers, such as convolutional layers, pooling layers, and fully connected layers. These layers are trained to extract features from the input image at different levels of abstraction. The features from each layer are then combined to form a single set of features.

Another method is to use feature pyramid networks (FPN), which combines features from different layers of a CNN in a pyramid-like structure. This allows for the extraction of features at different scales, which can be helpful in tasks such as object detection, where the size of the object can vary greatly.

A third method is to use an attention-based feature fusion technique, which uses an attention mechanism to selectively focus on different regions of the input image when extracting features. This allows the network to pay more attention to the regions of the image that are most important for the task at hand.

We propose using simple feature-fusion blocks that solely connect the output of the convolutional blocks directly to the top of the network (Figure 1). A feature-fusion block involves a batch normalization, dropout, and global average pooling layers. These layers do not extract additional features. The feature-fusion blocks are placed after the second, third, and fourth convolutional groups and are connected to the output of the max pooling layers that follow (Figure 1). In this way, the extorted image features of each convolutional group are connected directly to the classification layer at the top of the network, ensuring that no further processing is applied. Therefore, the feature-fusion blocks negate the hierarchical feature-extraction manner of the VGG19.

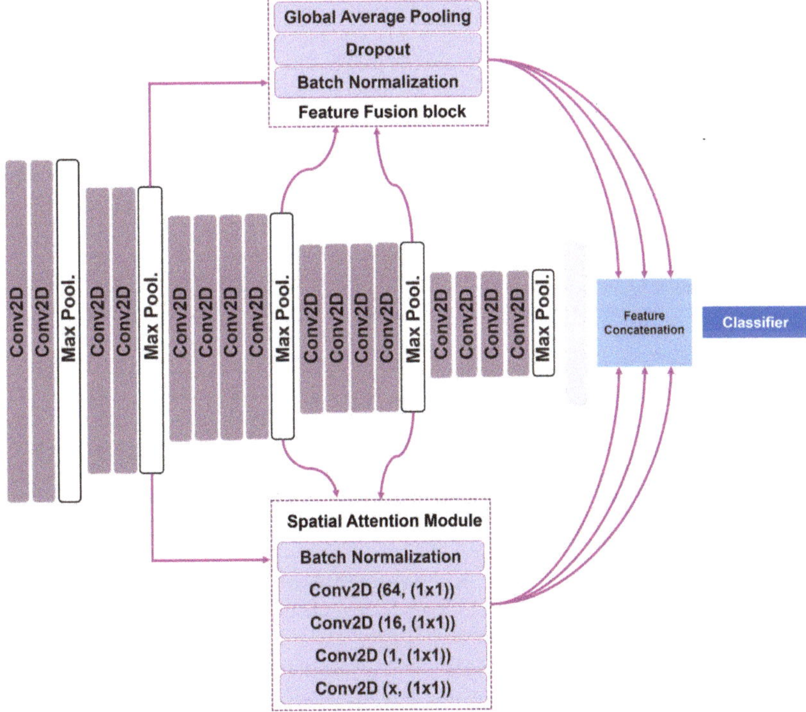

Figure 1. Attention Feature-Fusion VGG19 network.

3.2.3. Attention Mechanism

The attention mechanism is a technique used in convolutional neural networks (CNNs) to focus on the essential parts of an image when making predictions. It allows the network to selectively attend to different regions of the input image rather than treating the entire image as a single input.

The attention mechanism is typically implemented as an additional layer in the CNN, which is trained to learn the importance of different input image regions. This layer is often referred to as the attention layer. The attention layer is trained to learn a weighting for each region of the input image, which is used to determine the importance of that region when making predictions.

There are different attention mechanisms, but one of the most widely used is the "soft attention" mechanism. It is a different form of attention, which means it can be trained with backpropagation. This is accomplished by computing a set of attention weights for each region of the input image and then using these weights to weight the contributions of each region to the final prediction.

The attention mechanism can be applied differently in CNNs depending on the task. For instance, in image captioning, attention can focus selectively on different regions of an image when generating a text description of the image. In image classification, attention can be used to focus on specific regions of an image that are most relevant to the class of interest.

We propose using five-layered attention modules located after the second, third, and fourth convolutional groups and connected to the output of the following max pooling layers. The first layer is a batch normalization layer. The second, third, fourth, and fifth layers are convolutional operations utilizing 64, 16, 1, and x filters of 1×1 kernel size. The x number depends on the convolution group to which the attention module belongs. The attention module in the second group has an x of 128. Accordingly, the third and fourth groups have an x of 256 and 512, respectively. The extracted features are multiplied with the output of the convolutional group and connected to the network's top.

3.3. Datasets of the Study

The study enrols three classification datasets derived from four repositories, as presented in Table 1.

Table 1. Study's datasets overview.

Dataset	Source (s)	Number of Images
Brain Tumors Dataset	https://www.kaggle.com/datasets/adityakomaravolu/brain-tumor-mri-images (accessed on 24 November 2022). https://www.kaggle.com/datasets/roroyaseen/brain-tumor-data-mri (accessed on 24 November 2022).	7023 and 19,226. Total 26,249
Brain Disorders Dataset	https://www.kaggle.com/datasets/farjanakabirsamanta/alzheimer-diseases-3-class (accessed on 24 November 2022).	7756
Dementia Grading Dataset	https://www.kaggle.com/datasets/sachinkumar413/alzheimer-mri-dataset (accessed on 24 November 2022).	6400

3.3.1. Brain Tumors Dataset

Glioma is a type of brain tumor that arises from glial cells, which are the supporting cells of the nervous system [28]. They can be benign or malignant, and the malignant forms can be very aggressive. Symptoms can include headaches, seizures, and changes in cognitive function. Meningioma is a type of brain tumor arising from the meninges, the protective membranes covering the brain, and spinal cord [28]. They are generally benign tumors, but can cause symptoms such as headaches, seizures, and changes in cognitive function. Pituitary tumors develop in the pituitary gland, a small gland located at the base of the brain that produces hormones that regulate growth, metabolism, and other bodily functions [28]. Pituitary tumors can be benign or malignant and can cause symptoms such as headaches, vision problems, and changes in hormone levels.

In the present study, the final brain tumors dataset combines two publicly available datasets (Table 1). The datasets contain MRI images of glioma (8208 images), meningioma (7866 images), pituitary tumors (8175 images), and controls (2000). Henceforth, the classes are addressed as follows: glioma (G), meningioma (M), and pituitary (P).

The images are preprocessed and converted into JPEG format. The dataset is suitable for training DL networks for discriminating the classes.

3.3.2. Brain Disorders Dataset

AD is a progressive brain disorder that affects memory, thinking and behavior [29]. It is the most common cause of dementia, a general term for a decline in cognitive function severe enough to interfere with daily life. AD symptoms typically develop slowly and worsen over time, eventually leading to severe cognitive impairment and the inability to carry out daily activities [29]. The cause of AD is not fully understood. However, it is thought to be related to genetic, lifestyle, and environmental factors.

PD is a progressive nervous system disorder that affects movement [29]. It is caused by the loss of dopamine-producing cells in the brain, leading to symptoms such as tremors, stiffness, slow movement, and difficulty with balance and coordination. Parkinson's disease can also cause non-motor symptoms such as depression, anxiety, and cognitive impairment. The cause of PD is not fully understood. However, it involves a combination of genetic and environmental factors. There is no cure for PD, but medications and other treatments can help manage symptoms.

The study's dataset contains 7756 MRI images separated into three classes: PD with 906 images, AD with 3200 images, and controls with 3650 images. All the images are 176×208 pixels in size and come in JPEG format. This dataset is used to teach the network to distinguish between these classes.

3.3.3. Dementia Grading Dataset

Dementia is a general term for a decline in cognitive function severe enough to interfere with daily life. The severity of dementia can be graded or classified in various ways, but the most widely used system is the clinical dementia rating (CDR) scale [30,31]. The CDR scale ranges from 0 to 3, with 0 indicating no dementia, 1 indicating very mild dementia, 2 indicating mild dementia, and 3 indicating moderate to severe dementia.

The CDR scale assesses six areas of cognitive function: memory, orientation, judgment and problem solving, community affairs, home and hobbies, and personal care. The score in each area is used to determine the overall CDR score.

The CDR scale is a widely used and accepted tool for evaluating the severity of dementia and tracking its progression over time. Other scales used to grade dementia are the global deterioration scale (GDS) and the global clinical impression of change (GCIC).

It is important to note that while grading or classifying the severity of dementia can be helpful for research and tracking the progression of the disease, it is not always a definitive measure of an individual's cognitive or functional abilities.

The dataset of the particular study is collected from several websites/hospitals/public repositories and consists of MRI images of 128×128 pixel size. The total number of images is 6400. The distribution between classes is as follows: mildly demented (896 images), moderately demented (64 images), non-demented (3200 images), and very mildly demented (2240 images). Henceforth, the above classes are addressed as: mildly demented (Mi), moderately demented (Mo), and very mildly demented (VMi).

The images are preprocessed and converted into JPEG format.

3.4. Experiment Setup

The experiments were conducted on a workstation featuring an 11th Gen Intel®Core™ i9-11900KF @3.50GHz processor, an NVIDIA GeForce RTX 3080 Ti GPU and 64 GB of RAM, running a 64-bit operating system. Tensorflow 2.9.0 and Sklearn 1.0.2 were used during the experiments, both written in Python 3.9.

The assessment of the models was conducted using 10-fold cross-validation. During each fold, the accuracy (ACC), sensitivity (SEN), specificity (SPE), positive predicted value (PPV), negative predicting value (NPV), false positive rate (FPR), false negative rate (FNR), F-1 score (F1) of the run were calculated based on the recorded true positives (T.P.), false positives (FP), true negatives (TN), and false negatives (FN) of each class. In addition, from the predicted probabilities, we computed the area under curve score (AUC).

4. Results

Section 4.1 describes the classification performance of the AFF-VGG19 network on the brain tumor dataset. Section 4.2. presents the results of the network when classifying the brain disorders dataset. Accordingly, in Section 4.3., the performance of AFF-VGG19 in the dementia grading dataset is presented. Finally, Section 4.4. presents comparisons between the proposed AFF-VGG19 network and alternative state-of-the-art networks.

4.1. Brain Tumor Classification

The proposed network achieves an aggregated accuracy of 0.9353, computed based on the total true positives between the classes. The model shows excellent performance in distinguishing between tumor and non-tumor MRI images. Specifically, the network exhibits an accuracy of 0.9795 in the control class, with a very small FPR (0.0219), as Table 2 presents.

Table 2. Performance metrics of AFF-VGG19 on the brain tumor dataset. G stands for glioma, M for meningioma, P for pituitary.

	ACC	SEN	SPE	PPV	NPV	FPR	FNR	F1	AUC
G	0.9505	0.9676	0.9427	0.8849	0.9846	0.0573	0.0324	0.9244	0.9552
M	0.9304	0.9062	0.9408	0.8675	0.9591	0.0592	0.0938	0.8864	0.9235
P	0.9572	0.9161	0.9758	0.9449	0.9626	0.0242	0.0839	0.9303	0.9460
Control	0.9795	0.9960	0.9781	0.7895	0.9997	0.0219	0.0040	0.8808	0.9871

For the G class, the network achieves an accuracy of 0.9505, a sensitivity of 0.9676, and a specificity of 0.9427. The AUC score reaches 0.9552. For the M class, the network achieves an accuracy of 0.9304, a sensitivity of 0.9062, and a specificity of 0.9408. The AUC score reaches 0.9235. Accordingly, for the P class, the network achieves an accuracy of 0.9572, a sensitivity of 0.9161, and a specificity of 0.9758. The AUC score reaches 0.9460.

The relatively large FNR (0.0938) and accuracy (0.9304) in the M class indicate that the network performs sub-optimally in the discrimination of the M class from the rest (M versus ALL classification).

4.2. Brain Disorders Classification

AFF-VGG19 achieves an aggregated accuracy of 0.9565. The model shows excellent performance in distinguishing between AD-PD and control MRI images. Specifically, the network exhibits an accuracy of 0.9621 in the control class, with a very small FPR (0.0375), as Table 3 presents.

Table 3. Performance metrics of AFF-VGG19 on the brain disorders dataset. AD stands for Alzheimer's disease and PD for Parkinson's disease.

	ACC	SEN	SPE	PPV	NPV	FPR	FNR	F1	AUC
AD	0.9409	0.9222	0.9541	0.9339	0.9458	0.0459	0.0778	0.9280	0.9382
PD	0.9489	0.9860	0.9160	0.9125	0.9866	0.0840	0.0140	0.9479	0.9510
Control	0.9621	0.9592	0.9625	0.7718	0.9944	0.0375	0.0408	0.8553	0.9608

For the AD class, the network achieves an accuracy of 0.9409, a sensitivity of 0.9222, and a specificity of 0.9541. The AUC score reaches 0.9382. For the PD class, the network achieves an accuracy of 0.9489, a sensitivity of 0.9860, and a specificity of 0.9160. The AUC score reaches 0.9510.

AFF-VGG19 yields a relatively high FPR in PD detection and a larger FNR in AD detection (0.0840 and 0.0778, respectively).

4.3. Dementia Grading

AFF-VGG19 achieves an aggregated accuracy of 0.9497. The model shows excellent performance in distinguishing between Mo-Mi-VMi and control MRI images. Specifically, the network exhibits an accuracy of 0.9769 in the control class, with a very small FPR (0.0394), as Table 4 presents. In addition, a very low FNR is recorded (0.0069).

Table 4. Performance metrics of AFF-VGG19 on the dementia grading dataset. Mo stands for moderate, Mi for mild, and VMi for very mild.

	ACC	SEN	SPE	PPV	NPV	FPR	FNR	F1	AUC
Mo	0.9670	0.9531	0.9672	0.2268	0.9995	0.0328	0.0469	0.3664	0.9601
Mi	0.9264	0.8281	0.9424	0.7007	0.9712	0.0576	0.1719	0.7591	0.8853
VMi	0.9539	0.9362	0.9635	0.9324	0.9656	0.0365	0.0638	0.9343	0.9498
Control	0.9769	0.9931	0.9606	0.9619	0.9929	0.0394	0.0069	0.9772	0.9769

For the Mo class, the network achieves an accuracy of 0.9670, a sensitivity of 0.9531, and a specificity of 0.9672. The AUC score reaches 0.9601. For the Mi class, the network achieves an accuracy of 0.9264, a sensitivity of 0.8281, and a specificity of 0.9424. The AUC score reaches 0.8853. For the VMi class, the network yields an accuracy of 0.9539, a sensitivity of 0.9362, a specificity of 0.9635, and an AUC of 0.9498.

Figure 2 summarizes the performance of AFF-VGG19. Figure 2 presents the ROC curve, and the training and validation accuracy and loss for the brain tumor and brain disorder datasets.

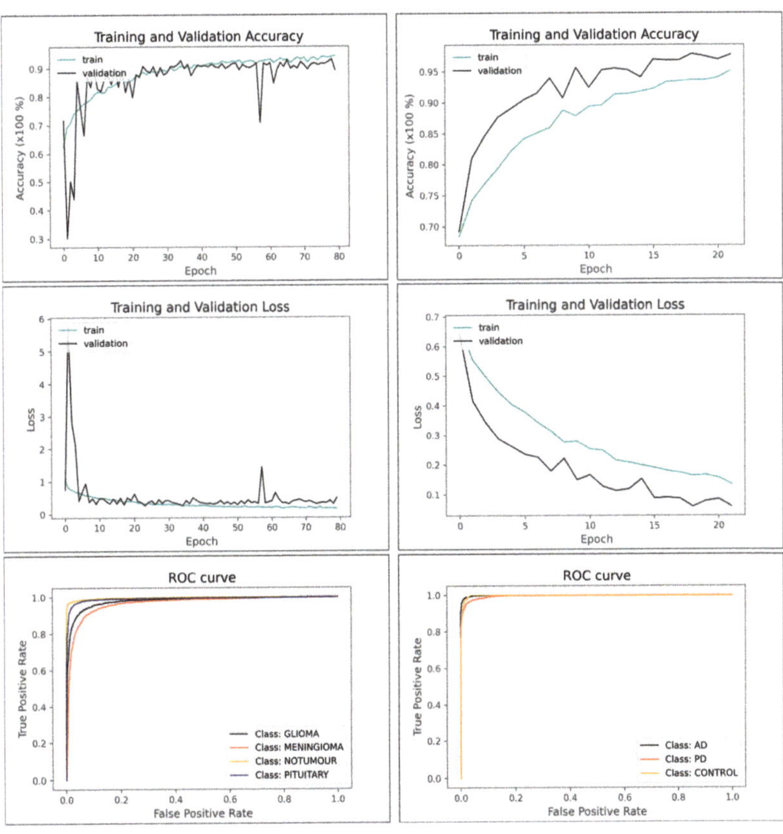

Figure 2. Training–validation accuracy–losses and ROC curve of AFF-VGG19.

4.4. Comparison with The State-Of-The-Art

AFF-VGG19 obtains the highest aggregated accuracy in every dataset (Table 5) compared to state-of-the-art CNNs, which do not use attention modules. In addition, the baseline VGG19 network stands out among its competitors when comparing non-attention-based models. Specifically, VGG19 exhibits the best accuracy in the brain tumor dataset (accuracy of 0.9108), the best accuracy in the brain disorder dataset (accuracy of 0.8981), and the fourth-best accuracy in the dementia grading dataset (accuracy of 0.9022).

Table 5. Accuracy of 17 state-of-the-art networks on the three datasets of the study.

	Brain Tumor	Brain Disorder	Dementia Grading
Xception	0.8850	0.8925	0.8786
VGG16	0.8897	0.8704	0.9088
VGG19	0.9108	0.8981	0.9022
ResNet152	0.8600	0.8646	0.8983
ResNet152V2	0.8801	0.8613	0.8984
InceptionV3	0.8713	0.8657	0.8961
InceptionResNetV2	0.8848	0.8605	0.8872
MobileNet	0.8689	0.8511	0.9091
MobileNetV2	0.8512	0.8312	0.9150
DenseNet169	0.8732	0.8442	0.9066
DenseNet201	0.8715	0.8495	0.8891
NASNetMobile	0.8594	0.8695	0.9011
EfficientNetB6	0.8734	0.8726	0.8975
EfficientNetB7	0.8606	0.8717	0.8892
EfficientNetV2B3	0.8804	0.8693	0.8814
ConvNeXtLarge	0.8732	0.8463	0.9095
ConvNeXtXLarge	0.8702	0.8422	0.8898
ATT-FF-VGG19	0.9353	0.9565	0.9497

The results of Table 5 justify the selection of the baseline VGG19 as the main component for an attention-based feature-fusion network.

The study compares the proposed method and methods presented by recent related works on similar datasets. Table 6 summarizes the results.

Table 6. Comparisons with related research.

First Author	Ref. No.	Test Data Size	Classes	Method	ACC	SEN	SPE
Sadad	[14]	612 slices	G-M-P	NASNet	0.996	-	-
Allah	[15]	460 slices	G-M-P	VGG19	G: 0.9854 M: 0.9857 P: 1	G: 0.9777 M: 0.9804 P: 1	G: 0.9914 M: 0.9871 P: 1
Rasool	[16]	692 slices	G-M-P-controls	Google-Net	0.981	G: 0.978 M: 0.973 P: 0.989 N: 0.987	
Kang	[17]	692 slices		DenseNet-169	0.9204	-	-
This study		26,249 slices	G-M-P-controls	AFF-VGG19	G: 0.9505 M: 0.9304 P: 0.9572	G: 9676 M: 0.9062 P: 0.9161	G: 0.9427 M: 0.9062 P: 0.9758
Bhan	[19]	1055 Slices	PD-controls	LeNet-5	0.9792	-	-
Sivaranjini	[18]	36 patients	PD-controls	AlexNet	0.889	-	-
Hussain	[20]	11 patients	AD-controls	CNN	0.9775	AD: 0.92 C: 1	-

Table 6. *Cont.*

First Author	Ref. No.	Test Data Size	Classes	Method	ACC	SEN	SPE
This study		7756 slices	PD-AD-controls	AFF-VGG19	PD: 0.9409 AD: 0.9489	PD: 0.9222 AD: 0.9860	PD: 0.9541 AD: 0.9160
Salehi	[21]	7635 slices	Mi-VMi-controls	CNN	0.99	-	-
Mohammed	[32]	6400 slices	Mi-VMi-Mo-controls	AlexNet	94.8	93	97.75
This study		6400 slices	Mi-VMi-Mo-controls	AFF-VGG19	Mo: 0.967 Mi: 0.9264 VMi: 0.9539	Mo: 0.9531 Mi: 0.8281 VMi: 0.9362	Mo: 0.9672 Mi: 0.9424 VMi: 0.9635

Compared to recent literature, the present study utilized large-scale data (brain tumor and brain disorders datasets). Still, the results are consistent with the literature and verify that the proposed methodology is robust for big-data classification.

4.5. Reproducibility

This section presents the results of statistical significance tests to verify the reproducibility of the experiments and the stability of the proposed approach. For this purpose, AFF-VGG19 was trained and validated under a 10-fold cross-validation on each dataset 20 times, and a T-test was performed. The results verify that the model produces consistent outcomes without statistically significant deviations from the initially reported accuracy (Table 7).

Table 7. Statistical significance test results.

	Brain Tumor	Brain Disorder	Dementia Grading
Mean	0.9355	0.9558	0.9491
Standard Deviation	0.002	0.002	0.001
t-statistic	0.4	−1.09	−1.9
Null Hypothesis	Mean = 0.9355	Mean = 0.9565	Mean = 0.9497
Result	At the 0.05 level, the population mean is NOT significantly different from the test mean (0.9353).	At the 0.05 level, the population mean is NOT significantly different from the test mean (0.9565).	At the 0.05 level, the population mean is NOT significantly different from the test mean (0.9497).

5. Discussion

DL will likely play an important role in disease diagnosis and classification from MRI images. With the ability to detect subtle changes in MRI images, accurately diagnose and classify diseases, detect and segment lesions, identify biomarkers, develop personalized medicine, and develop new diagnostic tools and therapies, deep learning has the potential to revolutionize medical imaging and improve patient outcomes.

The study proposed a modification of the baseline VGG19 network that improves its feature-extraction capabilities. Integrating the feature-fusion block and attention module avoided the hierarchical nature of the baseline model and improved the classification accuracy. The model was evaluated using three MRI datasets related to brain tumor discrimination, brain disorder classification, and dementia grading. The AFF-VGG19 network demonstrates superiority against state-of-the-art networks. It attains an accuracy of 0.9353 in distinguishing between three brain tumor classes, an accuracy of 0.9565 in distinguishing between AD and PD, and an accuracy of 0.9497 in grading cases of dementia. The high FPR in PD detection and high FNR in AD detection may have their cause in the fact that these two classes may give similar findings and patterns in specific parts of the image, such that they confuse the model as there are no distinct differences. For this purpose, it would be useful in future research to also consider clinical data that would probably help to better and more accurately determine the image classes.

Without integrating the attention and feature-fusion blocks, the baseline VGG19 network proved superior to the rest of the pretrained networks (Table 5). Therefore, the attention and feature-fusion blocks were implemented using the baseline VGG19 architecture as the main feature-extraction pipeline.

The study has limitations that the authors aim to tackle in the future. Firstly, the proposed network needs further evaluation using more MRI datasets to verify its effectiveness. Secondly, more fine-tuning and hyper-parameter tuning may be required depending on the particular dataset and classification task. In the present study, the same parameters were used for each dataset, which may decrease the efficiency. Thirdly, the attention modules can be further improved and attached to other state-of-the-art networks.

A CNN can be trained to identify features such as edges, textures, and shapes in an image associated with a particular disease or abnormality. By visualizing the intermediate representations of the network, medical experts can gain insights into which features the network is using to make its predictions. This can help to validate the network's predictions and provide additional information that can be used to support medical decision-making. In this context, the lack of post-hoc explainability-enhancing algorithms in the present study is a limitation and an opportunity for future studies.

Nevertheless, the proposed network achieves a top accuracy in every task. It proves to be superior to the baseline VGG19 model and other pretrained networks.

6. Conclusions

Recent advances in the architecture and functions of CNNs have allowed for a dramatic improvement in the performance of these models. The study proposes an innovative modification of VGG19 with integration of feature-fusion blocks and attention models to enrich the encapsulation of important image features that lead to more precise image classification. The AFF-VGG19 network demonstrated obtained an accuracy of 0.9353 in distinguishing between three brain tumor classes, an accuracy of 0.9565 in distinguishing between AD and PD, and an accuracy of 0.9497 in grading cases of dementia from MRI images. Future research should focus on evaluating the network using more datasets, enhancing the explainability of the framework, and tuning the attention modules to obtain more precise results.

Author Contributions: Conceptualization, I.D.A.; Methodology, I.D.A. and S.A.; software, I.D.A. and S.A.; validation, I.D.A. and M.T.; Resources, S.A. and M.T.; data curation, S.A. and M.T.; Supervision, I.D.A.; Writing—original draft, I.D.A. and M.T.; Writing—review and editing, S.A. and M.T. All authors have read and agreed to the published version of the manuscript.

Funding: This research received no external funding.

Data Availability Statement: The data presented in this study are openly available.

Conflicts of Interest: The authors declare no conflict of interest.

References

1. Goodfellow, I.; Bengio, Y.; Courville, A.; Bengio, Y. *Deep Learning*; MIT Press: Cambridge, MA, USA, 2016; Volume 1.
2. Plewes, D.B.; Kucharczyk, W. Physics of MRI: A Primer. *J. Magn. Reson. Imaging* **2012**, *35*, 1038–1054. [CrossRef]
3. Lundervold, A.S.; Lundervold, A. An Overview of Deep Learning in Medical Imaging Focusing on MRI. *Z. Med. Phys.* **2019**, *29*, 102–127. [CrossRef] [PubMed]
4. Turkbey, B.; Haider, M.A. Deep Learning-Based Artificial Intelligence Applications in Prostate MRI: Brief Summary. *Br. J. Radiol. BJR* **2022**, *95*, 20210563. [CrossRef] [PubMed]
5. Noor, M.B.T.; Zenia, N.Z.; Kaiser, M.S.; Mamun, S.A.; Mahmud, M. Application of Deep Learning in Detecting Neurological Disorders from Magnetic Resonance Images: A Survey on the Detection of Alzheimer's Disease, Parkinson's Disease and Schizophrenia. *Brain Inf.* **2020**, *7*, 11. [CrossRef] [PubMed]
6. Mostapha, M.; Styner, M. Role of Deep Learning in Infant Brain MRI Analysis. *Magn. Reson. Imaging* **2019**, *64*, 171–189. [CrossRef]
7. Akkus, Z.; Galimzianova, A.; Hoogi, A.; Rubin, D.L.; Erickson, B.J. Deep Learning for Brain MRI Segmentation: State of the Art and Future Directions. *J. Digit. Imaging* **2017**, *30*, 449–459. [CrossRef]
8. Tao, Q.; Lelieveldt, B.P.F.; van der Geest, R.J. Deep Learning for Quantitative Cardiac MRI. *Am. J. Roentgenol.* **2020**, *214*, 529–535. [CrossRef]

9. Schelb, P.; Kohl, S.; Radtke, J.P.; Wiesenfarth, M.; Kickingereder, P.; Bickelhaupt, S.; Kuder, T.A.; Stenzinger, A.; Hohenfellner, M.; Schlemmer, H.-P.; et al. Classification of Cancer at Prostate MRI: Deep Learning versus Clinical PI-RADS Assessment. *Radiology* **2019**, *293*, 607–617. [CrossRef]
10. LeCun, Y.; Bengio, Y.; Hinton, G. Deep learning. *Nature* **2015**, *521*, 436–444. [CrossRef]
11. Vaswani, A.; Shazeer, N.; Parmar, N.; Uszkoreit, J.; Jones, L.; Gomez, A.N.; Kaiser, L.; Polosukhin, I. Attention Is All You Need. *arXiv* **2017**. [CrossRef]
12. Le, Q.V.; Ngiam, J.; Coates, A.; Lahiri, A.; Prochnow, B.; Ng, A.Y. On optimization methods for deep learning. In Proceedings of the ICML, Bellevue, WA, USA, 28 June–2 July 2011.
13. Simonyan, K.; Zisserman, A. Very Deep Convolutional Networks for Large-Scale Image Recognition. *arXiv* **2015**, arXiv:1409.1556.
14. Sadad, T.; Rehman, A.; Munir, A.; Saba, T.; Tariq, U.; Ayesha, N.; Abbasi, R. Brain tumor detection and multi-classification using advanced deep learning techniques. *Microsc. Res. Tech.* **2021**, *84*, 1296–1308. [CrossRef]
15. Gab Allah, A.M.; Sarhan, A.M.; Elshennawy, N.M. Classification of Brain MRI Tumor Images Based on Deep Learning PGGAN Augmentation. *Diagnostics* **2021**, *11*, 2343. [CrossRef] [PubMed]
16. Rasool, M.; Ismail, N.A.; Boulila, W.; Ammar, A.; Samma, H.; Yafooz, W.M.S.; Emara, A.-H.M. A Hybrid Deep Learning Model for Brain Tumour Classification. *Entropy* **2022**, *24*, 799. [CrossRef] [PubMed]
17. Kang, J.; Ullah, Z.; Gwak, J. MRI-Based Brain Tumor Classification Using Ensemble of Deep Features and Machine Learning Classifiers. *Sensors* **2021**, *21*, 2222. [CrossRef] [PubMed]
18. Sivaranjini, S.; Sujatha, C.M. Deep learning based diagnosis of Parkinson's disease using convolutional neural network. *Multimed. Tools Appl.* **2020**, *79*, 15467–15479. [CrossRef]
19. Bhan, A.; Kapoor, S.; Gulati, M.; Goyal, A. Early Diagnosis of Parkinson's Disease in brain MRI using Deep Learning Algorithm. In Proceedings of the 2021 Third International Conference on Intelligent Communication Technologies and Virtual Mobile Networks (ICICV), Tirunelveli, India, 4–6 February 2021; IEEE: Tirunelveli, India, 2021; pp. 1467–1470.
20. Hussain, E.; Hasan, M.; Hassan, S.Z.; Hassan Azmi, T.; Rahman, M.A.; Zavid Parvez, M. Deep Learning Based Binary Classification for Alzheimer's Disease Detection using Brain MRI Images. In Proceedings of the 2020 15th IEEE Conference on Industrial Electronics and Applications (ICIEA), Kristiansand, Norway, 9–13 November 2020; IEEE: Kristiansand, Norway, 2020; pp. 1115–1120.
21. Salehi, A.W.; Baglat, P.; Sharma, B.B.; Gupta, G.; Upadhya, A. A CNN Model: Earlier Diagnosis and Classification of Alzheimer Disease using MRI. In Proceedings of the 2020 International Conference on Smart Electronics and Communication (ICOSEC), Trichy, India, 10–12 September 2020; IEEE: Trichy, India, 2020; pp. 156–161.
22. Alloghani, M.; Al-Jumeily, D.; Mustafina, J.; Hussain, A.; Aljaaf, A.J. A Systematic Review on Supervised and Unsupervised Machine Learning Algorithms for Data Science. In *Supervised and Unsupervised Learning for Data Science*; Unsupervised and Semi-Supervised Learning; Berry, M.W., Mohamed, A., Yap, B.W., Eds.; Springer International Publishing: Cham, Switzerland, 2020; pp. 3–21. ISBN 978-3-030-22474-5.
23. Apostolopoulos, I.D.; Papathanasiou, N.D. Classification of lung nodule malignancy in computed tomography imaging utilizing generative adversarial networks and semi-supervised transfer learning. *Biocybern. Biomed. Eng.* **2021**, *41*, 1243–1257. [CrossRef]
24. Deng, J.; Dong, W.; Socher, R.; Li, L.-J.; Li, K.; Fei-Fei, L. Imagenet: A large-scale hierarchical image database. In Proceedings of the 2009 IEEE Conference on Computer Vision and Pattern Recognition, Miami, FL, USA, 20–25 June 2009; IEEE: New York, NY, USA; pp. 248–255.
25. Huh, M.; Agrawal, P.; Efros, A.A. What makes ImageNet good for transfer learning? *arXiv* **2016**, arXiv:1608.08614.
26. Weiss, K.; Khoshgoftaar, T.M.; Wang, D. A Survey of Transfer Learning. *J. Big Data* **2016**, *3*, 9. [CrossRef]
27. Apostolopoulos, I.D.; Pintelas, E.G.; Livieris, I.E.; Apostolopoulos, D.J.; Papathanasiou, N.D.; Pintelas, P.E.; Panayiotakis, G.S. Automatic classification of solitary pulmonary nodules in PET/CT imaging employing transfer learning techniques. *Med. Biol. Eng. Comput.* **2021**, *59*, 1299–1310. [CrossRef]
28. Falkenstetter, S.; Leitner, J.; Brunner, S.M.; Rieder, T.N.; Kofler, B.; Weis, S. Galanin System in Human Glioma and Pituitary Adenoma. *Front. Endocrinol.* **2020**, *11*, 155. [CrossRef] [PubMed]
29. Coskun, P.; Wyrembak, J.; Schriner, S.E.; Chen, H.-W.; Marciniack, C.; LaFerla, F.; Wallace, D.C. A Mitochondrial Etiology of Alzheimer and Parkinson Disease. *Biochim. Biophys. Acta BBA-Gen. Subj.* **2012**, *1820*, 553–564. [CrossRef]
30. O'Bryant, S.E.; Lacritz, L.H.; Hall, J.; Waring, S.C.; Chan, W.; Khodr, Z.G.; Massman, P.J.; Hobson, V.; Cullum, C.M. Validation of the New Interpretive Guidelines for the Clinical Dementia Rating Scale Sum of Boxes Score in the National Alzheimer's Coordinating Center Database. *Arch. Neurol.* **2010**, *67*, 746–749. [CrossRef] [PubMed]
31. Coley, N.; Andrieu, S.; Jaros, M.; Weiner, M.; Cedarbaum, J.; Vellas, B. Suitability of the Clinical Dementia Rating-Sum of Boxes as a Single Primary Endpoint for Alzheimer's Disease Trials. *Alzheimer's Dement.* **2011**, *7*, 602–610.e2. [CrossRef]
32. Mohammed, B.A.; Senan, E.M.; Rassem, T.H.; Makbol, N.M.; Alanazi, A.A.; Al-Mekhlafi, Z.G.; Almurayziq, T.S.; Ghaleb, F.A. Multi-Method Analysis of Medical Records and MRI Images for Early Diagnosis of Dementia and Alzheimer's Disease Based on Deep Learning and Hybrid Methods. *Electronics* **2021**, *10*, 2860. [CrossRef]

MDPI
St. Alban-Anlage 66
4052 Basel
Switzerland
www.mdpi.com

Information Editorial Office
E-mail: information@mdpi.com
www.mdpi.com/journal/information

www.ingramcontent.com/pod-product-compliance
Lightning Source LLC
LaVergne TN
LVHW071241170726
843515LV00006BA/1297

9 783725 805556